Future Horizons

Future Horizons

—

Canadian Digital Humanities

Edited by
PAUL BARRETT AND SARAH ROGER

University of Ottawa Press

2023

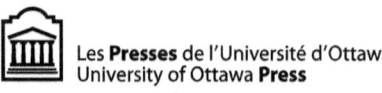

Les **Presses** de l'Université d'Ottawa
University of Ottawa **Press**

The University of Ottawa Press (UOP) is proud to be the oldest of the francophone university presses in Canada and the oldest bilingual university publisher in North America. Since 1936, UOP has been enriching intellectual and cultural discourse by producing peer-reviewed and award-winning books in the humanities and social sciences, in French and in English.

www.Press.uOttawa.ca

Library and Archives Canada Cataloguing in Publication

Title: Future horizons : Canadian digital humanities / edited by Paul Barrett and Sarah Roger.
Names: Barrett, Paul, 1979- editor. | Roger, Sarah Rachelle, 1981- editor.
Description: Includes bibliographical references.
Identifiers: Canadiana (print) 20220451737 | Canadiana (ebook) 20220451745 | ISBN 9780776640051 (softcover) | ISBN 9780776640068 (PDF) | ISBN 9780776640075 (EPUB)
Subjects: LCSH: Digital humanities—Research—Canada. | LCSH: Digital humanities—Study and teaching—Canada.
Classification: LCC AZ105 .F88 2023 | DDC 001.30285—dc23

Legal Deposit: Second Quarter 2023
Library and Archives Canada
© Paul Barrett and Sarah Roger 2023
All rights reserved.

Production Team

Copy editing	Robbie McCaw
Proofreading	Michael Waldin, Crystal Chan, Céline Parent
Typesetting	Transforma
Cover design	Lefrançois Agence B2B

Cover Image

Radiant Loop, generative art by Andrew James MacDonald, 2022.

The University of Ottawa Press gratefully acknowledges the support extended to its publishing list by the Government of Canada, the Canada Council for the Arts, the Ontario Arts Council, the Social Sciences and Humanities Research Council and the Canadian Federation for the Humanities and Social Sciences through the Awards to Scholarly Publications Program, and by the University of Ottawa.

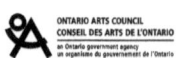
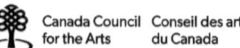
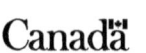

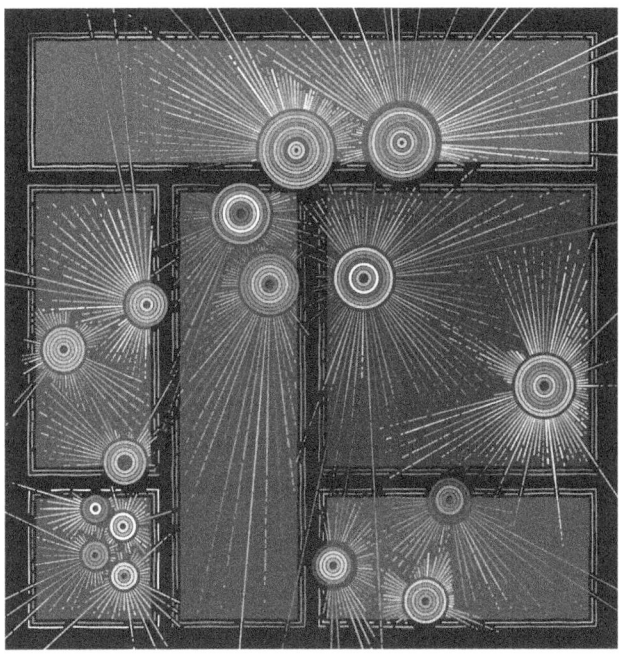

Radiant Loop, 2022, Generative Art

The artwork that features on the cover is a recent piece that was born out of pandemic living. It explores themes of connection, segregation, and isolation. Andrew James MacDonald is a Montréal-based generative artist who creates art using math and code. Born and raised in Hamilton, he studied multimedia at McMaster University and proceeded to forge a career as a web developer.

He has been involved in several digital humanities projects, most notably Voyant Tools. By combining his programming skills with his keen eye for colour and space, Andrew has been creating works in the field of generative art for over a decade.

Contents

Abstract

Across more than twenty chapters, *Future Horizons* explores the past, present, and future of digital humanities research, teaching, and experimentation in Canada. Bringing together work by established and emerging scholars, this collection presents contemporary initiatives in the digital humanities alongside a reassessment of the field's legacy and conversations about its future potential. It also offers a historical view of some important, yet largely unknown, digital projects in Canada.

Future Horizons offers deep dives into projects that enlist a wide range of approaches—from digital games to makerspaces, sound archives to born-digital poetry, visual arts to digital textual analysis—and that work with both historical and contemporary Canadian materials. The chapters in this collection demonstrate how these diverse approaches challenge disciplinary knowledge by enabling humanities researchers to ask new questions.

The writings here challenge the idea that there is either a single definition of digital humanities or a collective national identity. By looking to digital engagements with race, Indigeneity, gender, and sexuality—not to mention history, poetry, and nationhood—this volume expands what it means to work at the intersection of digital humanities and humanities in Canada today.

List of Figures

Acknowledgements

F uture Horizons proposes alternative visions and methods for recording and interpreting national identity and culture. Although the collection explores digital humanities (DH) in Canada, it rejects the idea of a singular, agreed-upon, or even wholly acceptable definition of Canada as nation. Many of the collection's contributors grapple with defining nationhood and related questions, and in various ways the chapters foreground the traditional, ancestral, unceded, and treaty territories of many Indigenous peoples on which the authors live and work.

The editors of this collection are grateful to be able to work on Anishinaabeg, Haudenosaunee, and Huron-Wendat territory, which is covered by the Dish with One Spoon Treaty, Treaty 13, and the Williams Treaties. The University of Ottawa Press, where the collection was prepared and published, is located on unceded Algonquin territory.

It takes many people to make an edited collection, and even more people to make an edited collection about digital humanities. Many of the essays in Future Horizons reference the vast number of collaborators and contributors required to keep a DH project afloat. Although we do not have space here to name every project mentioned in this collection—let alone name every person who has contributed to these projects—we want to acknowledge those whose labour underpins the vibrant DH community in Canada.

At the University of Ottawa Press, we have many people to thank: series editor Dean Irvine for seeing the potential in this collection even before a single contributor signed on; acquisitions editors Caroline Boudreau and Laurence Sylvain for guiding the book through publication; director Lara Mainville for her support of this collection and for recognizing the benefits of open access for publications in digital humanities. For all the behind-the-scenes work that made this collection possible, thank you to Maryse Cloutier, managing editor; Martin Llewellyn, production coordinator; Othmane El Mortaji, marketing manager; and Benoit Deneault, digital marketing and production coordinator. Thank you also to

Robbie McCaw for copyediting and Michael Waldin, Crystal Chan, and Céline Parent for proofreading.

Thank you to artist and DH programmer Andrew MacDonald for allowing us to feature one of his works on our cover. Thank you to Ray Siemens for his early review of the collection, and to the two anonymous reviewers whose feedback helped us to hone it.

We are grateful to Sandra Djwa for sharing her archive of punch cards and related outputs with us, and for agreeing to be interviewed for this collection. Our appreciation also to *Canadian Literature* for allowing us to reprint Djwa's article "Canadian Poetry and the Computer" (1970).

Much credit goes to Kiera Obbard, our colleague and research assistant extraordinaire, who has worked with us at the University of Guelph for many years—on this collection and several other projects.

This collection was made possible thanks to financial support from the Department of English and Cultural Studies at McMaster University, the School of English and Theatre Studies at the University of Guelph, and grants from the Social Sciences and Humanities Research Council of Canada.

Above all, thank you to our colleagues who generously collaborated on this collection, which was first conceived of pre-pandemic, when all our *future horizons* looked rather different. We are grateful to you for sharing your work—work that in some cases captures projects right before the pandemic began, in others actively engage with the pandemic's impact on digital humanities in Canada, and in all looks ahead to what the future of our field may hold.

CHAPTER 1

Digital Canadas? Transforming the Nation

Sarah Roger and Paul Barrett

Future Horizons: Canadian Digital Humanities takes the idiosyncratic approach of situating the digital humanities (DH) within a national context in order to reterritorialize DH research and teaching. Given the often-assumed extraterritoriality of digital scholarship (setting aside the platitudes of how, in a post-pandemic world, we have all had to learn to live and work digitally) and the perceived incompatibility of national framings and DH work, this collection argues for the particular import of Canadian DH research and teaching, and it rethinks the place of the digital in relation to the humanities. Canada has long had an oversized presence in the digital humanities, thanks in no small part to the work of researchers and advocates to harness the capacities of federal research-funding programs in the creation of individual projects, digital research networks, research infrastructure, and educational opportunities. In this sense, locating the digital humanities in national and spatial frameworks allows us to render visible these networks of funding, power, information, and collaboration that make DH scholarship—in Canada and elsewhere—possible.

This is not a gesture toward the particular concern with place that has animated Canadian cultural criticism for more than half a century (articulated most famously in Northrop Frye's argument that Canadians are primarily concerned with the question of "Where is here?") but rather an attempt to understand how locating DH practices within particular places and national frameworks—even in opposition to those frameworks—allows us to pose new questions. In what ways is DH research in Canada engaged with research questions and paradigms that matter here? Indeed, as the growing chorus of post-colonial DH scholars insist, claims of digital extraterritoriality are often guises that conceal how Western, often American, forms of research and scholarship present themselves as universal. This collection thus asks how the national space and culture of Canada continue to matter to contemporary DH researchers.

Marshall McLuhan famously identified Canada as a "borderline case," wherein the country is composed of "multiple borderlines, psychic, social, and geographic." Within the maelstrom of his trademark language of "vortices of energy and power," McLuhan links the particularity of the Canadian borderline experience to "[t]he vast new borders of electric energy and information created by radio and television" (1977, 241). Fittingly, he delivered his speech on the Canadian borderline condition via the technology of radio during Canada's centennial celebrations. These multiple borderlines are, for McLuhan, spaces of contested identities, disagreement, fragmentation; new technologies exacerbate, or transform, the Canadian borderline experience. Where a more conservative generation of critics—Harold Innis, George Grant, J. B. Brebner, and others—lamented technology's transformation of society, for others this transformation reinvigorates notions of citizenship, community, and nation.[1] For A. J. M. Smith, this experience of life on the borderline of empires (British and American) and languages (French and English) enables the Canadian critic to engage in a form of "eclectic detachment" wherein the critic is "immersed both in the European and the North American cultural tradition [...] but he is not *of it*" (1961, 23; emphasis in original). The position of the borderline critic is thus reinterpreted from one of lack or uncertainty to one of productive in-betweenness.[2] Caught between a history born of negotiation with both European forebears and contemporary America, Canadians speak from a novel position of syncretic invention.

Smith, McLuhan, and other early critics of Canadian cultural production sought to understand the apparently unique position of the Canadian critic in relation to an inherited European tradition, the Indigenous peoples and cultures the nation was aggressively colonizing, and the global hegemon to the south. Pheng Cheah's identification of the import of the "organismic metaphor of the political body" for ideas of nation finds something of its negative presence in Canada: the national *Bildung* as an already fragmented, perpetually delayed story recycled from other sources. For George Grant, "the impossibility of Canada" (2005, 67) results from an enforced internationalism alongside "the profundity with which technological civilization enfolds us as our destiny" (1976, 294). McLuhan's argument that "Canadians never got 'delivery' on their first national identity image in the nineteenth century and are the people who learned to live without the bold accents of the national ego-trippers of other lands" (1997, 227) finds echoes in Frye's "Haunted by a Lack of Ghosts" (1976), Sylvia Söderlind's "Ghost-National Arguments" (2006), and Jonathan Kertzer's "Worrying the Nation" (1998). Yet this Canadian assertion of a post-national identity (or is it a post-identity nation?) remains indebted to, even as it rejects, a nineteenth-century vision of nation that represses its colonial histories and present.[3]

While the trope of the borderline and the idea of eclectic detachment are evident in the essays collected in *Future Horizons*, they can only take us so far; mining these essays and projects for evidence of some unalloyed expression of

Canadian digital humanities will lead to predictable results. Perhaps the most valuable function of panacea formulations of national identity is to reveal not the borderlines but the emerging fissures that promise to swallow the project whole. The authors in this collection invoke the language of border, disruption, transgression, and transformation—and, noteworthy in a DH context, the attempts to archive or otherwise snapshot these shifting borders—in order to show how the failure of these ur-statements of national culture prompt new, more interesting questions. Technology and the digital humanities have played significant roles in both framing and rebutting this national hand-wringing over Canada's lack of identity, as well as offering alternative visions and methods for recording and interpreting national identity and culture—thus the multiple "Canadas" in our title. This is evident in the earliest digital projects in Canada as well as in more recent Canadian articulations of the digital humanities. For John Bonnett and Kevin Kee, "the digital humanities bear the 'thematic stamp [...] of transition'" (2006, 10), and for Dean Irvine, Vanessa Lent, and Bart Vautour, early DH work in Canada evinces a "renovatory act of editing" (2017, 6); our authors extend this project into a transformative act that challenges staid ideas of national culture.

The earliest example of this (in this collection, and perhaps even in the landscape of Canadian digital humanities) is Sandra Djwa's largely unknown digital concordance of Canadian poetry. Djwa's work rejects the Canadian cultural establishment's vision of Canadian literary culture and employs distant reading and traditional close reading to offer an alternative view of Canadian poetry. It is stunning that Djwa's early research and essay were completed in the 1960s and remain largely absent from discussions of digital humanities in Canada. Djwa's work is significant not merely because it is one of the earliest DH projects in Canada but also because it sets the pattern for the use of digital methods to reassess national paradigms. It is a demonstration of the kinds of projects we collect here: work that is often illegible within dominant frameworks of humanities research that also challenges definitions of the digital and the humanities alike. This is evident in Susan Brown, Kim Martin, and Asen Ivanov's conception of digital research infrastructure as a series of "boundary objects" that place distinct fields into productive, critical dialogue. David Gaertner tempers DH researchers' call for boundary transgression, taking, instead, "boundary work as a means to better articulate the places where DH practitioners should stop," particularly when working with Indigenous knowledge and cultural objects. Deanna Fong and Ryan Fitzpatrick extend Djwa's use of computational systems to interrogate the boundaries of textuality: "No longer can we simply question how the book will act upon, or transmit to, the reader, we must also consider the ways the book is, even at the moment of its inception, already enmeshed in a social and spatial production." *Future Horizons* demonstrates the ways in which DH work in Canada is "enmeshed in a social and spatial production," of communities of practice, methods of research, and complex engagement with a history of digital humanities in Canada.

Yet, the essays in this collection have not been assembled in the interest of reaffirming some vision of Canadian identity, nor of articulating an authoritative history.[4] Rather, this book aims to link the histories of the digital humanities in Canada to contemporary digital projects and research that may embrace or resist being labelled as digital humanities. Reterritorializing the digital humanities in Canada means grappling with the funding networks, bilingual research frameworks, and research questions that animate these projects while also thinking about their points of continuity and distinction from American and European DH work. For Andrea Zeffiro, ambivalence toward the digital humanities and the nation alike enables a critical "staying with the trouble" of digital humanities (Bucher quoted in Zeffiro)—a concept Zeffiro engages with in this collection as she explores "'routing' as a trope through which we can conjure DH in Canada." Routing is particularly suited to Canadian digital humanities as it "encourages diversions and deviations" in order to ruminate on digital "scholarship in its current state but also as something that does not yet exist." Similarly, Dani Spinosa's analysis of electronic literature brings "to the surface a digital humanities that could be [...] interested in opening, blurring, or breaking down national borders." Essays including Zeffiro's and Spinosa's cut a series of interconnected routes through digital humanities that foreground the future possibilities for the digital humanities in Canada.

Similarly, *Future Horizons* avoids venturing into the ongoing debates about what constitutes the digital humanities and its distinctions from computational literary studies, cultural analytics, and stylometrics (the statistical analysis of linguistic or literary style). For an excellent overview of these debates, see the essays collected in Melissa Terras, Julianne Nyhan, and Edward Vanhoutte's *Defining Digital Humanities: A Reader* (2013), particularly Geoffrey Rockwell's "Is Humanities Computing an Academic Discipline?" and Julia Flanders's "The Productive Unease of 21st-century Digital Scholarship." Indeed, the field of digital humanities, insofar as there is one, might be best characterized by a recurring worry over whether the digital and the humanities can ever be peaceably united.

To the extent that this book does engage those debates, it does so inductively and implicitly to challenge what conceptions of the humanities are at stake in digital research and teaching. Too often in the debates over the merits of the digital humanities, the digital is the term that raises ire while the humanities are treated as settled and uncontested. It is surprising how non-conversant many of the most stringent critiques and best-intentioned defences of the digital humanities are with the contemporary criticisms of the humanities that have emerged from post-colonial studies, Black studies, feminist studies, Indigenous studies, and queer studies. These efforts to locate the digital humanities within the humanities often rely on archaic or nostalgic notions of the humanities, as though the author is pleading their case to Edmund Burke or, in our Canadian context, Frye. The principles upon which Frye bases his own "confession of faith as a humanist" (Good 1997, 75) remain implicit within much contemporary DH work. As Gaertner writes in this collection, drawing on the work of Elizabeth Losh and Jacqueline Wernimont, "digital humanities can no longer afford to uphold the appearance of 'friction-free,' neo-Cartesian engagements with technology and the digital." This "retro-humanist" (Bianco 2012) conception of the humanities neglects the numerous important challenges to the idea of the humanities in terms of its methods but also, more importantly, in its role in endorsing a particular vision of who gets to count as human.

The essays in this collection therefore use the unhomeliness of the digital and the humanities as a basis to challenge not merely humanities *methods* but also the broader political and social stakes of DH work. If Willard McCarty defines the digital humanities as a space in which "ideas and machines interact asynchronously to deepen the fundamental problems, rather than solve them" (2013), then the essays in this collection demonstrate that the humanities themselves, as well as their framing within particular national and cultural contexts, are part of this fundamental problem. Roopika Risam notes that these "new" digital methodologies "are built on the histories and traditions of humanities knowledge production that have been deeply implicated in both colonialism and neocolonialism" (2018). This is particularly pronounced in Canada and forms a component of our call for reterritorialization: how can the digital humanities offer an alternative mode of thinking about place that challenges colonial paradigms of knowledge and the material practices of ongoing colonization? We thus heed Risam's call for theorizing digital humanities as both methodology and praxis and foreground DH research that is equally concerned with reimagining both the digital and the humanities: for example, Kendra Cowley turns, in her chapter, to the digital humanities to develop a "cartographic practice that might destabilize colonial orientations to the land by tuning into the fluidity of sound that exceeds the silencing, stilling imperative of settler colonialism." Across the collection, essays respond to Gaertner's demand, that all of us in digital humanities, including those of us who have not traditionally grappled with legacies of power in our work, can do more to "disrupt" the conversations that shape the field. [...] As a field, we must foster an environment of broad accountability in which we are all responsible for amplifying marginalized voices and critiquing and dismantling power.

Djwa's digital concordance of Canadian poetry, Mark V. Campbell's analysis of the disruptive potential of the hip hop archive, Fong and Fitzpatrick's development of the Fred Wah archive, and Allan Cho and Sarah Zhang's reconsideration of Chinese-Canadian immigration data are all digital projects that call for new methods for engaging archives in a manner that challenges the underlying discourses of humanism within the digital humanities.

Djwa's work is also exemplary as one of the first instances of DH scholars drawing on institutional infrastructure to pursue cross-disciplinary digital projects. DH research in Canada has been supported with immense research infrastructure, ranging from Djwa's support from the Computing Centre at UBC and Carleton's Centre for Editing Early Canadian Texts to the contemporary infrastructural supports for the Digital Humanities Summer Institute, the Digital Research Alliance of Canada (and Compute Canada before it), the Canadian Writing Research Collaboratory, SpokenWeb, Editing Modernism in Canada, the Implementing New Knowledge Environments Partnership, and other large-scale DH projects and research. In this respect, the digital humanities are very much at home in Canada and have often represented the kinds of research that has attracted Canadian funding bodies. Indeed, Brown, Martin, and Ivanov trace the evolution of digital research infrastructure in Canada, calling for "a strong network of partners [who] would knit the scholarly community into closer collaboration with publishers, GLAM [galleries, libraries, archives, and museums] institutions, and information stakeholders, foster shared expertise and infrastructural costs, and produce a potentially ever-growing community of practice." As McCarty reflected in 2012: "The Canadians have done marvels in convincing the government and setting up structures of funding," and there are "[p]er capita more digital humanists in Canada than anywhere else in the world, I think"—a statement that may well still be true today.

Yet despite substantial institutional support, *Future Horizons* also highlights the kinds of digital projects that may not be legibly digital humanities and, perhaps for that reason, do not have the same public profile nor have received funding from major Canadian granting organizations. Projects such as the Fred

Wah Digital Archive, the Indigenous New Media Collective, native-land.ca, and the Northside Hip Hop Archive—not to mention creative interventions such as those by Cowley and Klara du Plessis, and those described by Spinosa and Julia Polyck-O'Neill—challenge ideas of what DH research looks like and provoke new ways of thinking about the relationship between the digital and the humanities. These projects follow Alan Galey's provocation: "Why speculate when we can prototype" (2013, 108), while also seeking to provoke new meanings for the digital humanities as they engage with a diverse range of scholars and critics, including Said, Fanon, Wynter, Spivak, Butler, and beyond—critics who notably examine the relationship between digital culture, definitions of the humanities, and the question of who counts as human. Polyck-O'Neill's examination of the relationship between visual arts and digital humanities demonstrates how installations, prototypes, and engagement with the visual archive, from W. E. B. Du Bois to Theaster Gates, can challenge humanist epistemologies—and how the Canadian approach to teaching overlooks the power of the visual in training digital humanities' future practitioners. These projects employ DH work to challenge the "story of humanism" as a "European coming-of-age story" (Scott 2000, 121), and redefine the digital humanities as a method ripe for subaltern and contrapuntal reimagining.

In some respects, Polyck-O'Neill's approach and our approach more generally follows that of Anne Burdick et al.'s 2012 introductory text *Digital Humanities*, wherein they argue: "Digital Humanities projects can be described by sketching their structure at several levels." By *structure*, the authors refer to design, computation, processing, digitization, classification, description, metadata, organization, navigation, curation, analysis, editing, modelling, and prototyping. This list of formal features and practices provides a series of keywords by which to identify and group digital research and teaching. Yet, to this structural approach with its list of formal features, we add the question of how the combination of the digital and the humanities calls for new approaches to both the *digital* and the *humanities* as structural and formal concepts.

This collection, therefore, draws together a series of Canadian digital projects in order to raise the particular question of how the digital humanities transform conceptions of the humanities in Canadian contexts. Of course, one border that remains securely intact within this collection is that of language: these essays address DH only in English and in anglophone contexts. We welcome a similar volume from our colleagues working in French and bilingual contexts. Yet across these chapters, this volume attends to thematic concerns and individual techniques, scholarly investigations and ongoing artistic interventions. It meditates on the successes of past projects, while also proposing ways in which the field can do better. In doing so, Future Horizons runs the gamut from federally funded projects to local, ephemeral contributions, from archiving Canadian literary history to fostering nuanced approaches to complex, emergent concerns.

SITUATING AND DISRUPTING DIGITAL SCHOLARSHIP

The book's first section, "Situating and Disrupting Digital Scholarship," opens with theoretical and practical provocations for digital humanities' place within the academy via Risam's exploration of nationhood in digital humanities and Zeffiro's roadmap for the future of digital humanities. In "Where Is the Nation in Digital Humanities?" Risam explores the tensions between the digital humanities as a theoretically transnational field and the practical realities of it as a discipline that is governed by anglocentric, hegemonic structures. Building on her definition of "postcolonial digital humanities" as "an approach to uncovering and intervening in the disruptions within the digital cultural record produced by colonialism and neocolonialism" (2018, 3), Risam identifies the need for a deeper engagement with this new type of digital humanities—one "at the borderlands [that] recognizes the limitations of the nation and offers frameworks for challenging the persistence of the nation." In conversation with Risam, Zeffiro's "Rerouting the Digital (Humanities) Scholarship in Canada" "speculate[s] on the possibilities for what exists beyond and adjacent to digital (humanities) scholarship in Canada." Zeffiro looks for a way to reframe the hegemonic structures at play in Canadian digital humanities, seeking a way to celebrate emergent "feminist, queer, anti-racist, and anti-colonial ways of knowing, producing and engaging in research and community organizing." As part of this process, which presages many of the interventions made by the authors in this collection, Zeffiro calls for a more sustainable, broader model of digital humanities that foregrounds currently invisible labour—a call that recurs throughout the collection, and which is often addressed by contributors' active recognition of their collaborators and colleagues in their chapters.

Building on Risam's and Zeffiro's challenges to the status quo, the remainder of this first section looks to areas where the disrupting potential of digital humanities has complicated the entrenched power dynamics via scholarship that is inclusive of, yet also holds appropriately distinct space for, fields that have been heretofore marginalized. In "Closed, Open, Stopped: Indigenous Sovereignty and the Possibility of Decolonial Digital Humanities," Gaertner proposes a dynamic of open, closed, and stopped as a respectful space-making approach to Indigenous materials in the digital humanities. He argues that the digital humanities' traditional push for openness, embodied by the open access and open educational resource movements, risks overriding Indigenous peoples' right to reclaim and redistribute knowledge outside of colonial-derived knowledge systems and Western epistemologies.

Where Gaertner pulls the digital archive towards the margins, Jon Saklofske, Polyck-O'Neill, and Martin and Rashmeet Kaur push back against some of the mainstays of the field—digital games, data visualizations, and makerspaces—in critical interrogations of tools that practitioners of the digital humanities often take for granted. In "This Game Needs to Be Made," Saklofske presents a compelling case for digital games as productive arguments, constructed with the same intellectual rigour and theoretical precision of standard modes of scholarship. He proposes that games have the potential to provide transformative critical experiences beyond what is possible via traditional, linguistically limited argumentation. There are echoes of Saklofske's view that games are a complex medium in Polyck-O'Neill's assessment of data visualizations. She argues that scholars tend to treat data visualizations as a flat source of facts when they should be regarded as nuanced interpretive structures. She highlights a need for better visual literacy among digital humanists (and for better visual literacy training in Canadian digital humanities and beyond), arguing for "shifts in the use and development of tools for the visualization of data that more accurately reflect scholarly findings," and "a necessary critical expansion of the practice and interpretation of data visualization as a rich visual medium." For Polyck-O'Neill, this expansion is essential to understanding how visual representation can challenge how we view categories of race. In "Making, Conversation: An Experiment in Public Digital Humanities," Martin and Kaur discuss their lived experience with makerspaces and their work with the "DIYversity Project." In their analysis they interrogate barriers within the maker community, and in their conversation they model a form of community building and a series of learned best practices that offers an alternative vision of maker culture within a DH context.

DIGITAL POETICS

The book's second section looks to one of the key areas into which digital humanities has—and continues—to engage, intersect with, and even contribute to primary source humanities outputs: poetry. It starts with a look back to digital humanities' Canadian origins, proposing that Djwa's computational concordances (generated in the 1960s using an IBM 7044 mainframe computer) offer insight into the questions that arise at the intersection of digital tools and concepts of nation in the humanities. Djwa's ambitious project, which used computational methods to map a Canadian poetic tradition, was unrecognized at the time and remains "largely forgotten" today despite the fact that it "marks mile zero in computer-assisted criticism in Canadian literature" (Irvine, Lent, and Vautour 2017, 10). Both the merits of her early DH work and the misrecognition of it within Canadian literature reveal a great deal about the perceived relationship between the digital and the humanities today.

Turning from analysis to creation, this section swerves to take in the works of poets such as bpNichol and moves through to contemporary artists such as W. Mark Sutherland. With contributions from Gregory Betts, Eric Schmaltz, Spinosa, and du Plessis, this section proposes an inextricable intertwining of the digital and the poetic in the works of Canadian poets. Extending McCarty's observation that "Computing machines and scholarly intelligence change each other recursively" (2013, 2), these authors demonstrate that just as the digital humanities can augment our readings of poetry, so too does the digital shape the humanities by intervening in how poetry is written and read.[5]

Betts's "'saga uv th relees uv huuman spirit from compuewterr funckshuns': Space Conquest, IBM, and the Anti-digital Anxiety of Early Canadian Digital Poetics (1960–1968)" describes an "anxiety felt by Canada's [1960s] avant-garde about the intersection of computers and poetry"—an anxiety mirrored by a fascination with the potential that the digital has to enhance the creation of and access to poetry. As Schmaltz shows in "The Digits in the Digital: Bodies in the Machines of Canadian Concrete Poetry," it is impossible to extract the digital from the poetic in late twentieth-century Canadian poetry; the very fact of the computer—of the role it plays in drafting poems, of preparing them for press, of disseminating them across Canada and beyond—has shaped how poets create and how readers interpret. In these regards, the humanities as a creative fount have been inescapably and indelibly imprinted upon by the digital. In "Nations of Touch: The Politics of Electronic Literature as Digital Humanities," Spinosa pulls this thread through to the present by exploring the relationship between digital humanities and (Canadian) electronic poetry, arguing that "both are primarily and strategically interested in disrupting borders—of genre, of nation, of the line between literary text and scholarship—and that disruption is enhanced when DH scholarship and electronic literary production work in tandem." Spinosa demonstrates how digital humanities and electronic poetry have unsettled the academy and upended the structures that have traditionally narrowed both the types of poetry to which readers have access and the theoretical frameworks into which their readings of these poems are forced. In "Stop Words," Canadian poet and DH scholar du Plessis offers a fresh perspective on computational textual analysis and digital poetics. She questions the convention of deeming meaningless the frequently occurring words (stop words) that are usually stripped from textual analysis, instead enlisting the tools of digital humanities to generate playful poetic meaning.

DIGITAL CANADIAN ARCHIVES

Future Horizons' final section turns to recent and ongoing projects in the digital humanities in Canada to showcase the ways in which the field is building on the ideological and theoretical ideas posited in the book's first two sections. These chapters function as a time capsule of Canadian digital humanities in the early 2020s. Because DH projects often work with ephemeral or deteriorating materials, they can be circumstantial or impermanent, dependent on funding and human resources, or designed to fill a specific role at a certain time, as the authors of these chapters acknowledge. In this regard, this section constitutes a contribution back to the archive, in both its analogue and digital forms. In concert with this, the relationship between DH projects and the archives on which they draw—particularly the potential of digital tools for archival preservation and subsequent interpretation—is a thread that connects the projects featured in this chapter.

In "Wages Due Both Then and Now: Labour and the Public Good," Pascale Dangoisse, Constance Crompton, and Michelle Schwartz situate the Lesbian and Gay Liberation in Canada (lglc.ca) archives within the digital humanities, proposing ways of applying the non-traditional approaches of the digital humanities to materials and subject matter that has been sidelined in—or does not fit within the traditional formats of—the archive. They argue that the digital humanities presents unique opportunities to empower people through research and information.

In "Analog Thrills, Digital Spills: On the Fred Wah Digital Archive Version 2.0," Fong and Fitzpatrick show a different side of the archival process. In their first-person account of building the Fred Wah Digital Archive (fredwah.ca), Fong and Fitzpatrick make a bid for reframing the approach to the labour underpinning the digital humanities—a call that echoes Zeffiro's argument in the book's first section. They highlight the precarious nature of building and maintaining digital archives: the tenuous employment of the people involved, and the ways in which labour is entangled with ethical questions of how to archive lives that are "still unfolding in the present." Many of the difficulties that Djwa describes having faced in the 1960s remain true today; Fong and Fitzpatrick tellingly assert that "we must not commit the mistake of using affect to mask unethical labour practices, assuming that participation in a community is reward enough in itself when the community in question may be subject to differences of experience and power."

In "Humanizing the Archive: The Potential of Hip-Hop Archives in the Digital Humanities," Campbell draws a parallel argument about the power embedded in digital archives, positing that the archive is not a neutral storehouse but can be a site of silence and oppression for Black people. He argues that incorporating into the archive the "Black lives that are often conspicuously absent or significantly underrepresented in the field" of digital humanities could engender a radical restructuring of the field. In particular, he looks to digital tools and techniques for archiving hip-hop—and the genre's attendant emphasis on sampling and remixing—to show how the digital humanities has the potential to showcase a more expansive Black life.

In "Sounding Digital Humanities," Katherine McLeod proposes ways of working with sidelined materials and offers compelling justifications for doing so. She looks at how the SpokenWeb archive of Canadian literary sound recordings (spokenweb.ca) handles audio materials, which are traditionally under-used and under-preserved. She suggests that working with non-traditional materials facilitates experimentation in ways of making and knowing. Similarly, Cowley's "Unsettling Colonial Mapping: Sonic-Spatial Representations of amiskwaciwâs-kahikan" looks at some of the problems that arise from working with under-utilized archival audio materials. Cowley's project is a sonic-spatial mapping of the lands on which the University of Alberta is located; her exploration of the university's archive of audio recordings is a way of acknowledging the problems inherent in the DH tools and the limitations of preservation.

Brown, Martin, and Ivanov's "Linking Out: The Long Now of Digital Humanities Infrastructures" recalls their collective experiences with the Orlando Project, the Canadian Writing Research Collaboratory, and the Linked Infrastructure for Networked Cultural Scholarship and reflects on transformations in Canadian digital research infrastructure and the challenges of large-scale projects such as these. Their chapter identifies the infrastructural turn in DH research, and they conceive of DH infrastructure as a diverse ecology of actors, frameworks, systems, and communities of practice. Graham H. Jensen follows suit with another vision of infrastructure informed by community labour and practice, arguing for the value of digitization for enhancing our ways of knowing in "Beyond 'Mere Digitization': Introducing the Canadian Modernist Magazines Project." He explains how the digitization process for the Canadian Modernist Magazines Project (modernistmags.ca) has enriched the material beyond the mere act of preserving it. In particular, he points to the ways in which connecting contemporary approaches to cataloguing are challenged when brought to bear on digitized, historical material. In doing so, Jensen also points to the labour this work involved (something that many other contributors also highlight), saying: "Understood as a form of intellectual as well as technical labour, digitization poses a problem to institutions that have been quick to stress the importance of collaborative, public-facing, open, and non-traditional forms of scholarly production, but slow to acknowledge it in ways that matter to everyone engaged in this work." While exploring a digital archive may conjure for its users a sense of engaged community, the act of creating it can be labour intensive, expensive, and challenging. Too often, the emphasis is on the end product alone, and the creative and technical labour required to build it go unacknowledged or are seen as mere technical supports and not important scholarly contributions.

Finally, and looking beyond gaps in the scholarship created by digital humanities to gaps in the archive itself that the digital can uncover, Cho and Zhang's "A Legacy of Race and Data: Mining the History of Exclusion" looks at Chinese-Canadian immigration data from 1885 to 1923, to show how DH methods can open the archive to new interpretive possibilities by way of what is missing as much as by way of what is there. Their projects demonstrate how novel digital approaches challenge both what it means to read the archive as well as the kinds of insurgent knowledge that can be produced in dialogue with a digital archival approach.

In her chapter, Zeffiro asks: "Do (or does) the digital humanities exist? This question is by no means a call for performing an inventory of the field, nor is it a plea to readjust the definitional enclosures of DH scholarship. As we already know, both are futile exercises. And besides, as this anthology demonstrates so vividly, the digital humanities in Canada are unbound, dynamic, and multidirectional." *Future Horizons* takes this unbound, dynamic, multidirectional digital humanities as its starting point, and therefore does not attempt to sum up the discipline's current state. Instead, across its many chapters it gestures toward the field's emerging concerns, new paradigms, and marginalized voices. By troubling received notions of the digital, the humanities, and the attendant discourses of humanism that animate both, the volume places digital humanities into a complex dialogue with notions of race, Indigeneity, history, gender, sexuality, and nation. *Future Horizons* weaves together incongruous ideas of national culture and digital humanities in order to understand the contemporary stakes of—and potential for—DH work in Canada.

NOTES

1. Harold Innis is representative of their view, characterized by Philip Massolin as the Canadian Tory tradition: "Modern civilization [...] characterized by an enormous increase in the output of mechanized knowledge [...] has produced a state of numbness [...] and self-complacency only equaled by laughing gas. [...] The demands of the machine are insatiable" (Innis 2017, 383).

2. Smith's conception predicts Abdul JanMohamed's notion of the "specular border intellectual" (1992) that he identifies in the work of Edward Said. By contrast, JanMohamed describes the "syncretic intellectual" as "able to combine elements of the two cultures in order to articulate new syncretic forms and experiences" (97).

3. McLuhan's proposition, for example, that "there are 250,000 unnamed lakes in Ontario alone," and his description of the "idyllic playgrounds of our largely unoccupied land of lakes and forests" (1977, 229), depend on a familiar conception of *terra nullius* that completely disavows Indigenous histories and contemporary presences and epistemologies. This romantic-nationalist vision of Canadian culture has been exposed by Indigenous writers and thinkers as an aesthetic-ideological justification for colonization and the theft of land. Jordan Abel's digital poetry is exemplary in its deconstruction of the discourses of *terra nullius* upon which romantic nationalism depends, and in speaking back to the colonial archive with an Indigenous vision of land and space.

4. For two effective overviews of that history, see Bonnett and Kee (2010) and Siemens and Moorman (2006).

5. For a more in-depth consideration of the intersection between poetic production, analysis, and the computer in Canada in the 1960s and 1970s, including the works of Jean A. Baudot, Stephen Scobie, Robert Ian Scott, Peter Stevens, and Charles Stock, see Irvine, Lent, and Vautour (2017, 12–13).

REFERENCES

Bianco, Jamie Skye. 2012. "This Digital Humanities Which Is Not One." In *Debates in the Digital Humanities*, edited by Matthew K. Gold, 96–112. Minneapolis: University of Minnesota Press. https://doi.org/10.5749/minnesota/9780816677948.003.0012.

Bonnett, John, and Kevin Kee. 2010. "Transitions: A Prologue and Preview of Digital Humanities Research in Canada." *Digital Studies / Le champ numérique* 1, no. 2. https://doi.org/10.16995/dscn.106.

Burdick, Anne, Johanna Drucker, Peter Lunenfeld, Todd Pressier, and Jeffrey Schnapp. 2012. *Digital_Humanities*. Cambridge, Mass.: The MIT Press. https://doi.org/10.7551/mitpress/9248.001.0001.

Cheah, Pheng. 2012. *Spectral Nationality Passages of Freedom from Kant to Postcolonial Literatures of Liberation*. New York: Columbia University Press.

Djwa, Sandra. 1970. "Canadian Poetry and the Computer." In "The Frontiers of Literature," Special Issue, *Canadian Literature* 46 (Autumn): 43–54.

Flanders, Julia. 2009. "The Productive Unease of 21st-century Digital Scholarship." *Digital Humanities Quarterly* 3, no. 3.

Frye, Northrop. 2006a. "Haunted by a Lack of Ghosts" In *Northrop Frye on Canada*, edited by Jean O'Grady and David Staines, 472–92. Vol. 12 of *The Collected Works of Northrop Frye*. Toronto, Ont.: University of Toronto Press.

———. 2006b. "Literary and Mechanical Models." In *"The Secular Scripture" and Other Writings on Critical Theory*, edited by Joseph Adamson and Jean Wilson, 451–66. Vol. 18 of *The Collected Works of Northrop Frye*. Toronto, Ont.: University of Toronto Press.

Galey, Alan. 2013. "The Human Presence in Digital Artefacts." In *Text and Genre in Reconstruction*, edited by Willard McCarty. Cambridge, U.K.: Open Book Publishers.

Good, Graham. 1995. "Northrop Frye and Liberal Humanism." *Canadian Literature* 148 (Spring): 75–91.

Grant, George. 1976. "The Computer Does Not Impose on Us the Way it Should Be Used." In *Beyond Industrial Growth*, edited by Abraham Rotstein. Toronto, Ont.: University of Toronto Press. https://doi.org/10.3138/9781487583460-008.

———. 2005. *Lament for a Nation: The Defeat of Canadian Nationalism*. Montréal, Que., and Kingston, Ont.: McGill-Queen's University Press.

Innis, Harold A. 2017. *Essays in Canadian Economic History*, edited by Mary Q. Innis. Toronto, Ont.: University of Toronto Press. https://doi.org/10.3138/9781487512590-001.

Irvine, Dean, Vanessa Lent, and Bart Vautour. 2017. "Introduction." *Making Canada New: Editing, Modernism, and New Media*, 3–30. Toronto, Ont.: University of Toronto Press. https://doi.org/10.3138/9781487511357.

JanMohamed, Abdul R. 1992. "Worldliness-without-World, Homelessness-as-Home: Toward a Definition of the Specular Border Intellectual." In *Edward Said: A Critical Reader*, edited by Michael Sprinker, 96–120. Oxford, U.K.: Basil Blackwell.

Kertzer, Jonathan. 1998. *Worrying the Nation: Imagining a National Literature in English Canada*. Toronto, Ont.: University of Toronto Press. https://doi.org/10.3138/9781442683693.

McCarty, Willard. 2013. "Introduction." *Text and Genre in Reconstruction*, 1–11. Cambridge, U.K.: Open Book Publishers.

McCarty, Willard, Julianne Nyhan, Anne Welsh, and Jessica Salmon. 2012. "Questioning, Asking and Enduring Curiosity: An Oral History Conversation between Julianne Nyhan and Willard McCarty." *Digital Humanities Quarterly* 6, no. 3.

McLuhan, Marshall. 1977. "Canada: The Borderline Case." In *The Canadian Imagination*, edited by David Staines. 226–48. Cambridge, Mass.: Harvard University Press.

Risam, Roopika. 2018. *New Digital Worlds: Postcolonial Digital Humanities in Theory, Praxis, and Pedagogy*. Evanston, Ill.: Northwestern University Press.

Rockwell, Geoffrey. 2013. "Is Humanities Computing an Academic Discipline?" In *Defining Digital Humanities: A Reader*, edited by Melissa Terras, Julianne Nyhan, and Edward Vanhoutte. Oxford, U.K.: Routledge.

Scott, David. 2000. "The Re-enchantment of Humanism: An Interview with Sylvia Wynter." *Small Axe* 8: 119–207.

Siemens, Raymond George, and David Moorman. 2006. *Mind Technologies: Humanities Computing and the Canadian Academic Community*. Calgary, Alb.: University of Calgary Press. https://doi.org/10.2307/j.ctv6gqqt9.

Smith, A. J. M. 1961. "Eclectic Detachment." *Canadian Literature* 9 (Summer): 6–14.

Söderlind, Sylvia. 2006. "Ghost-National Arguments." *University of Toronto Quarterly* 75, no. 2: 673–92. https://doi.org/10.3138/utq.75.2.673.

Terras, Melissa, Julianne Nyhan, and Edward Vanhoutte, eds. 2013. *Defining Digital Humanities: A Reader*. Oxford, U.K.: Routledge.

Part 1

Situating and Disrupting Digital Scholarship

CHAPTER 2

Where Is the Nation in Digital Humanities, Revisited

Roopika Risam

In 2016, I assembled a roundtable at the Modern Language Association convention to investigate the question, "Where is the nation in digital humanities?" The roundtable was inspired by the title of a post that Paul Barrett wrote for the Postcolonial Digital Humanities website. Barrett had noted that among the critiques of the relationship between colonialism and knowledge production raised by scholars of post-colonial digital humanities, a noticeable absence was "the continued salience of the nation as an organizing structure and category of analysis" (Barrett 2014). Very much of its time, the roundtable, which included Alex Gil, Sara Humphreys, Toniesha Taylor, and Dhanashree Thorat, took up the question of the nation through the relationship between the local and the global. Participants raised questions about the anglocentric nature of digital humanities research, methods, and organizational structures to consider how we might decentre the anglophone hegemony incipient in the digital humanities. These are issues that I later examined in my book *New Digital Worlds: Postcolonial Digital Humanities in Theory, Praxis, and Pedagogy* (2018). At the heart of this investigation is the understanding that the digital humanities operate through a dialectic of the global and the local. The overdetermining conceptual practices of the global digital humanities are, in fact, methodologies and epistemologies of the Global North—and a very narrow swath of the Global North (not the minoritized, not the Indigenous) at that. Yet, in practice, DH methods are decidedly and necessarily local, accounting for the economic, political, and cultural circumstances that shape the intellectual choices we make.

These claims about digital humanities continue to hold true, and there remains significant work to be done to ensure that local practices are not continually overwritten by putatively "global" ones. However, the contemporary world looked quite different when I began this essay in early 2020 (and continues to change as I complete my revisions in 2022) than it did at that moment, in early 2016, when we gathered to discuss the role of the nation in digital humanities. Then, it seemed that emphasizing the nation and national contexts was a feasible way of insisting on the primacy of the local over the global. But Barrett was, in fact, correct; we had not accounted for the limitations of the nation as an organizing principle. Moreover, we were not anticipating the resurgence of nationalism as a corrosive political force. We were not counting on public acceptance of xenophobic nationalism in the United States and other countries. We had not yet wrapped our minds around how effectively the anti-immigrant discourses that were circulating would bring populist governments into mainstream politics. We did not expect that Russia would, once again, invade Ukraine. Instead, we were relying on national identity and national contexts for the digital humanities in the Global South to supplant the overdetermining power of the Global North. In retrospect, we were woefully naïve and had failed to address Barrett's question: "Where is the nation in digital humanities?" Or we had put too much faith in the formation of national DH communities as the answer.

From a vantage point of the beginning of the 2020s, I briefly examine how the nation has emerged as an organizing principle within the DH scholarly community globally in the last half of the 2010s and look forward to scholarship better positioned to respond to the relationship between the local and global beyond nationalist frames. This scholarship, I suggest, may more fully realize the promises of post-colonial digital humanities.

THE UNITED NATIONS OF DIGITAL HUMANITIES

In 2005, the Alliance of Digital Humanities Organizations (ADHO) was formed, uniting two separate professional organizations: the U.S.-based Association for Computers and the Humanities and the Europe-based Association for Literary and Linguistic Computing. The goal of ADHO is to coordinate DH initiatives and promote digital research in the humanities around the world (ADHO, n.d.-a). The addition of other geographically based constituent organizations followed suit (the Canadian Society for Digital Humanities / Société canadienne des humanités numériques, the Australasian Association for Digital Humanities, and the Japanese Association for Digital Humanities). The initial exception to this geographical organizing principle was centerNet, an organization founded to represent DH centres around the world. The geographical trend was further bucked by the addition of Humanistica, a francophone organization, and Red de Humanidades Digitales, a Spanish-language organization with strong roots in Latin America. Additional national entities, such as the Taiwanese Association for Digital Humanities, as well as regional entities such as the Digital Humanities Association of Southern Africa, were admitted as well, as ADHO developed a series of procedures and revised governance structure to account for the admittance of new constituent organizations (ADHO, n.d.-b).

In 2012, a special-interest group of ADHO, Global Outlook::Digital Humanities (GO::DH), was formed in response to the ways that the Global North practitioners of the digital humanities—in the United States, Canada, and Europe—had largely positioned themselves as the sum total of the international reach of the digital humanities through ADHO and the international conference it organizes. Initially conceptualized as an organization driven by scholars in the United Kingdom, the United States, and Canada that undertook outreach to DH practitioners in areas not represented within ADHO constituent organizations, GO::DH was launched as a network intended to link practitioners around the world to one another (O'Donnell et al. 2015). With a founding executive board that included DH practitioners in the United States, Canada, Nigeria, Cuba, and Italy, its mission shifted away from the original idea of outreach towards breaking down barriers to collaboration among digital humanists around the world (GO::DH, n.d.).[1]

Early projects supported by GO::DH, such as *Around Digital Humanities in 80 Days* (*Around DH*), were our methods of building transnational relationships. Led by Gil, along with a team of editors (which included me), *Around DH* sought to identify and represent the diverse methods of digital humanities around the world and develop collaborations with practitioners. The primary issue in such a project is that what "counts" as digital humanities can be easily overdetermined by how "digital humanities" is defined by scholars of the Global North. To avoid this, we began by crowdsourcing projects from around the world that blended humanities research and digital methods, regardless of whether they fit the Global North model of the digital humanities or, indeed, whether their creators identified them as "digital humanities" projects. The result was an extensive data set of global digital scholarship. From this crowdsourced list, editors for geographical regions curated entries on the projects that were published daily during the summer of 2014. By promoting self-identification of digital scholarship and intentionally avoiding the dominant ways of defining DH methods, *Around DH* offered an early example of what a transnational DH project could look like.

At various points in the early years of GO::DH, we were encouraged by ADHO to facilitate the organization of DH practitioners in the Global South into national groupings. However, the majority of us on the executive board resisted this idea, arguing that how DH organizations were formed should not be determined by ADHO or by GO::DH. Rather, practitioners themselves were the best poised to recognize whether national, regional, linguistic, or other organizing logics made the most sense for their work and their constituencies. In this way, GO::DH resisted the default logic of nation as an organizing principle, rebuffing ADHO's attempts to organize the world of digital humanities from the top down. Rather, it supported the work of scholars and, instead, emphasized the importance of grassroots organization determined by those who are engaged in the work.

The addition of new organizations to ADHO necessarily posed a threat to the initial power that the Association of Computers and the Humanities and the Association for Literary and Linguistic Computing (now the European Association for Digital Humanities, or EADH), wielded in the formation of the organization. With the primary organization of ADHO originally designed as a steering committee with representatives from constituent organizations, the addition of new organizations introduced new voting members to decision-making processes within ADHO. It is, perhaps, no surprise that the introduction of new constituent organizations led to a reassessment of the organizational structures of ADHO, raising questions about whether each organization should have a single vote, whether ACH and EADH should have additional voting representation due to their status as founding organizations, and whether special-interest groups like GO::DH should have some form of representation as well (ADHO 2016). There was also, always, the implicit concern that the finances of ADHO were undergirded by subscriptions to the journal *Digital Scholarship in the Humanities*, which is linked to EADH. All told, the entire organizational structure of ADHO was revised, with a bipartite structure comprising an executive board and a constituent organization board. The organization board includes representatives from the constituent organizations and establishes policy, strategy, and vision for ADHO, and it appoints the executive board to enact their decisions (ADHO, n.d.-c). It is also, perhaps, no surprise that as the governance structure changed and the two founding organizations found their power and influence diluted by the admission of new organizations, a new crop of nation-based European organizations began making moves toward ADHO membership as well. This could have the effect of further increasing European power within what is ostensibly the United Nations of digital humanities. Indeed, the nation perhaps becomes a recourse for the consolidation of further power for Europe, increased representation for European interests, and a greater concentration of resources in Europe through ADHO's funding model.

REGIONAL AND NATIONAL LOGICS

New forms of organization emerging from the connections forged by GO::DH shed light on the challenges of both regional and national logics in the digital humanities. The case of the now-defunct organization the South Asian Digital

Humanities (SADH) network illustrates the power of the nation as an organizing principle and how it has, in fact, come to wield the power it does in the organizing logics of digital humanities. SADH was organized to promote digital scholarship by scholars in South Asia and abroad, bringing together a community of scholars working on South Asia as a region (Risam 2016). One might argue that SADH was trying to be *too* inclusive, *too* transnational, in its attempt to link scholars in South Asia and scholars of South Asia while also addressing attendant issues of infrastructure, access, and policy. Yet, in light of the history of colonial India before the partition of 1947, the regional SADH is a far more sensible framing mechanism than a national one. Despite the existence of DH practices in other South Asian countries, such as Pakistan and Nepal, India quickly became centred, and the idea of a "South Asian" organization was scrapped. In its place, the Digital Humanities Alliance of India (DHAI) was formed. In 2019, DHAI was registered as an organization and rebranded the Digital Humanities Alliance for Research and Teaching Innovations (DHARTI) to further refine its mission within India.

SADH was in many ways a missed opportunity to develop organizing logics that aren't national, to foster transnational collaboration, and to recognize that the nation itself is a recent construct. There is an implicit tension between the history and cultural heritage of the region, which prior to 1947 was not comprised of multiple nations in the sense of the contemporary nation-state. Moreover, stewardship of cultural heritage requires an approach that is not national but regional and transnational to facilitate the work most adequately. But that is, of course, a key tension in the digital humanities at the ADHO level as well; the predominantly geographic organizing principle has significantly more to do with the location of practitioners than the actual topic of their work. Therefore, SADH was a lost opportunity to imagine a different mode of organizing that could have been a model for fidelity to the scholarship it facilitates, rather than location of participants. One might argue that the transition from SADH to DHAI and DHARTI was a response to the fact that the people best positioned to get funding for digital humanities in South Asia were, in fact, in India, leading to the group's purview becoming de facto national. The nation won again.

DIGITAL HUMANITIES AT THE BORDERLANDS

While the nation-based logic and geographical affinities at the heart of ADHO remain the organizing principle for the digital humanities, it reaffirms—as the case of SADH shows—the limitations of the nation, particularly for facilitating transnational connections for scholarship. This is not to suggest that we should not have increased organization of the digital humanities among nations or regions, particularly those whose DH practitioners do not have representation within ADHO and see a value in it. However, we also need greater attention to alternate modes of organizing. This was the original goal of GO::DH, as a meeting spot that could create south-to-south solidarities and collaborations that could exist beyond and around the operations of ADHO power. However, the interest of many GO::DH members in using the special-interest group to gain individual power in ADHO made that challenging. In fact, recent discussions among board members of GO::DH about becoming a constituent organization itself speaks to the fact that those who have gained control of the special interest group are more interested in its (and their own) status within ADHO as an institution rather than how we originally imagined the group: practitioners from around the world undertaking collaborative projects to make connections and expand our network.

There are examples of larger- and smaller-scale projects that are exploring what digital humanities might look like in the borderlands, beyond the framework of nations. Borderlands are physical spaces where nations meet as well as liminal spaces of hybridity and cultural connection. Gloria Anzaldúa's *Borderlands / La Frontera* (1987) articulates the critical connection between the two: "The U.S.-Mexican border *es una herida abierta* [is an open wound] where the Third World grates against the first and bleeds. And before a scab forms, it haemorrhages again, the lifeblood of two worlds merging to form a third country—a border culture" (25). Digital humanities at the borderlands recognizes the limitations of the nation and offers frameworks for challenging the persistence of the nation. A borderlands approach to digital humanities is being enacted both at the level of digital projects themselves and, increasingly, through efforts to develop transnational DH organizations.

Through digital projects, the border and borderlands—and, thus, the nation—have become sites of inquiry. The Borderlands Archives Cartography project, directed by Sylvia Fernández and Maira Álvarez, is one such example. The project visualizes the geographic locations of newspapers published in the U.S.-Mexico borderlands between the late nineteenth and mid-twentieth centuries. Critically, the project is organized by historical periods, reflecting geographical and political transitions of the border over time and emphasizing that the current border is one incarnation of a shifting landscape of national boundaries. As Fernández and Álvarez (n.d.) note, the project is a response to "the constant and current aggressive, political rhetoric that displays the geographic and ideological border between the United States and Mexico as a threat." They push back against a border defined by political discourse, positioning the borderland as "a space where different cultures co-exist under strong political, economic, and social hegemonies; as well as a space where regions influence each other, but maintain their own identities." In this regard, Fernández and Álvarez's work brings Anzaldúa's insights on tensions between the political and cultural space of the border to bear on digital humanities—not only in content but in method, as their responsive approach to the shifting location of the border over time indicates.

The effort to create the US Latino Digital Humanities Center at the University of Houston is another example of a borderland-driven, transnational approach to digital humanities. Directors Gabriela Baeza Ventura and Carolina Villarroel received a $750,000 grant from the Mellon Foundation to establish the program, based on their previous work running the Recovering the US Hispanic Literary Heritage program at the University of Houston (Arte Público Press, n.d.). Their goal is to facilitate digital scholarship based on the hundreds of thousands of documents authored by Latinos from the colonial period through the mid-twentieth century. Critically, both programs recognized that accurately capturing Latino literary heritage requires an approach that does not delimit itself by nation. Instead, their efforts focus on the borderlands of the United States and Mexico, promoting transnational, multilingual, and transdisciplinary approaches to digital humanities.

The practices that emerge from projects like Borderlands Archives Cartography and from new forms of organization like US Latino Digital Humanities are forerunners of practices that can unsettle the nation. In particular, they challenge the fixity of the border by focusing on the hybrid cultures of the borderlands, on the cultural production of migrants, on the voices of those who are not often represented in the dominant cultures of nations. Their work offers important inspiration for the development of DH practices that resist the overdetermining force of the nation in scholarship through their focus on the hybridity, plurality, and polyvocality of the borderlands. Such examples provide models of how to consciously resist the reinscription of the nation in DH practices while still meeting the important goal of privileging the local over the global. Through the complex forms of representation they make possible, these initiatives demonstrate what we are at risk of losing when centring the national as an organizing principle: the voices of those excluded from dominant culture.

While the DH community has grown significantly through connections made by ADHO, it is critical that attention to the nation does not foster essentialist scholarship or reinforce the dominant cultures of those nations. The nation alone cannot be a solution to the problem of the global and the local in the digital humanities. Rather, the emphasis must be on the multiplicity of practices within diverse communities inside nations and on a better understanding of the borderlands as a framework for producing hybrid knowledges. Indeed, the answer is to not accept nationalism as the answer to the broader problem of diversity and representation in the digital humanities but to recognize that plurality and polyvocality from multiple locales are the only way to realize greater equity and justice in the digital cultural record.

NOTE

1. In full disclosure, I served on the founding executive board of GO::DH and subsequently served as vice-president.

REFERENCES

ADHO (Alliance of Digital Humanities Organizations). 2016. "ADHO Governance Proposals." *Alliance of Digital Humanities Organizations.* https://adho.org/administration/steering/adho-governance-proposals.

———. n.d.-a. "About." *Alliance of Digital Humanities Organizations.* http://adho.org/about.

———. n.d.-b. "Admissions Protocol." *Alliance of Digital Humanities Organizations.* http://adho.org/administration/admissions-committee/admissions-protocol.

———. n.d.-c. "New ADHO Governance and Leadership." *Alliance of Digital Humanities Organizations.* https://adho.org/announcements/2019/new-adho-governance-and-leadership.

Anzaldúa, Gloria. 1987. *Borderlands/La Frontera: The New Mestiza.* San Francisco, Calif.: Aunt Lute Books.

Arte Público Press. n.d. "Digital Humanities." *Recovering the US Hispanic Literary Heritage.* https://artepublicopress.com/digital-humanities/.

Barrett, Paul. 2014. "Where Is the Nation in Postcolonial Digital Humanities?" *Postcolonial Digital Humanities,* January 20, 2014. https://dhpoco.org/blog/2014/01/20/where-is-the-nation-in-postcolonial-digital-humanities/.

Fernández, Sylvia, and Maira Álvarez. n.d. "About." *Borderlands Archives Cartography.* https://www.bacartography.org/autoras.

GO::DH (Global Outlook::Digital Humanities). n.d. "About." *Global Outlook::Digital Humanities.* http://www.globaloutlookdh.org/.

O'Donnell, Daniel, Katherine L. Walter, Alex Gil, and Neil Fraistat. 2015. "Only Connect: The Globalization of the Digital Humanities." In *A New Companion* to *Digital Humanities*, edited by Susan Schreibman, Ray Siemens, and John Unsworth, 493–510. Malden, Mass.: Wiley. https://doi.org/10.1002/9781118680605.ch34.

Risam, Roopika. 2016. "Diasporizing the Digital Humanities: Displacing the Center and Periphery." *International Journal of E-Politics* 7, no. 3: 65–78. https://doi.org/10.4018/IJEP.2016070105.

———. 2018. *New Digital Worlds: Postcolonial Digital Humanities in Theory, Praxis, and Pedagogy*. Evanston, Ill.: Northwestern University Press.

CHAPTER 3

Rerouting Digital (Humanities) Scholarship in Canada

Andrea Zeffiro

reroute
verb
uk / ˌriː'ruːt/ us / ˌriː'raʊt/
   to change the route of something

In their introduction to this volume, Sarah Roger and Paul Barrett reflect on how "locating DH practices within particular places and national frameworks—even in opposition to those frameworks—allows us to pose new questions." The editors invite us to question how we think about the digital humanities, and also to trouble precisely *how* we think about it and to veer (i.e., reroute) from established histories, practices, and critiques toward *future horizons*.

My argument in this chapter responds by offering a series of provocations about the digital humanities in Canada, although the body of scholarship cannot be cleanly delimited from other national contexts, even when the organizing frame of analysis is a national one. Given this, my piece is in conversation with American DH scholarship, yet my discussion is rooted (or routed) in a Canadian context.

This perspective is also rooted in my own experience working in the Canadian academy in an administrative capacity. Since 2015, I have served as academic director for the Sherman Centre for Digital Scholarship, at McMaster University. Established in 2012, the Sherman Centre is a notable node in the Canadian DH network because it was the first digital scholarship centre to reside in a university library.[1] The Centre, as a place and as a community, is an always-present reference point throughout this piece because it has informed so much of my thinking about digital (humanities) scholarship. Yet what is reflected here is a culmination of observations I have been mulling over for the past 15 years, which is to signal how these issues and concerns extend beyond DH scholarship and envelop other contiguous fields like communication and media studies and the media arts. I employ the term "routing" as a trope through which we can conjure the digital humanities in Canada. To reroute means to change the direction of something; it insinuates, and possibly even encourages digressions and deviations. The provocations in this chapter are contained but tangential and seek to reroute readers through thinking about digital (humanities) scholarship that moves away from digital humanities specifically. Rather, each section—shortcut, detour, divergence, roadwork, and service road—ruminates on digital (humanities) scholarship in its current state but also as something that does not yet exist. Through the figurative lens of "rerouting," I abandon a central course of digital humanities, veer away from it, and come back again in order to speculate on the possibilities for what exists beyond and adjacent to digital (humanities) scholarship in Canada.

SHORTCUT

My experience of/in digital humanities is peculiar, I think, because it is channelled mostly through administration. Although my research trajectory bears a resemblance to DH research, it is the work I do in facilitating, supporting, championing, and administering digital humanities that I would claim as digital (humanities) scholarship. As a faculty member with a cross appointment with the library, my understanding of how to do research as gleaned through traditional scholarly contexts is *always* in negotiation with library practices and protocols, especially around notions of "service." I have found it challenging at times to deflect my perception of research—or what constitutes research—and submit to a kind of service ethic. I work in the service of digital (humanities) scholarship. This chapter, in part, considers how administrative labour is a "detour" from the proper work of scholarship. Service work in the academy is notoriously asymmetrical, with scholars from equity-seeking and -deserving groups experiencing heavier service burdens and spending a disproportionate amount of their time on the invisible work of sustaining academia (Brown 2018). Working in the library has reshaped my perception of labour conditions and precarity in academia. I am acutely aware of how my understanding of labour in an academic context is shaped by my experiences as a graduate student, postdoc, contract faculty, and, finally, as a tenure-track faculty member. And I carry with me a sensitivity to precarity, the mechanisms through which entitlement and privilege play out through power imbalances, and the disproportionate allocation of resources across academic units that exacerbates divisions of labour.

How do we do the work of digital humanities? For many years my answer to this question went something like "the digital humanities are collaborative because research happens when people with different skills and expertise work together." It works as a basic definition for a facet of digital (humanities) scholarship. Recently, however, I have started to think about "collaboration"—how the term is used to qualify partnerships across the university, but also how it can obfuscate the labour of producing DH research. In this respect, I am continuously assessing how we do the work of DH scholarship in relation to attribution and in/visible labour. How does service—and servitude—sustain the digital humanities? How do we do better at rendering transparent individual contributions that support and sustain DH initiatives? What are other forms of (non-academic) labour and labouring bodies that sustain digital (humanities) scholarship? Rerouting digital humanities requires diversions from well-trodden paths that replicate a white, masculine, middle-class, and Western bias in knowledge production (McPherson 2012), and that fetishize computational social-science methods that similarly normalize "gentlemanly" technical expertise (Savage 2013, 18) and continue to valorize the traditionally more prized work of intellectual expression.

My understanding of the digital humanities as a broader field is deeply informed by the local assemblages and the unique concerns that shape and sway digital (humanities) scholarship at the local level (Risam 2016). However, I am also acutely aware of what Roopika Risam identifies as the local-global quandary within digital humanities. One cannot make general assessments based on one's local context, especially those of us doing digital humanities in the Global North. Who are we to define digital humanities writ large?

Throughout this chapter I refer to digital humanities, DH, digital scholarship, and digital (humanities) scholarship somewhat interchangeably, but also with discretion. My play on digital (humanities) scholarship encompasses my day-to-day relationship to digital scholarship and to the digital humanities. I tend to favour the term "digital scholarship," not simply because of the work I do at the Sherman Centre but also because it is broad enough to envelop contemporary forms of scholarly inquiry that integrate digital methods and computational approaches in research and teaching. In the context of the Sherman Centre, I view the digital humanities as digital scholarship, hence digital (humanities) scholarship, but the digital humanities, as I argue, designates a scholarly field marked by particular research trends, interests, organizations, and events. That said, I am also wholly uninterested in any kind of project aimed at delimiting research contexts or cementing definitions. I would much rather leave it up to individuals to self-identify.

I arrived at digital humanities rather circuitously, perhaps even by accident, and most certainly with some resistance. I find myself still navigating contradictory and conflicting states in my positioning within/alongside/against digital (humanities) scholarship. In part, this tension has to do with trying to reconcile what feels like a split identity. In my administrative role, I am often advocating for the integration of computational tools into research and teaching and championing the ways in which these new approaches contribute to an expanding array of methods and sources of evidence. However, in my research and teaching, I am also wary of the ways in which many of the same tools and methods exacerbate inequities in the production of new knowledge about what sorts of research attract funding, about the inequitable distribution of resources across departments, and about how questions of power, equity, and positionalities are sidelined for tools training across multiple facets of collective life. This chapter is deeply immersed in the rhythms of navigating these contradictory stances and asks if it is possible to have both positive and negative sentiments about digital (humanities) scholarship?

I have come to explore this feeling of mis/alignment in relation to what Tania Bucher (2019) describes as ambivalence. As Bucher writes:

> Far from being agreeable or a cop-out, the ambivalent position means having to negotiate an ongoing tension without necessarily finding resolution. The kind of ambivalence I have in mind is not about occupying an indifferent position. It's not an "anything goes" attitude, nor does it involve compromise. Ambivalence isn't a lack of belief, but rather the ability to "stay with the trouble" of questioning basic assumptions and to be transparent about them. (3)

Bucher argues for the virtue of ambivalence in thinking about digital technologies because it allows us to perceive and comprehend conflicting and contradictory things. As an alternative mode of critical positionality, ambivalence makes it possible to observe the relative strengths and weaknesses of a range of positions, activities, and engagements within the digital humanities by making it impossible to ignore the infrastructure and hidden labour needed to support digital (humanities) scholarship.

In what follows, I practise ambivalence by assessing some of my basic assumptions about digital (humanities) scholarship in broad terms, but also channelled through my local DH context, and I do so by being transparent about my impressions. I "mind the gaps" (Brown 2011) between the celebratory proclamations about how the digital will amplify humanities research and the critical assessments of how new computation tools and sources of evidence produce new dilemmas. What does it mean to be ambivalent about digital (humanities) scholarship? Can one be ambivalent and still care? Can we feel ambivalence toward both the things we are concerned about and also the things we are most hopeful about? Who gets to inhabit or claim ambivalence? Is ambivalence a state to aspire to? Can we remain critical of some practices, agendas, or trajectories in digital (humanities) scholarship in Canada while also continuing to be optimistic about others?

DETOUR

Ambivalence, the capacity to see competing versions of things at once, can enable us to take stock of the existing forms of digital humanities, and the ways in which they can be said to exist, and how some versions of digital humanities become dominant and thus obscure others. Do (or does) the digital humanities exist? This question is by no means a call for performing an inventory of the field, nor is it a plea to readjust the definitional enclosures of DH scholarship. As we already know, both are futile exercises. And besides, as this anthology demonstrates so vividly, the digital humanities in Canada are unbound, dynamic, and multidirectional. In this respect, the digital humanities are alive and well. Rather, I pose the question rhetorically to disentangle the digital humanities as a field of cultural production (Bourdieu 1993). If we consider the digital humanities as a "site of struggles in which what is at stake is the power to impose the dominant definition" in order to "delimit the population of those entitled to take part in the struggle to define" it (Bourdieu 1994, 42), then we might begin to envision and destabilize what the digital humanities means and for whom. We can trace any number of genealogies that anchor the digital humanities in particular geographical contexts and scholarly disciplines. Recuperating the roots (or routes) of the digital humanities in Canada could mean that scholars acquire a sense of how their field of focus came into being. Origin stories anchor us to a thing and to a re/imagined group of people and allow us to have a sense of belonging. History creates community. But history privileges particular worldviews and approaches to knowledge production and preservation strategies. History excludes. By drawing out the tensions and contradictions of digital (humanities) scholarship, we can recognize how digital humanities is manufactured by struggles over the power to assert meaning. A scholarly field like digital humanities is not a definitive trajectory that moves from point A to point B. Instead, if we approach digital humanities as undisciplined and evolving, we can find ways to move through it, beside it, and against it. Rather than seek to resolve tensions, we embrace provocations.

Indeed, the digital humanities have long existed through the under-recognized historic use of technologies within the humanities in Canada before digital humanities was acknowledged more officially (Bowness 2013). We can look to established scholars who have worked in the digital humanities long before the digital humanities was the digital humanities, when the convergence between humanities research and emerging technologies was sometimes referred to as "humanities computing." Matthew Kirschenbaum (2012) and Adeline Koh (2015) have mapped the connection between digital humanities and humanities computing, citing the latter as an antecedent to the former. Both Koh and Kirschenbaum note the similarities carried forward from humanities computing, but even more crucially, they recognize how digital humanities evolved with more critical inflection (Smith 2007). To paraphrase Koh, digital humanities has encouraged scholars to delve into the theoretical underpinnings and social consequences of computational methods. And even then, digital (humanities) scholarship, like a lot of other academic domains, can do more still to engage with the ways technology mediates and re/produces race, class, gender, and sexuality (Gold and Klein 2018; Earhart 2012) by drawing out the politics and norms embedded within the hardware and software that sustain digital (humanities) scholarship.

The relationship between digital humanities and humanities computing—and digital scholarship—is accentuated here to underscore the emphasis on technology as an appendage to the humanities. Technology as an adjunct to humanities research insinuates that the union between the humanities as a field of study and technology as a tool in research is new. Even though we know it is not, we continue to perpetuate the myth, even more so now that universities in Canada are increasingly pressured to demonstrate direct links to employability while being confronted by the threat of performance-based outcomes for funding. Wendi Hui Kyong Chun and Lisa Marie Rhody (2014) have urged us to appraise "the general euphoria surrounding technology and education" by taking aim not at the humanities per se but, rather, "at the larger project of rewriting political and pedagogical problems into technological ones, into problems that technology can fix" (3). Indeed, improving our approaches to digital pedagogy and thoughtfully integrating classroom learning technologies into our teaching environments are possible ways to mitigate the pressures of needing to demonstrate "real world" applicability. As Safiya Noble (2019) reminds us, the alleged neutral stance asserted by many in information studies and digital humanities is demonstrative of how colonization is a process marked by forgetting: "It is through this stance of not being engaged with the Western colonial past," writes

Noble, "a past that has never ended, that we perpetuate digital media practices that exploit the labor of people of color, as well as the environment. If ever there were a place for digital humanists to engage and critique, it is at the intersection of neocolonial investments in information, communication, and technology infrastructures: investments that rest precariously on colonial history, past and present" (1).

What good is it to teach digital tools and approaches without balancing technology-enhanced learning with critical digital literacies engaged in drawing out colonial history, past and present? If students remain untrained to reflect on the approaches they use in consuming, producing, and sharing new knowledge, then we are reproducing models that privilege the power and mystique of the black box (Pinch 1992). Perhaps this is even more urgent given the dependency on technology companies during the global pandemic. The emergency shift to remote learning left no time to consider the long-term exploitative effects of relying on proprietary platforms and software for everything from delivering lectures to proctoring exams. Many educational technologies like Blackboard, Echo360, and Kaltura use Amazon Web Services, the same cloud computing platform used by intermediary companies like Palantir, which develops AI technologies for militaries and police departments.

We are taxed to ask questions about our technocentric humanities projects, whether categorized as digital humanities or not. What digital tools, approaches, and sources of evidence do we integrate into our research and teaching? How do these tools, approaches, and sources of evidence shape the production and representation of knowledge? Who will benefit from this knowledge? In what ways do we seek to draw out the contradictions and tensions of working with these tools, approaches, and sources of evidence? Who is excluded and how? How do our choices in hardware, software, and sources of evidence intersect with, reject, or critique issues of in/equity and in/justice? Where do we go from here? And who is issued a licence to move us along?

DIVERGENCE

My formative understanding of digital humanities came from a decade (2005–2015) of working within collaborative and interdisciplinary research networks engaged in the creation of technological artifacts and immersive experiences. These were large-scale research teams composed of media artists, interaction designers, software engineers, computer scientists, ethnographers, neuroscientists, and medical doctors. Research happened in physical "lab" environments, with projects spanning multiple institutions and partners. My experiences in these interdisciplinary research labs informed my understanding of academic research as an inherently collaborative and creative endeavour. However, these research configurations are not without challenges. The lab model forces us to explicitly raise the question of how research is practised and how research is supported differently in institutional contexts, which is rarely explicit in humanities work. Coming from a lab model into a centre *for* research, I have had to readjust my expectations for how a research entity like the Sherman Centre establishes a relationship to research activities. The prepositions "of" and "for" are important to bring into the fold. The Sherman Centre, for instance, is a centre *for* research; it is in support of or in service of research. Whereas a research entity that is *of* research—like a lab—is organized primarily around a research agenda that is often driven by a researcher's profile. Both models expose and centre the mode and process of collaboratively building ideas rather than just the built result of knowledge or singular expertise.

The humanities lab model is thriving because humanities research has been changing for almost forty years. The lab model responds to these transformations by effectively supporting interdisciplinary and collaborative research and organizing shared use of digital tools and methods (Arac 1997; Davidson 1999; Svenson 2016). However, the term "lab" in the context of humanities research is almost as contentious as the word "digital." Some scholars view the emergence of the lab model as the "scientification of the humanities" (Gottschall 2008). Particularly as the lab model emerged in part from the pressures imposed by the neo-liberal university, where "productivity,' 'economic efficiency,' and delivering 'value for money'" are dominant organizing principles (Shore 2008). A lab can serve as a measurement of research productivity and responds well to funding programs and performance metrics. For these reasons, a humanities lab can mimic the entrepreneurial impulse of the incubator model from Big Tech. This is not to suggest that humanities labs ignore innovative and inventive models. In fact, because much of the work we are increasingly asked to do as researchers and educators is to "train" students to acquire research skills to take with them into professional contexts, it makes sense if these experiential learning environments understand industry. All the same, if we are appropriating a model and training students in its image, it becomes necessary to question whom that model serves and how, and to find ways to make transparent how that model shapes the production of knowledge both within and outside the university. In my experience, the lab was for the most part a generative space that supported research collaborations that were defined by a particular project and granting cycle. As a doctoral and then as a postdoctoral student, it was in those research contexts that I was able to identify my strengths in large-scale project management and interdisciplinary research. I was also able to identify my boundaries as a collaborator and recognize the models for scholarship and collaboration that I was unwilling to perpetuate.

As Urszula Pawlicka-Deger (2019) has shown, the lab model materialized as a crucial element in humanities infrastructure to support new collaborative practices and methods at the intersection of digital technologies and tools. In this respect, the humanities lab has also come to signal a new way to engage research and knowledge production. These labs engage extensively with communities outside of the university and thoughtfully share resources. And even when a particular individual's research agenda directs research, the process itself is collaborative and funding is directed toward graduate and undergraduate training and professionalization, and community engagement.

Charting the evolution of the humanities lab in two distinct phases, from 1983–2010 to 2010–2018, Pawlicka-Deger (2019) suggests we move one step further in our configuration of labs as humanities infrastructure and consider how a lab can serve as an infrastructure for public engagement. Citing Humanities for All, an initiative of the National Humanities Alliance Foundation that showcases over 1,500 publicly engaged humanities initiatives, Pawlicka-Deger describes infrastructures for engagement as "structures that support engaged scholarship, including degree programs, centers, funding opportunities, digital technologies, and curriculum reorientation initiatives" (Humanities for All 2019). Pawlicka-Deger calls on researchers to envision a research model in the humanities that is a catalyst of intervention. In the final section of this chapter, "Service Road," I expand on the possibility offered by reading digital humanities through the concept of "infrastructure" to imagine how we might bake justice into (digital) humanities scholarship by interrogating what, how, and why we do what we do and for whom. My formative experiences in interdisciplinary research labs allowed me to navigate the "research centre" context to a certain extent. More than anything, those experiences have enabled me to consider the subtle distinction between spaces *of* research and spaces *for* research; however, the lab model I had experienced was not homogenous with a centre *for* research.

In a lab model that is *of research*, there is symmetry between a faculty member's research profile, their lab, and the kind of collaborations and programming activities that are carried out by it. When these labs function as infrastructures for engagement, activities and resources are shared in numerous ways, and research and administrative practices are deliberately and painstakingly transparent and deliberative. I am thinking of the Technoscience Research Unit at the University of Toronto, the Environmental Media Lab at the University of Calgary, the Civic Laboratory for Environmental Action Research at Memorial University, and the THINC Lab at the University of Guelph. These research units are stewarded by scholars who are deeply committed to feminist, queer, anti-racist, and anti-colonial ways of knowing, producing, and engaging in research and community organizing. Indeed, these are the models we want to learn from and emulate.

Comparably, the *for-research* model also offers training opportunities and engages in research collaborations; however, this model centres these resources across disciplines and hierarchies (Moritz et al. 2017). The *for-research* model is focused primarily on democratizing access to expertise, technological and knowledge infrastructure, and is committed to equitable and sustainable information dissemination and resource distribution for those seeking these things (Moritz et al. 2017). In Canada it is increasingly common to find these centres or units *for* digital (humanities) scholarship in university libraries.

As the first digital scholarship centre residing in a university library, the Sherman Centre has been in a unique position to reveal some of the dynamics at play in collaborations and how different forms of human infrastructure are made more or less visible in the academy. Since then, the centre has become a hub through which expertise and resources are shared with the campus community who are seeking to "do more with digital scholarship." The centre provides consulting, instruction, and technical support to faculty, staff, and students on any stage or aspect of a digital research or pedagogical project to help determine the approaches that best suit the project, whether big or small. Core programming includes multiple workshop series, a graduate residency program, a scholar-in-residency program, a visiting-scholar series, and an undergraduate course. The centre's physical resources include technical infrastructure, like high-performance workstations and a mini maker space, office space for the centre's staff, and shared meeting and workspace for library staff, faculty, and graduate students. The research infrastructure supported by the centre includes some physical and technical resources, but the focus is on shared knowledge and expertise and access to digital research services. Across all these collaborations and partnerships, Sherman Centre's staff play a vital un/official role as consultants and technical support (Poole and Garwood 2018). The centre is not just a space, nor a cache of resources. It is defined by the individuals constituting it. It is worthwhile to think about the centre, insofar as it holds all these various positions and activities, as a site of ambivalent relationship to and perspective on academe. The fact that the centre is situated within the library—a site of fluid, open, multidisciplinary inquiry that takes place alongside the traditional material means of research—is relevant to how we continue to perceive the library and academic labour. The centre supports a range of types of university work—service, administrative, professoriate, research—all of which are treated as being equally meaningful. I return to the complexities associated with the in/visbility of human infrastructure in digital (humanities) scholarship in the next section.

Following the Sherman Centre, several other Canadian institutions have since established similar units in their libraries, including the University of Alberta, the University of Victoria, the University of Saskatchewan, and Brock University. Indeed, this list of DH research centres, labs, and programs that exist in Canada is not exhaustive because my point is to stress the affinity between digital (humanities) scholarship and the library, as Vandegrift and Varner showed in 2015. As Spiro (2012) demonstrated, the digital humanities share many of the values of the library community: interdisciplinarity, openness, and collaboration. As Miriam Posner reminds us in a 2012 piece, "DH was being done in the library (and in the archive) well before it made its way into academic departments." Moreover, libraries and librarians continue to play a pivotal role in digital (humanities) scholarship. Reflecting on the U.S. context, Hswe and Varner explain how librarians were enthusiastic partners in early DH projects. "We have been such valuable collaborators over the years," they assert, "because the values of librarianship inform a deep interest in information access, a concern for information preservation, and a desire to make room for our diverse user communities" (Varner and Hswe 2016). Institutionally and professionally, the library is rooted in a deeply embedded practice of community building. Research communities can turn to the library not simply to resource or service digital (humanities) scholarship but also to take the lead in DH initiatives.

It is necessary to turn from how research is currently organized to imagine how we *could* organize these resources. The library is indispensable to any kind of re/envisioning of digital (humanities) scholarship. I am not suggesting that digital (humanities) scholarship happens exclusively within the library. I am advocating for a research infrastructure where DH resources and services are made accessible to anyone regardless of the focus of their research project, level of expertise, or discipline, and how these resources and services could be organized, managed, and sustained through the library.

How might digital (humanities) scholarship be not merely preserved but also provoked if the library stewarded its resource infrastructure? How might such a re/organization of resources enable the digital humanities to develop in divergent directions? Rather than corral digital (humanities) scholarship into a neatline (pun intended) we should aspire for divergence. Imagine, then, a campus with researchers equipped with labs and engaged in digital (humanities) scholarship reflective of their research agendas. Add to this research environment a central campus unit whose mandate is to evenly distribute a plethora of free DH resources and services to support researchers and student projects, engage in community-driven initiatives, and co-create, collaborate, and partner with the research community. Imagine, if you will, the infrastructure for engagement university campuses could sustain to educate students in experiential environments that reference industry, but more importantly, offer counterpoints to technology as a panacea for social and economic challenges. Engaging with complex issues demands ongoing collaborative and interdisciplinary partnerships that may or may not include digital approaches and computational resources.

ROADWORKS

A useful analogy for the hidden labour that sustains DH scholarship might be found in roadwork, the construction or repair work done to roads or utilities in proximity to transportation routes. Often, this work is done in the open and is therefore visible, albeit fleetingly, to those passing by. At other times, the bulk of the work happens discreetly, sometimes at night, when roads are less travelled. In these instances, the work can happen without appearing to happen at all. Roadwork happens both openly and in the background; it is work that repairs or augments infrastructure, but often without change to what is already there. I want to draw out the tension of in/visibility to explore how those of us who support digital (humanities) scholarship are visible but most often we direct, incite, consult, facilitate, manage, and observe in the background. This section provokes by asking, what does a sustainable and ethical labour model (Moritz et al. 2017) for digital (humanities) scholarship look like?

At the Sherman Centre, we offer digital scholarship services that are active forms of collaborative labour. We *consult* on grant applications by helping researchers refine their methodologies to include digital approaches. We *advise* those wishing to include digital projects in their courses by teaching tools and co-facilitating assignments. We *mentor* graduate students who are seeking to integrate digital methods and approaches into their research and require expertise outside of their department. We *teach* workshops that introduce the campus community to digital scholarship approaches. We *train* researchers and their research team in digital approaches to get projects off the ground. Four staff members constitute the Sherman Centre, and we can provide these resources because we can draw on the expertise and services from other units in the library.[2] Over the years, library staff external to the Sherman Centre have participated in many of the activities I list above, and we want to accentuate the connection publicly for them to claim as well. To that end, we formalized an affiliates program. Library staff affiliates contribute to the workshop series and participate in consultation meetings. Again, these activities were happening before, but organizing these collaborations in a formal way recognizes individual contributions and places the onus on the centre to make transparent the in/visible labour that sustains it.

In thinking through questions of labour/ing in digital (humanities) scholarship spaces, specifically within the library, I rely on the important work of others who have shared their reflections on the entanglements of individual acts that constitute DH activities. Bethany Nowviskie's now seminal reflection "A Skunk in the Library" (2011) considers how one of the most appealing qualities of library culture—its service ethic—obfuscates much of the work in/of the library and consigns library staff to the sidelines when it comes to collaborating on digital research and pedagogical projects. In turn, librarians are restricted from participating as true intellectual partners. Nowviskie's observations on the restrictions imposed by the library's organizational service mentality were echoed recently in a 2019 essay by Bobby Smiley, who notes that when the librarian relinquishes input in the shaping of a research project it relegates them to only handling technical support. Smiley urges us to recover the ways in which librarians have been—and continue to be—intrinsic in digital humanities *as* digital humanists.

What stands out for me in Nowiskie's and Smiley's evaluations of the library's role in scholarly research is the need to better communicate that role. In part, it demands a divergence in the perception of the library from a purely transactional or service model to one of a full collaborator and partner in research and teaching, as Christopher Millson-Martula and Kevin Gunn (2017) have shown. Even then, how would collaboration evolve in practice if our collective understanding of the library shifts from fleeting and extractive encounters to more prolonged engagement? For instance, in "Miracle Workers," a paper presented at an Alliance of Digital Humanities Organizations conference, Posner (2018) reminds us how a reconfiguration of the library's role in research can slide into what Alex Gil has called the "miracle work/er." The shift from viewing the librarian as technical support to project collaborator could mean that the librarian takes on more responsibility without adjusting the level of resources, institutional support, and compensation. This would run the risk of using new words to describe existing configurations of in/visible labour.

Complicating matters further is the issue of how we can reconfigure perceptions of the library when the makeup of that community is also evolving. Centres for digital (humanities) scholarship, for instance, integrate postdoctoral fellows and faculty as staff. In her 2018 paper, Carrie Johnston described this configuration as "flipped mentorship," a situation in which an early-career scholar is responsible for supporting multiple digital scholarship endeavours that are independent of their research and advising more advanced scholars in their research and teaching. Giving a name to an institutional relationship or practice means that we can better assess how to make the interaction work for all those involved and find ways to make collaborative configurations less amorphous. Naming things that were previously unnamed is resonant with Bourdieu's understanding of cultural production as a site of struggle and the framing of ambivalent positioning within academia: holding ambivalence might result in seeing multiple things at the same time, thereby allowing relationships to be named and recognized in a more transparent and equitable way.

In addition to drawing out the conditions of in/visible labour in our professional contexts, we need to ask those same questions about the in/visible labour that supports digital (humanities) scholarship, as a number of scholars have discussed (Gillespie 2018; Hicks 2018; Nakamura 2014; Roberts 2019; Noble 2019; Zeffiro 2020). Who and what contributes to the material sustenance of our research practices and processes of knowledge production? How are these labouring bodies implicated in our research endeavours? How might we do a better job at surfacing and foregrounding in/visible labouring bodies? Those of us who do the work of digital scholarship need to refocus our attention to the invisible supply chain that fuels digital culture writ large and draw out the material conditions of oppression, or we risk continuously reproducing processes and structures of hegemony, imperialism, and power (Risam 2019; Zeffiro 2020).

SERVICE ROAD

Paola Ricaurte Quijano (2018) has written about how, in Latin America, the digital humanities—in its capacity to offer alternatives to dominant systems of knowledge production—can recuperate some of the collaborative activities and creative engagements common to citizen laboratories. In particular, Quijano considers how the digital humanities can help cultivate research spaces attentive to collaborative formations that support open and innovative forms of knowledge production. Quijano's articulation of the digital humanities' capacity to facilitate new models for collaborative scholarship provides an opportunity to consider what such a model might look like in Canadian spaces of/for research. Moreover, Quijano's assessment of a local DH context might allow us to contemplate the forms of engagement we care most about recovering within our own locales.

We may look to the model Quijano describes, one in which newness and big-ness become unsustainable features rather than celebrated perks. In order for alternative models for digital (humanities) scholarship to flourish, institutions will need to reframe models of/for research that fetishize "big" technological infrastructure. As a starting point, we need to undermine the value we ascribe to expansion or, more fittingly, what Anna Lowenhaupt Tsing describes in a 2012 study as scalability, which is "the ability to expand—and expand, and expand—without rethinking basic elements" (505). Tsing tracks our enchantment with scalability in research contexts, describing how we have come to rely on and reward scalability. We see value in projects that can become bigger without hav-ing to change the frame. In turn, Tsing advocates for a "nonscalability theory that pays attention to the mounting pile of ruins that scalability leaves behind" (506). Non-scalability is a counterweight to scalability.

In the context of digital (humanities), we can challenge the assumption that bigger is always better. Non-scalability is an opportunity to co-create with oth-ers in ways that are thoughtful and attentive to sustainability (the environmental impact of software and hardware), equity (who is being asked to do what and by whom, and how are these relationships recognized?), deliberation (who has an opportunity to shape the project and in what ways?), and temporality (can we slow aspects of the research process in order to engage meaningfully with all those we anticipate will collaborate in the work?). As I intimated above, one possible way to describe all the services and technologies available to us is as an infrastructure. What do we mean by infrastructure? How do we envision such an organizational structure for the digital humanities?

In the spirit of non-scalability, I titled this section "Service Road" to invoke those passages that are most often adjacent to highways and allow for local traffic to gain access to properties. How might a detour to the service road disrupt and destabilize the processes through which scalability has come to hold the prom-ise of transformation of the humanities? What if we reroute digital (humanities) scholarship through the service roads? What if we focused on building infra-structures that are less like highways and more like service roads?

Indeed, as Risam affirms, "the centers of digital humanities produce their own margins" (2018, 6). How can centres (i.e., DH trends and space of/for research) recover alternative models for digital (humanities) scholarship? How can they dislodge long-standing regressive assumptions about the relationship between technology and humanities research and pedagogy? How can we engage with and critique, through our research and teaching, the ways in which our local and national economies of exploitation (past and present) link to global information, communication, and technology infrastructures?

Many of us are already well versed in the warranted critiques leveraged at the digital humanities of the Global North. Even as I write this, I am wondering how much of my commentary merely echoes those critiques without doing the work of offering meaningful alternatives. Perhaps continuous efforts aimed at inciting provocations, destabilizing seemingly stable agendas, and challenging neutral stances are in effect rerouting digital (humanities) scholarship. It is more that the larger project aimed at transforming the digital humanities is never complete because the field itself is continuously evolving. What aspects of digital (humanities) scholarship requires decentring? We share a responsibility to continuously decentre and reroute our ways of doing, knowing, and critiquing.

NOTES

1. I wish to acknowledge the following individuals whose visioning, energy, and creativity have shaped the Sherman Centre since its founding in 2012: Dale Askey, Matthew Davis, Jeff Demaine, Blake Dillon, Melissa Elliot, Danica Evering, John Fink, Michael Gallant, Cathy Grisé, Emily Goodwin, Myron Groover, Sil Hamilton, Katie Harding, Christine Homuth, Vivek Jadon, Krista Jamieson, Mica Jorgenson, Simran Kaur, Isaac Kinley, Sandra LaPointe, Debbie Lawlor, Allison Leanage, Veronica Litt, Matt McCollow, Sam McEwan, Gabriela Mircea, Devon Mordell, Paige Morgan, SM Mukkaram Nainar, Chris Myhr, Gil Niessen, Clementine Oberst, Isaac Pratt, Olga Perkovic, Alexander Schaaf, Subhanya Sivajothy, Vange Holtz Schramek, and Sarah Whitwell.

   Additionally, I wish to thank the following individuals who were part of the Sherman Centre's graduate residency program 2014–2022, and whose work has challenged and shaped the contours of digital (humanities) scholarship at McMaster University and beyond: Cameron Anderson, Emma Croll-Baehre, Marley Beach, Mark Belan, Helen Benny, Maddie Brockbank, Raquel Burgess, Alexis Carlota-Cochrane, Samantha Clarke, Linzey Corridon, Deena Abul Fottouh, Katherine Eaton, Jantina Ellens, Duygu Ertemin, Kristine Germann, Hayley Goodchild, Emily Goodwin, Samantha Stevens-Hall, Rudaina Hamed, Chris Handy, Arun Jacob, Adan Jerreat-Poole, Shaila Jamal, Michael Johnson, Mica Jorgenson, Melda Coskun Karadag, Theresa Kenney, Kellen Kurschinski, Kelsey Leonard, Melissa Marie (emmy) Legge, Angelo Mateo, Marrissa Mathews, Adrianna Michell, Brianne Morgan, Luis Navarro, Gloria Park, Jeremy Parsons, Sarah Paust, Shalen Prado, Akacia Propst, Channah Fonseca-Quezada, Jess Rauchberg, Mackenzie Salt, Daniel Schmidtke, Melodie Song, Bryor Snefjella, Stephen Surlin, Joann Varickanickal, Sarah Whitwell, and Tina Wilson.

2. At the time of writing this chapter, the Sherman Centre consisted of four staff members, myself included. An environmental scan of digital scholarship and DH centres performed in July 2020 indicated a trend toward larger staffing complements with expertise in research data management, data visualization, geospatial and statistical data, and bibliometrics and research impact analysis. With the support and guidance from the university librarian, Vivian Lewis, and the library's leadership group, the Sherman Centre underwent a reorganization. By early fall 2021, some existing library staff were reassigned to the centre, along with the addition of new staff positions.

REFERENCES

Arac, Jonathan. 1997. "Shop window or laboratory: Collection, collaboration, and the humanities." In *The Politics of Research*, edited by E. Ann Kaplan and George Levine, 116–26. New Brunswick, N.J.: Rutgers University Press.

Bourdieu, Pierre. 1993. *The Field of Cultural Production*. Edited by R. Johnson. Oxford, U.K.: Polity Press.

———. 1994. *Language and Symbolic Power*. Oxford, U.K.: Polity Press.

Bowness, Suzanne. 2013. "Parsing the Humanities." *University Affairs*, March 13. https://www.universityaffairs.ca/features/feature-article/parsing-the-digital-humanities/.

Brown, Susan. 2011. "Don't Mind the Gap: Evolving Digital Modes of Scholarly Production Across the Digital-Humanities Divide." In *The Culture of the Humanities*, edited by Daniel Coleman and Smaro Kamboureli, 203–31. Edmonton: University of Alberta Press.

———. 2018. "Delivery Service: Gender and the Political Unconscious of Digital Humanities." In *Bodies of Information: Intersectional Feminism and the Digital Humanities*, edited by Elizabeth Losh and Jacqueline Wernimont. 261–86. Minneapolis: University of Minnesota Press.

Bucher, Taina. 2019. "Bad Guys and Bag Ladies: On the Politics of Polemics and the Promise of Ambivalence." *Social Media + Society*. https://doi.org/10.1177/2056305119856705.

Chun, Wendi Hui Kyong, and Lisa Marie Rhody. 2014. "Working the Digital Humanities: Uncovering Shadows between the Dark and the Light." *differences* 25, no. 1: 1–25. https://doi.org/10.1215/10407391-2419985.

Davidson, Cathy. 1999. "What if Scholars in the Humanities Worked Together, in a Lab?" *Chronicle of Higher Education*. https://www.chronicle.com/article/What-If-Scholars-in-the/24009.

Earhart, Amy E. 2012. "Can Information be Unfettered? Race and the New Digital Humanities Canon." In *Debates in the Digital Humanities*, edited by Matthew K. Gold. Minneapolis: University of Minnesota Press. https://dhdebates.

gc.cuny.edu/read/untitled-88c11800-9446-469b-a3be-3fdb36bfbd1e/section/
cfoafo4d-73e3-4738-98d9-74c1ae3534e5#ch18.

Gillespie, Tarleton. 2018. *Custodians of the Internet: Platforms, Content Moderation, and the Hidden Decisions that Shape Social Media*. New Haven, Conn.: Yale University Press.

Gold, Matthew K., and Lauren F. Klein. 2019. Introduction to *Debates in the Digital Humanities 2019*, edited by Matthew K. Gold and Lauren F. Klein. Minneapolis: University of Minnesota Press. https://dhdebates.gc.cuny.edu/read/untitled-f2acf72c-a469-49d8-be35-67f9ac1e3a60/section/ocd11777-7d1b-4f2c-8fdf-4704e827c2c2#intro.

Gottschall, Jonathan. 2008. *Literature, Science, and a New Humanities*. New York: Palgrave Macmillan.

Hicks, Marie. 2018. *Programmed Inequality: How Britain Discarded Women Technologists and Lost Its Edge in Computing*. Cambridge, Mass.: The MIT Press.

Humanities for All. 2019. "About." https://humanitiesforall.org/about.

Johnston, Carrie. 2018. "Flipped Mentorship." Paper presented at ADHO DH 2018. Mexico City, Mexico. June 21, 2018. https://dh2018.adho.org/precarious-labor-in-the-digital-humanities/.

Kirschenbaum, Matthew. 2012. "What is Digital Humanities and What's it Doing in English Departments?" In *Debates in the Digital Humanities*, edited by Matthew K. Gold. Minneapolis: University of Minnesota Press. https://dhdebates.gc.cuny.edu/read/untitled-88c11800-9446-469b-a3be-3fdb36bfbd1e/section/f5640d43-b8eb-4d49-bc4b-eb31a16f3d06#ch01.

Koh, Adeline. 2015. "A Letter to the Humanities: DH Will Not Save You." *Hybrid Pedagogy*. April 19, 2015. https://hybridpedagogy.org/a-letter-to-the-humanities-dh-will-not-save-you/.

McPherson, Tara. 2012. "Why are the Digital Humanities So White? or Thinking the Histories of Race and Computation." In *Debates in the Digital Humanities*, edited by Matthew K. Gold. Minneapolis: University of Minnesota Press. https://dhdebates.gc.cuny.edu/read/untitled-88c11800-9446-469b-a3be-3fdb36bfbd1e/section/20df8acd-9ab9-4f35-8a5d-e91aa5f4a0ea#ch09

Millson-Martula, Christopher, and Kevin Gunn. 2017. "The Digital Humanities: Implications for Librarians, Libraries, and Librarianship." *College & Undergraduate Libraries* 24, no. 2–4: 135–39. https://doi.org/10.1080/10691316.2017.1387011.

Moritz, Carolyn, Rachel Smart, Aaron Retteen, Matthew Hunter, Sarah Stanley, Devin Soper, and Michah Vandegrift. 2017. "De-centering and Re-centering Digital Scholarship: A Manifesto." *Journal of New Librarianship* 2, no. 2: 102–9. https://doi.org/10.21173/newlibs/3/2.

Nakamura, Lisa. 2014. "Indigenous Circuits: Navajo Women and the Racialization of Early Electronic Manufacture." *American Quarterly* 66, no. 4: 919–41.

Noble, Safiya Umoja. 2019. "Toward a Critical Black Digital Humanities." In Gold and Klein 2019. https://dhdebates.gc.cuny.edu/read/4805e692-0823-4073-b431-5a684250a82d/section/5aafe7fe-db7e-4ec1-935f-09d8028a2687#en35.

Nowviskie, Bethany. 2011. "A Skunk in the Library." June 28, 2011. http://nowviskie.org/2011/a-skunk-in-the-library/.

Pawlicka-Deger, Urszula. 2019. "A Laboratory as Critical Infrastructure in the Humanities." *Humanities Laboratories: Critical Infrastructures and Knowledge Experiments*. Department of Digital Humanities, King's College. May 23, 2019. https://blogs.kcl.ac.uk/ddh/files/2019/06/2Pawlicka-Deger_WM_Fellow_lecture.pdf).

Pinch, Trevor J. 1992. "Opening Black Boxes: Science, Technology and Society." *Social Studies of Science* 22, no. 3: 487–510. https://doi.org/10.1177/030631279202200300.

Poole, Alex, and Deborah Garwood. 2018. "'Natural Allies': Librarians, Archivists, and Big Data in International Digital Humanities Project Work." *Journal of Documentation* 74, no. 4: 804–26. https://doi.org/10.1108/JD-10-2017-0137.

Posner, Miriam. 2012. "What Are Some Challenges to Doing DH in the Library?" *Miriam Posner's Blog*. August 10, 2012. http://miriamposner.com/blog/what-are-some-challenges-to-doing-dh-in-the-library/.

———. 2018. "Miracle Workers." Paper presented at ADHO DH 2018. Mexico City, Mexico. June 21, 2018. https://dh2018.adho.org/precarious-labor-in-the-digital-humanities/.

Quijano, Paola. 2018. "Citizen Laboratories and Digital Humanities." *Digital Humanities Quarterly* 12, no. 1. http://www.digitalhumanities.org/dhq/vol/12/1/000352/000352.html.

Risam, Roopika. 2016. "Navigating the Global Digital Humanities: Insights from Black Feminism." In Gold and Klein 2016. https://dhdebates.gc.cuny.edu/read/untitled/section/4316ff92-bad0-45e8-8f09-90f493c6f564.

———. 2019. "Navigating the Global Digital Humanities: Insights from Black Feminism." In Gold and Klein 2019.

Roberts, Sarah. 2019. *Behind the Screen: Content Moderation in the Shadows of Social Media*. New Haven, Conn.: Yale University Press.

Savage, Mike. 2013. "The 'Social Life of Methods': A Critical Introduction." *Theory, Culture & Society* 30, no. 4: 3–21.

Shore, Criss. 2008. "Audit Culture and Illiberal Governance: Universities and the Politics of Accountability." *Anthropological Theory* 8, no. 3: 278–98. https://doi.org/10.1177/1463499608093815.

Smiley, Bobby L. 2019. "From Humanities to Scholarship: Librarians, Labor, and the Digital." In Gold and Klein 2019. https://dhdebates.gc.cuny.edu/read/untitled-f2acf72c-a469-49d8-be35-67f9ac1e3a60/section/bf082d0f-e26b-4293-a7f6-a1ffdc10ba39#ch35.

Smith, Martha Nell. 2007. "The Human Touch Software of the Highest Order: Revisiting Editing as Interpretation." *Textual Cultures: Texts, Contexts, Interpretation* 2, no. 1: 1–15.

Spiro, Lisa. 2012. "'This Is Why We Fight': Defining the Values of the Digital Humanities." In *Debates in the Digital Humanities*, edited by Matthew K. Gold. Minneapolis: University of Minnesota Press. https://doi.org/10.5749/9781452963754.

Svenson, Patrik. 2016. *Big Digital Humanities. Imagining a Meeting Place for the Humanities and the Digital.* Ann Arbor: University of Michigan Press.

Tsing, Anna Lowenhaupt. 2012. "On Nonscalability: The Living World Is Not Amenable to Precision-nested Scales." *Common Knowledge* 18, no. 3: 505–524.

Vandegrift, Micah, and Stewart Varner. 2015. "Evolving in Common: Creating Mutually Supportive Relationships between Libraries and the Digital Humanities." *Journal of Library Administration* 53, no. 1: 67–78. https://doi.org/10.1080/01930826.2013.756699.

Varner, Stewart, and Patricia Hswe. "Special Report: Digital Humanities in Libraries." *American Libraries* January 4, 2016. https://americanlibrariesmagazine.org/2016/01/04/special-report-digital-humanities-libraries/

Zeffiro, Andrea. 2020. "Digitizing Labour in the Google Books Project: Gloved Fingertips and Severed Hands." In *Humans at Work in the Digital Age: Forms of Digital Textual Labor*, edited by Shawna Ross and Andrew Pilsch, 133–54. London, U.K.: Routledge.

CHAPTER 4

Closed, Open, Stopped: Indigenous Sovereignty and the Possibility of Decolonial Digital Humanities

David Gaertner

When we disappear Indigenous presence from our intellectual endeavors, our movement building, and our scholarship, we not only align ourselves with the wrong side of history, we necessarily negate any form of solidarity and become actors in the maintenance of settler colonialism.

—Glen Coulthard and Leanne Betasamosake
Simpson (2016, 255)

What happens when the outcome is a sustainable practice, a sustainable self in academia, a lifeline to others as a way of imagining a future together?

—Fiona Barnett (2014, 74)

In what has been labelled the "revolution" of the open access movement (Oberländer and Reimer, n.d., 1), is it possible for the digital humanities to articulate closure as a methodology? At the time of writing, open access (OA) and open educational resources (OER) have been firmly entrenched as galvanizing movements arisen from activist methodologies and a healthy disrespect for gatekeeping and the ivory tower. Peter Suber writes,

OA benefits literally everyone, for the same reasons that research itself benefits literally everyone. OA performs this service by facilitating research and making the results more widely available and useful. It benefits researchers as readers by helping them find and retrieve the information they need, and it benefits researchers as authors by helping them reach readers who can apply, cite, and build on their work. (2012, iv)

As Suber helps to illustrate, the availability of information is vastly uneven. Working in community contexts, access to research hidden behind paywalls is often a significant part of the value that my students and I can bring to a collaboration. Part of what the digital humanities can and should do is to resist and dismantle the systems that restrict knowledge dissemination to a select few. OA provides a clearly defined avenue for doing this work and it is laudable. To cite Suber once again, "digital technologies have created more than one revolution. Let's call this one [OA] the access revolution" (1).

However, without dismissing the benefits of OA noted by Suber, we must also be mindful that the logic of OA does not apply equally in all directions. For many communities, OA replicates colonial methodologies that posit extraction as a public good. Amy Earhart notes that OA risks obfuscating the historical and cultural nuances of knowledge mobilization and argues that "we are at a moment where we need to think about how the exploitation of data is related to historical exploitation of people(s), to reconnect the digital with the embodied experience" (2019, 371). This is perhaps particularly true in Indigenous contexts. Information expropriated from Indigenous communities includes everything from health data to membership, materials, housing, lands and resources, and traditional knowledge, among others. The costs of this extraction include diminished access to ancestral knowledge, loss of control over cultural heritage items, the appropriation and commercialization of cultural practices, as well as threats to authenticity and livelihood (Brown and Nicholas 2012, 309). When we examine the distribution of knowledge through the pipelines of power and privilege, we are also confronted with the many reasons why, for certain communities, the "open" distribution of knowledge has deeply material, affective, and situated resonances and consequences.

With Indigenous contexts in mind, where should the digital humanities locate itself in relation to the "access revolution" and, within that revolution, to social justice and resistance? How can the digital humanities and Indigenous studies work collaboratively and reciprocally toward critiquing and expanding institutional definitions of OA—and to what ends? Clearly, there is much to be gained from both fields via thoughtful relationships between Indigenous and DH organizations. Indeed, powerful reciprocal partnerships have already been documented in the literature (McMahon, LaHache, and Whiteduck 2015; Winters 2018; Guiliano and Heitman 2019). That said, despite remarkable interventions by some of our colleagues, I stand with Dorothy Kim and Jesse Stommel when I suggest that all of us in the digital humanities, including those of us who have not traditionally grappled with legacies of power in our work, can do more to "disrupt" the conversations that shape the field. The work of carving space for Indigenous scholarship in the digital humanities cannot be the sole responsibility of Indigenous scholars. As a field, we must foster an environment of broad accountability in which we are all responsible for amplifying marginalized voices and critiquing and dismantling power. Drawing on Kim and Stommel, this chapter is therefore a call for digital humanists "to resist, to hope, to protest, to play slant, to create communities, to demand change. Together" (2018, 33).

It is necessary to emphasize the last word in the above phrases from Kim and Stommel, "together." It is emphasized with the preceding full stop for good reason. Learning to do DH work with and through critiques of power and colonialism is something we can and should be moving toward as a community, not out of a sense of guilt or even of obligation, *but because doing so is a matter of professional competence.* In order for our students to succeed today, they must be prepared to grapple with power as it implicates their work with digital tools and platforms. At a bare minimum, they need to learn to build systems that do not reproduce or reinforce harm and identify the potential for harm in existing systems. Done well, broad engagement with legacies of power and accessibility across the digital humanities will equip our students to navigate an increasingly complex world using tools and ideas that foster critically informed engagement with the digital. Digital humanities can no longer afford to uphold the appearance of "friction-free," neo-Cartesian engagements with technology and the digital (Losh and Wernimont 2019, xii). Mistaken or not, these appearances spell the beginning of the end for the digital humanities. Other fields and disciplines are taking direct and specific action to address colonial violence, environmental justice, and white supremacy (see Earhart 2019, 367). We thus require evidence that "the digital humanities are finally maturing from their critically naïve beginnings" (Losh and Wernimont 2019, xii). Contending with legacies of power in the digital humanities not only offers a potentially transformative learning experience for our students, it also represents an opportunity for us, as a community of practice, to collectively address problems endemic in our society and in our education system. Doing so together has material consequences. When issues of power are taken up by instructors across the wide range of academic perspectives that claim space in the "big tent" of digital humanities, we provide students with options to understand core issues as they relate to their specific academic and career goals. In the tent, they can see the necessary work of disruption that Kim and Stommel advocate for, not as an ideological imposition but as the practical and critical knowledge necessary to meaningfully acknowledge and redress inherited legacies of violence as they are embodied in data and the digital. This is the work that is required of all of us now. Together.

Taking inspiration from collections such as Kim and Stommel's *Disrupting the Digital Humanities* and Elizabeth Losh and Jacqueline Wernimont's *Bodies of Information*, it is my hope that this chapter makes a modest contribution to the social justice infrastructure of the digital humanities. As a non-Indigenous researcher who works across the fields of new media and Indigenous studies, I have had the very good fortune to work closely with Indigenous communities, students, and colleagues on a range of digital and community-based projects, including video-game production, CMS development, and podcasting. This chapter is not a secret pass to working with Indigenous content or with Indigenous communities. That pass doesn't exist—if only because the diversity of perspectives and ontologies between and even within Indigenous communities is far too vast for any one way of conducting engagement. However, this chapter may function as a roadmap, developed from a long legacy of Indigenous scholarship on knowledge production/mobilization that illustrates how we can think more critically about the power dynamics that shape our relationships with marginalized groups. This roadmap builds from Earhart's work in contributing to a core of best practices that can be mobilized by DH practitioners in their work with community (2019, 369). However, it also acknowledges a long history of Indigenous scholarship on knowledge dissemination and the politics of infrastructure, tracing Vine Deloria's (1978) reflections on "wise and substantial" knowledge mobilization to Deanna Reder's expansive relational Indigenous literatures database, The People and the Text (n.d.).

With the aim of reorienting Indigenous knowledge dissemination toward self-determination and sovereignty, the question I pose in this chapter is, how can the idea of "openness" be reoriented so as to facilitate relationships between digital humanities and Indigenous studies? Building out of the lessons learned in community-engaged research, I posit closure—as opposed to openness—as a social-justice-oriented DH methodology. In this, I conceptualize closure not in the sense of shutting down but as a willingness to cede control. This is a difficult lesson for many of us. Earhart argues that "academics working on projects must be willing to cede control from the individual and the academic institution and position the project within a community or activist site" (2019, 372). By closure, I mean the regulated movement of data and knowledge based in informed ongoing consent, reciprocity, and culturally specific protocols around data dissemination. Warren Cariou (Métis) articulates these boundaries—and our responsibility to them as researchers of Indigenous knowledge—as "critical humility": "I propose...we approach this work with an understanding that we can't know it completely, and perhaps that there are even aspects of it that some or all of us *shouldn't* know. That kind of humility could help to create a more ethical kind of reading, one that is less appropriative and more sensitive to the cultural roles of these works" (2020, 8).

I see critical humility as an articulation of closure and as a place to begin relationships between digital humanities and Indigenous studies. By this I mean that, via closure, the digital humanities can more precisely articulate what it means to work *with* Indigenous communities and against legacies of settler colonialism. In this sense, closure privileges and amplifies Indigenous ways of knowing and acknowledges the ongoing impacts of colonialism on knowledge mobilization. Like Deb Verhoeven, who similarly problematizes OA through the portmanteau "clopenness," I see closure as a means to take up OA more robustly, and more ethically, as a methodology—by which I mean a critical space through which to interrogate how knowledge moves (or how it doesn't). Verhoeven illustrates how introducing closure to OA might also "offer insights into the ways academic infrastructures (as iterations of patriarchy/capitalism/neo-liberalism) apply an 'openness penalty' that works to obstruct new players (minorities) from entering" (2021). Where I expand on Verhoeven's work is in what closure, as an acknowledgement of sovereignty and self-determination, might mean in terms of building strong, sustainable community relationships in the digital humanities as a result of interrogating openness. In sum, my argument is that the digital humanities can, and should, be a doubly oriented practice: researching and advancing new digital practices for the collection, analysis, and dissemination of data while simultaneously incorporating social justice into these practices via boundary setting.

At its base, the closure I am attempting to articulate here is a matter of sovereignty, self-determination, and boundaries. I argue for a digital humanities that can avail itself of the difficult labour of resistance and protest, working *with* Indigenous communities and often outside of the usual DH comfort zones, while also acknowledging, as Cariou suggests, that there are limits to what we can know. For me, this means embracing what maker theorists identify as "boundary work" (Williams and Willet 2019, 801). According to Rachel D. Williams and Rebekah Willet, boundary work "refers to the activities in which individuals engage to situate their domains of knowledge; it involves processes of delineating what one does and what one does not do" (802). In the digital humanities, boundary work means bringing different communities together around an idea to generate digital objects that interrogate meaning and proliferate ideas via material culture. This work is done via the lens of cultural safety, namely ongoing conversations on positionality, consent, and intellectual reciprocity. At stake here is a relationship in which collaborators know they are free to say no or to withdraw consent. Specifically, I take boundary work as a means to better articulate the places where DH practitioners should stop. As academics, operating against a publish-or-perish mandate, "go" is a survival language, so "stop" can be a hard word to hear. However, as Earhart argues, the publish-or-perish narrative "also contributes to exploitation and abuses of the communities" that we work with (2019, 375). By "stopped", I mean community-based DH research that readily acknowledges and respects boundaries between Indigenous and settler communities. By acknowledging boundaries, we can more ethically identify spaces where the digital humanities can amplify Indigenous voices without appropriating them; where tools and infrastructure can be co-developed while upholding community intellectual property, training community members, and mobilizing community assets toward community ends. Acknowledging boundaries, in this sense, means being explicit about what we, as academics, can and cannot (or should not) engage with; it means identifying protocols for stewardship and establishing clear plans for knowledge transfer; it means acknowledging that not all knowledge should be open and understanding why. In respecting boundaries and building "stopped" research into our practices, we begin to develop a model of "collaboration that positions the academic as an equal, or even lesser, partner in the relationship, which is the only model that will begin to balance inequity" (Earhart 2019, 375).

Below, I attempt to unpack boundary work through one of Indigenous studies' great thinkers Vine Deloria (Standing Rock Sioux). Via Deloria, who in 1978 was loudly calling for innovative models to facilitate Indigenous knowledge mobilization, I look specifically at Indigenous sovereignty and self-determination as they apply to data collection and dissemination. In understanding sovereignty as a boundary, I argue we can better articulate what a healthy relationship between digital humanities and Indigenous studies looks like and make space for reciprocal research collaborations in which we work together against settler colonialism and white supremacy. From Deloria, I move into analysis of a case study, Reder's (Cree-Métis) The People and the Text, built in collaboration with Susan Brown's Canadian Writing Research Collaboratory (CWRC), which I suggest provides a cogent model for rigorous and ethical collaboration across digital humanities and Indigenous literatures. Looking at Reder's engagement with intellectual sovereignty (which, as I will illustrate, has its own lineage in Deloria's work), I attempt to illustrate how large-scale DH infrastructure can be shaped to facilitate radically relational spaces of textual engagement.

CLOSURE AS SOVEREIGNTY

My articulation of closure, as boundary work, is developed out of a rich tradition of Indigenous scholarship that challenges the ubiquity of openness, both

IRL (in real life) and URL (virtual). Specifically, I define closure in relation to Indigenous sovereignty, building out of Deloria's path-clearing work. Deloria offers a nuanced place to ground this conversation inasmuch as his thinking brings sovereignty and knowledge organization/dissemination together in productive and, in my opinion, critically under-researched ways. Deloria's notion of sovereignty provides a solid foundation from which to imagine how Indigenous–DH relations might operate, emphasizing non-interference, and what he identifies in his work on knowledge mobilization as "wise and substantial" (1978, 17) approaches to information science.

In many ways, Deloria's work sets the bar for the theorization of sovereignty in Indigenous studies. Robert Warrior (Osage) argues that "Deloria's consistent discussion of sovereignty [...] has not been often paralleled among contemporary American Indian intellectuals" (1992, 6). For Warrior, Deloria's contribution to Indigenous sovereignty remains a vital cornerstone of Indigenous studies because of his ability to conceptualize it as an "open-ended" process: that is, beyond the essentializing tropes that relegate Indigenous thought and culture to the past, moving instead toward future-oriented and nation-specific self-determination. That Deloria considered sovereignty as a relationship, and thus something that is continually growing and unfolding out of community, also leaves ample room to consider that concept in future-oriented spaces, including the digital. I will return to intellectual sovereignty and its articulation in the digital humanities momentarily, but for now it is enough to know that both Warrior and Deloria operate out of principles of Indigenous-led knowledge production and dissemination, not with the intent of ignoring ideas from outside of community, but rather of first amplifying and bolstering nation-specific knowledge and ways of knowing.

Broadly speaking, sovereignty as a political concept has clear roots in global political theory and Cold War politics, which worked to draw the indelible boundaries around nation-states that charted out systems of "war, deterrence, decision making, trade, monetary relations, and so on" (Philpott 2010, 297). In Indigenous contexts, these principles, which speak specifically to how a nation protects and maintains its borders, still apply. But via thinkers like Deloria and Warrior, sovereignty is also closely linked to "cultural integrity" and "intellectual sovereignty," which are specific to traditional knowledges and their persistence in the deleterious face of settler colonialism. According to Deloria, "sovereignty can be said to consist more of continued cultural integrity than of political powers. And to the degree that a nation loses its sense of cultural identity, to that degree it suffers a loss of sovereignty" (1999, 27). Deloria's definition of

Indigenous sovereignty is firmly rooted in the maintenance and protection of a community's self-determined interests, which necessitates that the people have the power to control the production and distribution of their knowledge according to their own principles and within the contexts of their own territory, without outside interference. In other words, "cultural integrity" refers to an active articulation of sovereignty that privileges tradition, culture, place, and continuity across past, present, and future tenses.

The assimilative imperatives of colonial nation-states, which balk at the prospect of sovereign borders *within* the bounded territory of colonial nation-states, is a clear threat to Indigenous sovereignty. However, perhaps more pertinent to the digital humanities, technology also poses a threat to Indigenous cultural integrity. Translating Deloria's ideas into the OA movement, it is evident that his definition of cultural integrity, while anachronistic, is at odds with the ways in which digital archives are conceptualized in open spaces, beyond the boundaries of a tribal community or First Nation, and thus accessible from any point outside of it. Without the contexts of place and people, Deloria argues, this knowledge is stripped of a significant part of what it is and what it is meant for. Indigenous sovereignty, in Deloria's conception, is upheld by the land and the nation and maintained by Indigenous stewards toward Indigenous ends. Settler colonialism, however, functions at least in part by deracinating Indigenous knowledge and extracting data for study outside of Indigenous boundaries (land-based, methodological, and otherwise). Removing Indigenous knowledge from sovereign boundaries thus risks stripping it of the contexts critical to its ongoing integrity. The ways in which the public typically (and ultimately incorrectly) conceptualizes open as "barrier-free" access (Suber 2012, 4) risks dissimulating the boundaries and structures that Deloria establishes as necessary to cultural integrity while recapitulating and reinforcing settler colonial politics that appeal to the public good—without acknowledging who that public is and how they came to be.

Thinking against the logics of open technologies and the imperative of "barrier-free" knowledge, at the centre of Deloria's definition of sovereignty is Indigenous control over anti-colonial reform, which he argues will lead to a unified vision and community strength. According to Deloria, Indigenous "governments must be allowed to structure [...] activities according to traditional precepts rather than being required to follow rules, regulations, and eligibility standards established for" colonial societies (1999, 27). This is to say that Deloria's sovereignty is grounded in self-determination and what Audra Simpson (Mohawk) identifies as the "refusal to recognize" colonial impositions (2014, 128), including assimilative and boundary-effacing initiatives positioned as a "public good."

Sovereignty, as the right to refuse non-Indigenous intervention, even when it is coded as "help" or "allyship," is about facilitating the space for Indigenous knowledge to flourish in the strength of its own contexts and without the burden of imposed infrastructures. It is Indigenous-centred, Indigenous controlled, and mobilized toward maintaining and proliferating Indigenous integrity.

While Deloria's conception of sovereignty should give us pause when considering the implications of Indigenous–settler collaborations in a field like digital humanities (namely how and if settler DH scholars can collaborate without imposing deeply embedded, and sometimes unrecognized, colonial infrastructures), it does not preclude technological intervention. In fact, Deloria's definition of sovereignty is at least partly driven by technical advancement, particularly in the prescient ways in which he imagined information repatriation and organization. Inasmuch as Deloria's sovereignty is about refusing interference, it is also about innovating substantial approaches to cultural integrity through the reclamation and dissemination of ancestral knowledge and Indigenous data that is currently managed by the settler state. According to Jennifer O'Neal, Deloria's decolonial politics are enveloped in his "recognition that information and knowledge are critical to the sovereignty and self-determination of Native nations" (2015, 3). In this sense, there is a clear connection between knowledge management and the cultural integrity that informs sovereignty in Deloria's framework.

Deloria's thinking in these regards is centred firmly in the repatriation of Indigenous knowledge, facilitated by the advancement of information organization and the strategic use of digitization. In a paper he prepared for the White House Preconference on Indian Library and Information Services, Deloria ties Indigenous sovereignty specifically to what he identifies as "the right to know": the right of Indigenous peoples to reclaim, use, and pass on their knowledge, including their data.[1] While access to personal information is considered a right in most modern democracies, until the United Nations Declaration on the Rights of Indigenous Peoples, adopted by the UN General Assembly in 2007, Indigenous data were (and in some cases still are) collected, maintained, and appropriated by colonial governments, in university and state archives, and museums. This was done under the auspices of "preservation" (a symptom of a colonial ideology that erroneously insists that Indigenous culture is dead or dying). That stolen data were subsequently mobilized by the state, through non-Indigenous research and state policy, as a means to maintain or recover control over Indigenous populations. In Deloria's formulation, the right to know is thus an insistence that Indigenous peoples regain control over the knowledge that has been and continues to be extracted from them as part of the colonial project.

It is also a call to action for information scientists (and, I would add, DH practitioners) to strategically organize and disseminate that data effectively *against* the colonial project. It is a refusal of settler stewardship, in archives, repositories, and databases, and an affirmation of the presence and future of Indigenous peoples articulated through dynamic, living relationships with data and knowledge.

While it is information-based, decolonization is not a metaphor in Deloria's conception of knowledge mobilization.[2] It is an argument—against settler colonialism's ongoing disruptions—for facilitating deeper connections to the land and to community through the endurance of Indigenous ways of knowing. Specifically, the "right to know" is the right "to know the past, to know the traditional alternatives advocated by their ancestors, to know the specific experiences of their communities, and to know about the world that surrounds them in the same intimate manner they once knew the plains, mountains, deserts, rivers, and woods" (Deloria 1978, 13). In other words, Deloria's information organization and knowledge mobilization is about reclaiming and redistributing Indigenous knowledges so as to better connect communities to the land, thus providing the groundwork, so to speak, for reclamation and resurgence.

Of course, contemporary thinkers in Indigenous studies—see, for example, Glen Coulthard and Leanne Betasamosake Simpson (2016)—have illustrated that, despite the best efforts of colonizers, Indigenous connections to land have never been entirely severed. Much of that history has been preserved due to the efforts of elders and knowledge keepers who put their lives at risk to protect and pass knowledge even as the state worked to ban, extract, and archive it. However, whether through the Indian Act, title extinguishment, forced removal, or legal measures levied to repress and eradicate Indigenous knowledges, settler colonialism, at its base, works to make land-based knowledges as tenuous as possible because it facilitates *terra nullius* and legitimizes colonial occupation.[3] With the "right to know," Deloria directly connects the extraction of Indigenous knowledge under colonization to experiences of alienation and isolation in the colonial environment. Returning knowledge and repatriating data is thus much more than an intellectual exercise for Deloria; it is about fortifying sovereignty by deepening relationships to the land via Indigenous knowledges and, therefore, amplifying what Coulthard and Simpson identify as "grounded normativity": the reciprocal practices, process and relationships "that are inherently informed by an intimate relationship to place" (254).

So where does technology sit here and where might the digital humanities intervene as a means to facilitate Deloria's vision? Inasmuch as Deloria's conception of sovereignty is closely intertwined with the movement and reclamation of Indigenous knowledge, thoughtfully deployed technology can be a means of repatriating knowledge and developing new, living engagements between documentation, stories, and data, and the communities to which they belong. Indeed, what remains of critical importance in Deloria's work is the forward thinking he applied to the right to know, particularly in terms of technological advancement. O'Neal (2015, 2–3) outlines the steps that Deloria sees as essential to facilitating the right to know. I've numbered and incorporated them here as a cursory roadmap for ways in which the digital humanities might contribute to sovereign agendas. I've also adapted some of O'Neal's language to reflect the Canadian contexts:

1. Inventory and catalogue existing records in federal possession.
2. Duplicate and make accessible pertinent Indigenous historical records.
3. Develop information services customized for First Nations, Métis, and Inuit.
4. Develop library and information science education for First Nations, Métis, and Inuit.
5. Provide digitization capabilities for Indigenous cultural resource centres.
6. Establish regional research centres, cataloguing, and stewardship.
7. Appropriate acquisition funding for repatriation.

Of particular significance to the digital humanities is the foresight Deloria had in terms of the reproduction, mobilization, and even digitization of Indigenous knowledge, which he saw as a means to repatriate information to communities whose membership was widely distributed as a result of legislation such as the Indian Removal Act in the United States. In sum, part of what I want to suggest is that, when taken up in response to the legacies of power that restrict knowledge mobilization, the digital humanities can play an important role in the contemporary development of the right to know, articulated specifically as Indigenous knowledge repatriation. Making data collected on Indigenous peoples readily available to the communities it belongs to facilitates the right to know, but it also makes technology legible as a potentially decolonial intervention.

That said, what "wise" might mean in the contexts of organization and dissemination is a thorny issue, particularly given that European and Indigenous library and information science cataloguing systems (for example the Brian Deer Classification System) are designed for very different audiences and use cases. Reclaiming knowledge does not mean subsequently subjecting it to the same old tired systems of colonial classification. In this sense, being attentive to Deloria's wise and substantial practices means supporting, signal-boosting, and implementing the Indigenous knowledge organization systems that already exist, while being attentive to the colonial ideologies that abide in the systems that ground our research.[4]

The work of dismantling colonial cataloguing systems is already well underway, led by Indigenous scholars and librarians. Kim Lawson (Heiltsuk) writes about Indigenous information organization systems that have been in place for thousands of years:

> First Nations are not only sources and consumers of knowledge, but also have their own knowledge systems rooted in complex oral cultures. These deep and often unrecognized differences lead to significant difficulties in communicating between indigenous and non-indigenous people. These difficulties are compounded in conflict-driven situations. There are not only disagreements about what is true but also disagreements about what makes information reliable or credible—what makes people trust it enough to act on it. (2004, 1)

What Lawson points to in her research is that information organization tacitly frames how knowledge is received and acted upon. Being attentive to how information is delivered provides deeper insight into what is defined as knowledge and what is not. Cataloguing is one of those delivery systems. My colleagues at X̱wi7x̱wa Library, the only Indigenous branch of an academic library in Canada, demonstrate the colonial ideologies that sit at the root of systems such as the Library of Congress Subject Headings (LCSH). They illustrate that one of the primary challenges for Indigenous information organization is that the literary warrant of materials is most often based on European literatures and nineteenth-century Western epistemologies. Embedded within these knowledge systems are colonial classification and description procedures that reproduce themselves by ignoring "Indigenous contexts due to historicization, omission, marginalization, lack of recognition of sovereign nations, lack of specificity, and lack of relevance" (Doyle, Dupont, and Lawson 2015, 111).

Christine Bone and Brett Lougheed illustrate that systems such as the LCSH homogenize Indigenous peoples across North America and obfuscate community-specific, locally meaningful points of access to Indigenous knowledge and data (2017, 83). This is perhaps nowhere more evident than in the ongoing use of the word "Indian" as a subject heading in LCSH, a term that carries with it a racist and violent history of colonization. According to Bone and Lougheed, the persistence of "Indian" as a subject header "is the problem in LCSH most often mentioned by reference librarians and library users [...] particularly Native Studies students and professors" (86). Undoing these classification systems necessitates wise, community-engaged research, but it also requires technical, financial, and administrative support for the Indigenous librarians and information scientists, such as Lawson, who have already cleared the path for the resurgence of Indigenous knowledge organization in libraries such as X̱wi7x̱wa. Thanks to efforts by Lawson, Bone and Lougheed, Sarah Dupont, Kayla Lar-Son, and many more, we are just now starting to see the fruit of this labour. Since I wrote the first draft of this chapter, the Canadian Research Knowledge Network (CRKN) has acted on the recommendations of X̱wi7x̱wa staff and other Indigenous librarians and gallery, library, archive, and museum workers to replace the subject heading "Indians of North America" with "Indigenous peoples" in the Canadiana collections. On January 25, 2022, CRKN publicly acknowledged "the need for national solutions to harmful and inappropriate subject headings and resource descriptions, and for respectful terminology for Indigenous peoples and all who have been marginalized or inappropriately represented in the history of Canada" (CRKN 2022). CRKN also produced a spreadsheet that outlines interim subject headings that can be used as a stopgap while a national vocabulary is being established for subject headings and resource descriptions.[5]

Deloria's work provides us with a rigorous, two-faceted approach to closure, which (1) defines that concept in relation to cultural integrity and the right to refuse non-Indigenous interference (even when that interference is coded as "help") and (2) grounds the resurgence of this sovereignty in the wise dissemination of knowledge and the decolonization of existing cataloguing systems. As O'Neal outlines, contemporary efforts to "decolonize" archives are thus related to bringing dissemination and access together in exacting and exciting ways; that is, they aim to "replac[e] Western ways of managing tribal archives with those rooted in the Indigenous epistemological traditional ways of knowing and stewarding collections" (2015, 2). Practically, this means being attentive to the ways in which colonial cataloguing systems reproduce and reinforce settler colonialism, while directing technical, financial, and administrative support toward existing systems of Indigenous knowledge organization.

What is at stake, then, in an equitable relationship between Indigenous studies and digital humanities is a sovereign approach to information organization and knowledge mobilization *to the benefit of both parties*. On the one hand, the digital humanities can support Indigenous knowledge organization with financial, technical, infrastructural, and administrative support that facilitates the development of wise and substantial digitization practices as determined by specific communities. On the other hand, the digital humanities is supported, in the sense that Kim and Stommel advocate for in *Disrupting the Digital Humanities*, by gaining a clearer understanding of the colonial architecture that undergirds the information technologies and infrastructures we traffic in as researchers and teachers of the digital. The radical, unrealized potential of the digital humanities, as Miriam Posner (2016) puts it, therefore lies in our willingness to go about "ripping apart and rebuilding the machinery of the archive and database so that it doesn't reproduce the logic that got us here in the first place." Indigenous knowledge systems represent a significant step in this direction, if we approach them with critical humility and a willingness to learn and respect community-specific boundaries informed by intellectual sovereignty and cultural integrity.

THE PEOPLE AND THE TEXT: "RECONNECTING KINSHIP"

In the final section of this chapter, I include an example of an Indigenous DH project that models what Deloria calls "wise and substantial" knowledge dissemination. I do so in part because of the respect that I have for this project and the researchers involved, and because it illustrates what I was proselytizing above: collaboration within and beyond the big tent of the digital humanities that facilitates Indigenous research via thoughtful development of DH infrastructure. That project is Deanna Reder's (Cree-Métis) The People and the Text (TPatT), which not only demonstrates a laudable collaboration between Reder and Susan Brown but also significantly contributes to Indigenous literary studies by reformulating how we think about our relation to digital text "as an encounter with a living being" (Cariou 2020, 8), an idea I will expound upon a little further on.

Reder's intervention into digital humanities via Indigenous literatures is ambitious and profound. The website/archive/pedagogical agenda of the project aims to "bring together scholars in collaboration with communities to establish ethical guidelines to train a new generation of scholars and to make best use of new web technologies to open up the literary past of Indigenous writing" (Reder, n.d.). Taking Warrior's work as a starting point (which, as noted above, is itself derived out of Deloria's thinking on sovereignty and cultural integrity), TPatT mobilizes Indigenous intellectual sovereignty as a modality through which to imagine our relationship to story and the digital. Building from Warrior, Reder notes that while other disciplines have devised research ethics and protocols to foreground Indigenous intellectual sovereignty in their work, literary studies—and, by proxy, digital humanities—remains frustratingly far behind. The pervasive notion of the literary scholar as a single, autonomous author whose engagement is with the text and the text alone, often stands in the way of literature scholars imagining community-based methodologies, including collaborative interpretation, place-based reading, and community-led research. Via the TPatT website and database, Reder and her team are working to reframe literary scholarship within an ecosystem of relationships and responsibilities, which do not so much decentre the text as they invite community and relations into how we understand it relationally.

To be clear, reimagining how Indigenous literary criticism can be rescoped to include community is not simply a theoretical or even methodological intervention. It is also a necessary response to our changing legal landscape. Reder points out that Indigenous peoples should have "the right to participate in and benefit from research" (Reder and Brown 2022) on texts that come from their communities. The United Nations Declaration on the Rights of Indigenous Peoples, which is now law in Canada, agrees:

> Indigenous peoples have the right to practise and revitalize their cultural traditions and customs. This includes the right to maintain, protect and develop the past, present and future manifestations of their cultures, such as archaeological and historical sites, artefacts, designs, ceremonies, *technologies* and visual and performing arts and *literature*.[6]

In the sense that the declaration conveys, TPatT mobilizes the digital humanities to open literary interpretation to Indigenous communities in ways that respond directly to legal circumstances and the changing nature through which we understand research in Indigenous literatures: not as a practice conducted *on* Indigenous peoples and texts but as relationships we build with and for Indigenous communities around their stories.

What is being closed in this formulation then? In the terms of intellectual sovereignty, closure, as defined by Warrior's work and extended into TPatT, is the sound of interpretive doors being shut. These might be the doors of deconstruction, psychoanalytic theory, historical materialism, etc. These doors are shut not for the sake of locking Western criticism out of the conversation forever but for making the pathway to local, community-specific critical approaches easier to find and follow. For Warrior, Indigenous intellectual sovereignty is concerned with moving critics away from the panacea of Western literary criticism toward nation-specific literary tools. However, he argues at the same time that critics must resist the idea that Indigenous peoples "need nothing outside of ourselves and our cultures in order to understand the world and our place in it" (Warrior

1992, 18). He continues to nuance the definition of intellectual sovereignty (and, I would argue, closure) by identifying it as an exercise in affirming the importance and impact of community-based intellectual practices: "[T]he struggle for sovereignty is not a struggle to be free from the influence of anything outside of ourselves, but a process of asserting the power we possess as communities and individuals to make decisions that affect our lives" (19). In other words, literary sovereignty is about drawing clear links between texts and ideas to specific Indigenous intellectual traditions with minimal background noise.

Handily, the internet is able to draw connections fairly well. TPatT is built on CWRC infrastructure, which itself has been developed from the Orlando project, which Brown developed to "harnesses the power of digital tools and methods to advance feminist literary scholarship" (Brown, n.d.). Key to this infrastructure is how CWRC uses links as "pathways out into multiple perspectives" (Brown, n.d.). Working with Brown and CWRC, Reder has shaped connections between people and text via the digital to generate a nexus of story and community-specific knowledge that mutually inform one another. TPatT enacts intellectual sovereignty by shepherding scholars toward engagements with Indigenous literatures that decentre the literary critic as the prime mover. As certain interpretive doors close, the website opens collaborative, digital spaces that, through CWRC infrastructure, associate literary works and the scholars that study them with the texts' communities of origin. The website builds a scholarly community by gathering Indigenous texts and supplementary materials and then making them available, for the most part via an OA model, to the public. At the same time, deploying what Verhoeven might call clopenness, it builds capacity in Indigenous intellectual sovereignty by providing training, resources, and nation-specific case studies for literary scholars to learn to work ethically and productively with specific communities and their literatures.

As I have gestured toward already, Reder's work with TPatT is made possible, in large part, because of the support offered by Brown and CWRC. In fact, the collaboration between Brown and Reder models how DH infrastructure can be effectively and ethically mobilized to amplify Indigenous voices and research— and also to demonstrate broad accountability. In leveraging CWRC's considerable resources and research networks, TPatT has a secure, stable, and sustainable home, and the technology to generate the constellation of links and resources that inform the site's intellectual sovereignty. In providing infrastructure and support, Brown amplifies Indigenous voices and facilitates Indigenous-led research creation while Reder maintains sovereign, self-determined representation and deployment of her project.

To be sure, CWRC operates in close relation to Canadian national discourses and the frameworks of the Western literary canon, which, understandably, may raise some concerns for scholars looking for Indigenous-made digital infrastructure (see, for instance, Duarte 2017). That said, *infrastructure is labour*, and for scholars such as Reder, whose plate is already overflowing from the demands placed on Indigenous scholars in academia, support is necessary for the success of the project if it can be leveraged toward the needs of the researcher. Further, CWRC is built with equity, diversity, and inclusion in mind. Building out of lessons learned from the Orlando project, CWRC generates constellations of meaning via pathways of linked data and resources—"drawing on the link's infrastructure and agenda, [the CWRC backend] use[s] linked data to provide context, situate knowledge, and advance diversity" (Brown, n.d.). Brown is able to build this infrastructure because of her experience as a DH scholar, and also because of her privileged position in the academy. As she makes plain in her presentations about CWRC, digital infrastructure is deeply political. The resources required to develop full-stack support for searchable, dynamic, and sustainable digital tools necessitates intervention from researchers that are deeply embedded in their institutions. Building robust digital infrastructure within the university system requires not only technical wherewithal (and deep and abiding patience) but also the capacity to apply for and manage large grants, hire and maintain servers, build accessible portals, develop a research data management plan, etc. All of which means employing staff, training students, liaising with industry and IT, and, often, working with an advisory. From Brown's perspective, the DH scholars that have the privilege to work with this type of institutional support also have a responsibility to leverage those resources for marginalized communities and researchers. In this, she cites Posner, who writes that:

> It's incumbent upon all of us (but particularly those of us who have platforms) to push for the inclusion of underrepresented communities in digital humanities work, because it will make all of our work stronger and sounder. We can't allow digital humanities to recapitulate the inequities and underrepresentations that plague Silicon Valley; or the systemic injustice, in our country [the United States] and abroad, that silences voices and lives. (2016)

With TPatT, Brown and her team are developing an ongoing, reciprocal relationship with Indigenous literature scholars that facilitates the development of Indigenous intellectual sovereignty alongside CWRC infrastructure. For instance, Brown and Reder joined the 2022 gathering of the Indigenous Literary Studies Association at the Gabriel Dumont Institute to present on the TPatT/ CWRC collaboration. I was also on that panel, presenting work from a video-game-development initiative, so I had a front-row seat for the conversation. What was evident in the TPatT/CWRC presentation, and this came from both Brown and Reder, was that, within CWRC, TPatT is afforded the space to develop the relational, community-oriented infrastructure that Reder needs for the project to succeed. IRL, Reder has a long-standing and broad career in community-engaged literary studies, including the publication of deeply sensitive work with Maria Campbell, who was also in the audience for the discussion.[7] However, how to translate the relationality of Reder's work into digital infrastructure—an infrastructure which, as authors such as Duarte have illustrated, is haunted by the legacies of data extraction and colonial cataloguing systems—was a significant concern for Reder, but it was also an issue that Brown was invested in addressing.

Citing Duarte and Miranda Belarde-Lewis as a means to open up the problem, Brown spoke to the ways in which CWRC has adapted to be mindful of "how cataloguing and classification practices become techniques of colonization" (Duarte and Belarde-Lewis 2015, 682), deploying tools such as the Traditional Knowledge (TK) labels created by Local Contexts. TK labels "allow communities to express local and specific conditions for sharing and engaging in future research and relationships in ways that are consistent with already existing community rules, governance and protocols for using, sharing and circulating knowledge and data" (Local Contexts, n.d.), and have been meaningfully deployed in large-scale community projects such as digitalsqewlets.ca. TK labels help to steer the ways in which knowledge moves online in accordance with community-specific protocols, but they are not an infrastructural intervention per se. Infrastructure is the digital technologies and protocols that provide the foundation for a platform's information technology and operations.

For CWRC, holding space for Indigenous initiatives on the platform means digging into the foundations of the project, which are broadly based on FAIR (findable, accessible, interoperable, reusable) data principles. From the onset, CWRC grounded its infrastructure on OA, interoperability, and preservation—"it aims to connect data by promoting and providing means to employ standards and best practices that make data shareable, interoperable, and preservable" (Brown, n.d.). However, as I have gone to some lengths to describe above, OA principles are often at odds with how Indigenous peoples want to control and benefit from their knowledge. As the Global Indigenous Data Alliance (GIDA) puts it, "the emphasis on greater data sharing alone creates a tension for Indigenous Peoples who are also asserting greater control over the application and use of Indigenous data and Indigenous Knowledge for collective benefit" (GIDA, n.d.). Addressing this tension within the CWRC infrastructure means harmonizing FAIR with what GIDA identifies as CARE principles (collective benefit, authority to control, responsibility, and ethics). In the language I am evoking here, the integration of CARE into a platform such as CWRC is an articulation of closure, particularly inasmuch as principles such as OA are framed as universally beneficial and necessary. According to GIDA:

> The current movement toward open data and open science does not fully engage with Indigenous Peoples' rights and interests. Existing principles within the open data movement (e.g., FAIR: findable, accessible, interoperable, reusable) primarily focus on characteristics of data that will facilitate increased data sharing among entities while ignoring power differentials and historical contexts.

The CARE principles, designed as a compliment to FAIR, "are people and purpose oriented, reflecting the crucial role of data in advancing Indigenous innovation and self-determination." By integrating CARE into a system that claims OA and interoperability as bedrocks of its infrastructure, Brown and her team generate a technical and epistemological space "to create value from Indigenous data in ways that are grounded in Indigenous worldviews."

Further, CARE also provides for relationality. CWRC's deliberate digital architecture, which is working toward building Indigenous data protocols into its infrastructure, provides the foundation necessary for Reder to conceptualize and build a database rooted in her practices as a Cree-Métis literary scholar. Since its inception, TPatT has been constructed as more than a static database, but rather as a constellation of dynamic relationships between texts and people that challenges how scholars think about the relationship between animate and inanimate entities. In Reder's conception, which she draws from Cariou, TPatT is conceptualized as a means to rethink our relationship to text, "as an encounter with a living being, or perhaps with a spirit" (Cariou 2020, 8). Toward this end, Reder and her team have imagined their digital archive as a living sanctuary for texts, complete with the community-oriented organizational structure necessary for these stories to nourish (and be nourished by) their communities of origin. The centrality of relationality in TPatT cannot simply be drawn out in the space between storyteller and audience, or audience and end user. It also takes shape in the intimate spaces that unfold around story as it is brought into being in community, with that story functioning as what the Stó:lō musicologist Dylan Robinson identifies as a "nonhuman ancestor" (2021, 91). In the way that TPatT gathers Indigenous stories, breathing new life into archival materials through CWRC's relational infrastructure, the website functions as a conduit for what Robinson calls "reconnecting kinship," a process of liberating and reviving material forms that have been held in archives and repositories (87). Robinson illustrates that the sustenance that these materials represent for community is significantly muted when the object is removed from its kin networks to be pre-served in archives, repositories, display cases, and even "locked into hard drives" (Brown, n.d.). Robinson goes so far as to suggest that what archivists call preser-vation is, in many Indigenous contexts, akin to incarceration (2021, 91).

For Robinson, Cariou, and Reder, cultural objects—including stories—that have been extracted from Indigenous communities and placed in archives need to be understood not as stagnant documents, or lifeless relics, but as relations: ancestors who hold a deep and powerful potential to nourish and sustain com-munity. These relations are not static or historical. They are not inanimate but dynamic and adaptive pieces of living culture that both give life to, and receive life from, kinship networks and community. Finding opportunities to be in relation to stories—outside of the archive—is thus not only a matter of repatriation but, also, more pointedly, a matter of rehabilitation, care, and intellectual sovereignty. Robinson points to the moments when non-human ancestors are reintroduced into the community, for instance when museum objects are repatriated, giving their relations the "opportunity to be fed and to feed [their] ancestors"—through cultural engagement based in reciprocity (92).

A non-textual example helps to illustrate how "reconnecting kinship" functions. Robinson cites a performance in which Mike and Mique'l Dangeli (Nisga'a, Tlingit, and Tsimshian) danced with an "amhalaayt ancestor" (frontlet of headdress), which had been freed from a plexiglass display case at the Agnes Etherington Arts Centre. By drawing the *amhalaayt* into their performance by dancing with and for it—and thus recontextualizing it within community cul-tural praxis—the Dangelis demonstrated how that relation, treated as a static and historical object in the arts centre, could be reintegrated into the community as a living entity, as part of an ongoing cultural practice. Recontextualized as such, Robinson writes that "it is life-giving and itself has life," meaning the *amhalaayt* acts in reciprocity with the community, much like a human relation would (2021, 92). The Dangelis are nourished by the cultural history contained in the *amha-laayt*, shaping their performance in relation to it, and the *amhalaayt* is awoken into contemporary contexts through the love and care of the performance and the performers. According to Mique'l Dangeli, "it is important that they know we acknowledge them, and that we love them" (Robinson 2021, 92). In this gesture of radical relationality, which Karyn Recollet identifies as a distinctly Indigenous "technolo[gy] of worlding," what settler audiences might see as static cultural heritage objects "are activated in communion with extensions of kinships" (2019, 91). Built as a means to connect story to community, TPatT enacts processes of "reconnecting kinship" that centre engagement with Indigenous literature as a deeply relational connection built out of dynamic, living relationships with story and text. As such, these stories are realized and activated as time-travelling bundles with the power to galvanize ancestors across past, present, and future—and CWRC provides resources and infrastructure to facilitate that.

CONCLUSION

What is closure as a DH methodology? As Verhoeven puts it, closure "enables us to be open to whatever we are not, that opens us to being challenged, and most importantly to change" (2021).

Measured equally against the legacies of data extraction that make up the archive and the path-clearing interventions being made by Indigenous scholars such as Reder, closure is a means of taking stock in digital humanities. Clearly defining the value that the digital humanities could bring to a community project and, just as importantly, identifying the harms and risks it carries is an essential part of building digital infrastructure with and for community and against colonial archival and cataloguing processes.

In this sense, closure as boundary work is also just good scholarly practice. For settler scholars, closure means not getting in the way; it means deconstructing one's own authority, expertise, and desire to control and, in doing so, holding space for community knowledge keepers, who, under the logic of OA qua the public domain, have had their knowledge appropriated, decontextualized, and monetized (to someone else's benefit) for over 150 years. What Indigenous studies throws into relief for digital humanities are questions of expropriation and enrichment. When knowledge is digitized and mobilized through digital infrastructure, who stands to benefit from those practices? Who is the work in relation with? And what is the work as a relation?

As Coulthard and Simpson insist in the passage that serves as the epigraph for this chapter, disappearing Indigenous presence from any academic discipline negates solidarity and tacitly contributes to the maintenance of settler colonialism. This is perhaps doubly true in the realm of the digital humanities, given the ways in which those same colonial systems work to position Indigeneity and technology as mutually exclusive terms and how, as Duarte has illustrated, technology was developed, at least in part, as a means to surveil and displace Indigenous peoples (Duarte 2017, 10). Guided by closure, non-Indigenous DH scholars should be supporting Indigenous interventions into the digital and working in relation with Indigenous peoples, not only because they need our support but also because working reciprocally and ethically with Indigenous communities and ideas to reconfigure our infrastructure will make DH stronger and more relevant to our students and our community partners. Indigenous interventions into data and the digital facilitate improved conditions for women and minorities in digital spaces; they foster critical thinking and engagement with technology, not only as a tool but also as a social practice; they compel us to seriously consider best practices for health, safety, and the environment in our labs and classrooms. By holding up sovereignty as it stems from foundational work in Indigenous studies, the digital humanities can position itself to foster the solidarity, community, and mutual support that ambitious, path-clearing projects such as The People and the Text take on.

Closure is not the end of a conversation. It is a beginning. The call for closure is a call to action in response to the imperative of OA, but it is simultaneously a call that unpacks and unsettles what the digital humanities is and who it serves.

NOTES

1. The right to know is now also affirmed in the United Nations Declaration on the Rights of Indigenous Peoples, particularly in articles 11–13.
2. Here I am referring to Eve Tuck and K. Wayne Yang's foundational essay "Decolonization Is Not a Metaphor" (2012). According to Tuck and Yang, decolonization must be directly connected to returning land to Indigenous peoples. The decolonial work we do in the digital humanities is often at risk of being a metaphor because it is abstracted from place through the "placeless" terrain of cyberspace (Barlow 1996). Deloria's work, however, makes explicit the connection between data, data management, and land.
3. The potlach ban (1885–1951), which made the practice of Indigenous ceremony a criminal offence, is just one example of the federal government's specific attack on Indigenous cultural integrity. For more on *terra nullius* and the connection between Indigenous knowledge and land, see Cariou (2014).
4. For an excellent summary of written resources, see Bone and Lougheed (2017).
5. Access CRKN's spreadsheet at https://docs.google.com/spreadsheets/d/1uPI55rpGE QT7OP3uJWVm2KWZoRpaznaW/edit#gid=584739606; "this spreadsheet is a living document and should not be considered exhaustive or definitive. It will be updated

as language changes, or as an updated national vocabulary is created. We encourage input and feedback on this document."

6. United Nations Declaration on the Rights of Indigenous Peoples Act, SC 2021, c. 14. https://laws-lois.justice.gc.ca/eng/acts/u-2.2/FullText.html; emphasis mine.

7. Probably the most prominent example of the compassionate and thoughtful community-engaged research Reder has done in community is the work that she and Alix Shield did in locating and repatriating sensitive passages that the original publisher had removed from Maria Campbell's *Halfbreed*. How Reder and Shield worked with Campbell to reincorporate these "lost" passages into a new edition of the novel is documented in Reder and Shield (2019).

REFERENCES

Barlow, John Perry. 1996. "A Declaration of the Independence of Cyberspace." EFF (blog). 1996. https://www.eff.org/cyberspace-independence.

Barnett, Fiona M. 2014. "The Brave Side of Digital Humanities." *differences* 25, no. 1: 64–78. https://doi.org/10.1215/10407391-2420003.

Bone, Christine, and Brett Lougheed. 2017. "Library of Congress Subject Headings Related to Indigenous Peoples: Changing LCSH for Use in a Canadian Archival Context." *Cataloging & Classification Quarterly* 56, no. 1: 83–95. https://doi.org/10.1080/01639374.2017.1382641.

Brown, Susan. n.d. "Canadian Writing Research Collaboratory." Canadian Writing Research Collaboratory. Accessed July 3, 2022. https://cwrc.ca/about.

Brown, Deidre, and George Nicholas. 2012. "Protecting Indigenous Cultural Property in the Age of Digital Democracy: Institutional and Communal Responses to Canadian First Nations and Māori Heritage Concerns." *Journal of Material Culture* 17, no. 3: 307–24. https://doi.org/10.1177/1359183512454065.

Cariou, Warren. 2014. "Edgework: Indigenous Poetics as Re-Placement." In *Indigenous Poetics in Canada*, 31–38. Waterloo, Ont.: Wilfrid Laurier University Press.

———. 2020. "On Critical Humility." *Studies in American Indian Literatures* 32, no. 3–4: 1–12. https://doi.org/10.1353/ail.2020.0015.

Coulthard, Glen, and Simpson, Leanne. 2016. "Grounded Normativity/Place-Based Solidarity." *American Quarterly* 86, no. 2: 249–55. https://doi.org/10.1353/aq.2016.0038.

CRKN (Canadian Research Knowledge Network). 2022. "Decolonizing Canadiana Metadata: An Overdue Step in Removing Harmful Subject Headings." January 25, 2022. https://www.crkn-rcdr.ca/en/decolonizing-canadiana-metadata-overdue-step-removing-harmful-subject-headings.

Deloria, Vine. 1978. *The Right to Know*. Washington, D.C.: Office of Library and Information Services, U.S. Department of the Interior.

———. 1999. "Self-Determination and the Concept of Sovereignty." In *Native American Sovereignty*, 107–14. New York and London, U.K.: Garland Publishing.

Doyle, Anne M., Sarah Dupont, and Kim Lawson. 2015. "Indigenization of Knowledge Organization at X̱wi7x̱wa Library." *Journal of Library and Information Studies* 13, no. 2: 107–34.

Duarte, Marisa, and Miranda Belarde-Lewis. 2015. "Imagining: Creating Spaces for Indigenous Ontologies." *Cataloging & Classification Quarterly* 53, no. 5–6: 677–702. https://doi.org/10.1080/01639374.2015.1018396.

Duarte, Marisa. 2017. *Network Sovereignty: Building the Internet Across Indian Country*. Seattle: University of Washington Press.

Earhart, Amy. 2019. "Can We Trust the University? Digital Humanities Collaborations with Historically Exploited Cultural Communities." In Losh and Wernimont 2019, 369–90. Minneapolis: University of Minnesota Press. https://doi.org/10.5749/j.ctv9hj9r9.23.

GIDA (Global Indigenous Data Alliance). n.d. "CARE Principles for Indigenous Data Governance." Accessed July 3, 2022. https://www.gida-global.org/care.

Guiliano, Jennifer, and Carolyn Heitman. 2019. "Difficult Heritage and the Complexities of Indigenous Data." *Journal of Cultural Analytics* 4, no. 1. https://doi.org/10.22148/16.044.

Kim, Dorothy, and Jesse Stommel. 2018. *Disrupting the Digital Humanities*. Goleta: Punctum Books. https://doi.org/10.2307/j.ctv19cwdqv.

Lawson, Kim. 2004. "Precious Fragments: First Nations Materials in Archives, Libraries and Museums." (master's thesis) Victoria, B.C.: University of Victoria.

Local Contexts. n.d. Accessed July 3, 2022. https://localcontexts.org/.

Losh, Elizabeth M., and Jacqueline Wernimont. 2019. Introduction to *Bodies of Information: Intersectional Feminism and the Digital Humanities*, edited by Elizabeth M. Losh and Jacqueline Wernimont, ix–xxv. Minneapolis: University of Minnesota Press. https://doi.org/10.5749/j.ctv9hj9r9.

McMahon, Rob, Tim LaHache, and Tim Whiteduck. 2015. "Digital Data Management Resurgence in Kahnawà:ke." *The International Indigenous Policy Journal* 6, no. 3: 1–19. https://doi.org/10.18584/iipj.2015.6.3.6.

Oberländer, Anja, and Torsten Reimer. n.d. "Open Access and the Library." In *Open Access and the Library*, 1–3. Basel, Switzerland: Multidisciplinary Digital Publishing Institute. http://doi.org/10.3390/publications7010003.

O'Neal, Jennifer. 2015. "'The Right to Know': Decolonizing Native American Archives." *Journal of Western Archives* 6, no. 1: 1–17. https://doi.org/10.26077/fc99-b022.

Philpott, Daniel. 2010. *Revolutions in Sovereignty: How Ideas Shaped Modern International Relations*. Princeton, N.J.: Princeton University Press. https://doi.org/10.1515/9781400824236.

Posner, Miriam. 2016. "What's Next: The Radical, Unrealized Potential of Digital Humanities." In *Debates in the Digital Humanities*, edited by Matthew K. Gold and Lauren F. Klein. Minneapolis: University of Minnesota Press. https://dhdebates.gc.cuny.edu/read/untitled/section/a22aca14-0eb0-4cc6-a622-6fee9428a357##ch03.

Recollet, Karyn. 2019. "Choreographies of the Fall: Futurity Bundles & Land-Ing When Future Falls Are Immanent." *Theatre* 49, no. 3: 89–105. https://doi.org/10.1215/01610775-7856639.

Reder, Deanna. n.d. "The People and the Text." The People and the Text. Accessed July 3, 2022. http://thepeopleandthetext.ca/.

Reder, Deanna, and Susan Brown. 2022. "Creating Collections in a Good Way: Incorporating Indigenous Research Ethics into a Database." Paper presented at Visiting to Tell Stories, 2022 Gathering of the Indigenous Literary Studies Association, Saskatoon, June 17.

Reder, Deanna, and Alix Shield. 2019. "'I Write This for All of You': Recovering the Unpublished RCMP 'Incident' in Maria Campbell's Halfbreed (1973)." *Canadian Literature* 237: 13–25.

Robinson, Dylan. 2021. *Hungry Listening: Resonant Theory for Indigenous Sound Studies.* Minneapolis: University of Minnesota Press. https://doi.org/10.5749/j.ctvzpv6bb.

Simpson, Audra. 2014. *Mohawk Interruptus: Political Life Across the Borders of Settler States.* Durham, N.C.: Duke University Press. https://doi.org/10.2307/j.ctv1198w8z.

Suber, Peter. 2012. *Open Access.* Cambridge, Mass.: The MIT Press. https://doi.org/10.7551/mitpress/9286.001.0001.

Tuck, Eve, and K. Wayne Yang. 2012. "Decolonization Is Not a Metaphor." *Decolonization: Indigeneity, Education & Society* 1, no. 1: 1–40. https://doi.org/10.25058/20112742.n38.04.

Verhoeven, Deb. 2021. "Scholarship in a Clopen World." Presented at the Putting Open Social Scholarship Into Practice, University of Victoria. https://inke.ca/putting-open-social-scholarship-into-practice/abstracts/#Verhoeven-Deb. https://doi.org/10.54590/pop.2022.002.

Warrior, Robert Allen. 1992. "Intellectual Sovereignty and The Struggle for An American Indian Future." *Wicazo Sa Review* 8, no. 1: 1–20. https://doi.org/10.2307/1409359.

Williams, Rachel D., and Rebekah Willet. 2019. "Makerspaces and Boundary Work: The Role of Librarians as Educators in Public Library Makerspaces." *Journal of Librarianship and Information Science* 51, no. 3: 801–13. https://doi.org/10.1177/0961000617742467.

Winters, Jasmin, and Justine Boudreau. 2018. "Supporting Self-Determined Indigenous Innovations: Rethinking the Digital Divide in Canada." *Technology Innovation Management Review* 8, no. 2: 38–48. https://doi.org/10.22215/timreview/1138.

CHAPTER 5

"This Game Needs to Be Made":
Playable Theories ⇌ Virtual Worlds

Jon Saklofske

In August 2018, the satirical news website the Hard Times featured a parodic article entitled "New Video Game 'Douche Debater' Lets You Play as Jordan Peterson," which humorously reported that a "spiritual sequel" to the popular videogame series *Phoenix Wright: Ace Attorney* would let players roleplay as Jordan Peterson, controversial University of Toronto psychology professor emeritus, as he uses "intelligence and assholery training to take on a variety of opponents" in hyperbolic debating scenarios (Amory 2018; see figure 5.1).

New Video Game 'Douche Debater' Lets You Play as Jordan Peterson

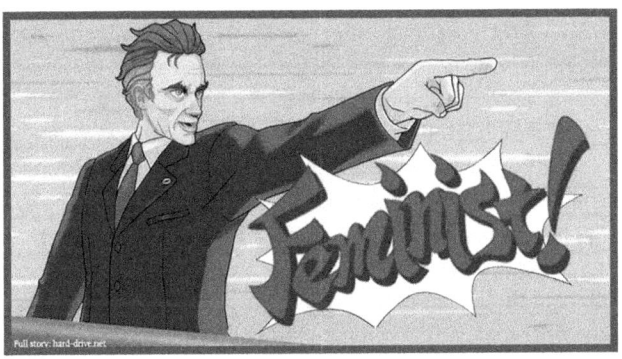

Figure 5.1. *Douche Debater, a parody of Phoenix Wright, Ace Attorney.*
Source: Amory 2018.

That article imagines Peterson's questionable positions and tactics as playable strategies and game mechanics to poke fun at him. While sharing this piece with a colleague, I commented, "this game *needs* to be made," recognizing and acknowledging the rhetorical power and critical effect that such a satirical and playable representation of Peterson would have on people's perspectives regarding him and his ideas.

Despite an increasing amount of excellent critical scholarly work, digital games are still often dismissed by academics, the press, and the general public as escapist and possessing little value in terms of critical thinking or theoretical engagement. However, games resembling the one imagined for the Hard Times article do exist and function quite differently than those designed for amusement. Experiences such as those offered through the Everyday Arcade and Molleindustria websites are designed as rhetorical tools, as interruptions and interventions, as surprising and unexpected design mods to familiar digital-game conventions, and also as a challenge to beliefs that such interactive participatory experiences are neutral, disengaged, leisure pastimes. Everyday Arcade describes itself as an "Emmy-nominated game company that makes playable news," offering reconfigured parodies of more well-known games such as The Voter Suppression Trail (based on the Oregon Trail educational game developed by Don Rawitsch, Bill Heinemann, and Paul Dillenberger in 1971), which critiques voter-suppression techniques experienced by many in the 2016 American presidential election, and Angry Olds (based on Angry Birds, Rovio's 2009 puzzle game), in which the player literally fires "old white men at the pillars of American society," knocking down edifices such as immigration, civil rights, and healthcare (Everyday Arcade 2022a, 2002b). With games such as The McDonald's Videogame (which exposes and critiques the exploitative corporate culture of the McDonald's fast-food chain), and the extremely controversial Operation Pedopriest game, which confronted the "code of silence" and self-protection practised by the Catholic Church in response to sex-crime accusations, Molleindustria defines its purpose as "the reappropriation of videogames" and the "radicalization of popular culture" through "satirical business simulations," "meditations on labour and alienation," "playable theories" and "politically incorrect pseudo-games" (Molleindustria 2003).

As Miguel Sicart affirms in Play Matters, games are political, but the "true political effect of these objects takes place when we occupy them, that is, when they become instruments for political expression" (2017, 73). The games mentioned above are simple rhetorical statements, political expressions, and/or propaganda of one type or another. In other words, they do not function all that differently from many popular commercial games (which rhetorically reinforce normative social values and hyperbolic masculine power fantasies), but they do offer an alternative perspective and a more lucid exposure of such functionality. According to Sicart, these kinds of games "are not played; we perform operations in order to activate and configure their [pre-programmed] messages. That is hardly a creative, appropriative activity. In fact, it is a guided activity through power structures toward purposes dictated beforehand" (73).

Instead of seeing digital games as mindless distractions, as invitations to perform pointed, prescriptive, and already-embedded declarations, or even as "aesthetic forms of rationalization," this chapter suggests an additional opportunity for their literal and conceptual employment in humanities scholarship and theory (Pedercini 2014). For this purpose, it is necessary to move beyond some of the conceptual baggage and traditional limitations associated with the term "game." While there are many competing and restrictive definitions for games, they often involve rule-based systems that promote algorithmic mastery and performative efficiency via competition, exploration, and acquisition.[1] However, the more conventional ideas of what constitutes a game can be considered a subset of the larger category of virtual worlds. Virtual worlds are engineered environments, defined by systems that create complexity through their parallel and conditional operations. Uncritically asserted, virtual worlds can replicate, reinforce, and even idealize conventional narratives and systems of perception, belief, and practice. However, their inherent complexity also enables opportunities for critical and transformative play within such constraint.

Some virtual-world environments have the potential to function as "playable theory," a phrase initially used by Paolo Pedercini to describe Molleindustria's *Free Culture* game in 2008 and subsequently used as a subtitle to the Molleindustria game *Leaky World* (2010), which offers an "interactive interpretation of the essay 'Conspiracy as Governance' by Julian Assange" (Molleindustria 2010). *Leaky World*, which both critiques some of the argumentative shortcomings of Assange's essay while also supporting transparency and whistleblowing overall, requires the player to literally connect the dots between nodes on a world map to establish information networks between "political elites" while also managing (by cutting and reconfiguring) leaky or insecure connections (see figure 5.2).

The simple mechanics involved in this endless process illustrate the conceptual complexities missing from Assange's account of "the drama of transnational power in the information age" (Molleindustria 2010). This attempt to create an experience that adapts, critiques, and expands conventional textual arguments via a particular set of representative mechanics in a virtual environment is a good first step toward the acknowledgement of the relationship between virtual worlds, digital-game experiences, and doing theoretically situated work. I'm inspired by this example and others, such as Jason Helms's metacritical "Play Smarter Not Harder" playable scholarly article (2019), which asks the reader to play a choice-based game about producing a scholarly publication from within Helms's self-conscious and self-reflexive publication. But I also want to note that these conceptualizations and applications are part of a small and relatively

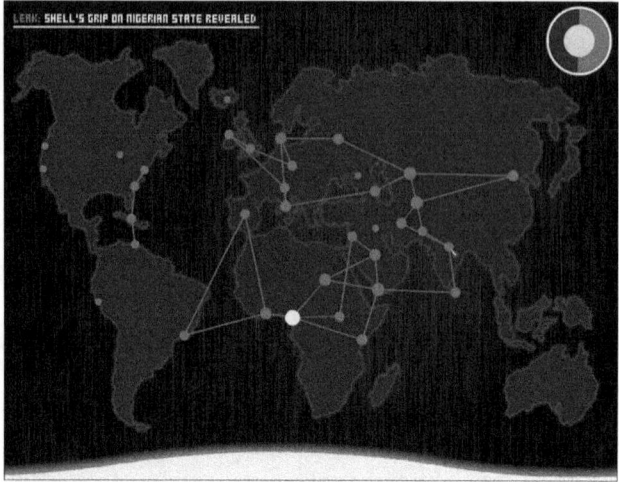

Figure 5.2. Molleindustria's *Leaky World: A Playable Theory Game*.

Source: Molleindustria 2010.

undeveloped approach to seeing the design of and engagement with virtual worlds as ways of constructing, testing, iterating, and communicating theoretical ideas and perspectives. In response, this chapter promotes an extension and further development of the idea of games as playable or experiential theory, as a unification of theory and praxis, and as humanities labs. As humanities labs, games (and their playable virtual worlds/theories) have the potential and flexibility to exist independently from extant real-life systems and imagined collectives (including nationalist fictions and economic metaverses).

Alan Galey and Stan Ruecker equated prototypes with arguments and theories, particularly as both involve a set of ideas, an explanation, and a formulation of principles (2010). This connection is useful in that it establishes and contributes to discourses around critical making and research-creation as alternative ways to model ideas and engage with research questions. It also calls attention to the inherent politics of such work: the prototypes themselves—the ways that they function and are structured—are inherently biased and meaningful, something that Mark Marino has powerfully reasserted in relation to programming code in his 2020 book *Critical Code Studies*. Galey and Ruecker's particular assertion playfully resonates with a programming definition for "argument" that has philosophical implications: "Argument" in programming is "a value passed to a function"—a functional value (Python.org 2022). Similarly, any philosophical argument involves values expressed as functions, expressions, purposes. Many social and political theories are imaginative, creative, speculative, and hypothetical "what if" thought experiments that often isolate processes, functions, or arguments. This what-if type of theorizing is a specific form of prototyping akin to the above conceptualization of virtual world building, to the creation and communication of an imagined, inhabitable, and interactive mental model. In other words, speculative theory prototypes virtual worlds. From within such worlds, we can tell stories, establish and validate narratives, and test characters and characteristics.

Theories thus enable certain stories to be told, emphasized, and promoted, and theoretical world building is about the construction of potential perspectives, the prototyping of interactive theoryscapes. Following from Jesper Juul's assertion that "rules themselves create fictions" (2011, 13), this chapter amalgamates and expands the connections made by Galey and Ruecker between theory, prototype, and argument to argue that digital games and—by extension—virtual worlds are theories and, as indicated by the equilibrium sign in the title of this chapter, that theories are virtual worlds. In essence, then, this kind of speculative theorizing is virtual world building, the communication of an imagined and interactive mental model. To inhabit the worlds imagined and prototyped by theory is to inhabit virtual environments in which particular narrative potentials have been foregrounded. Such speculative simulations model imaginative identifications of, interpretations of, and alternatives to social and political systems. Some examples of theories that involve speculative world building include Michel Foucault's world of power, Jacques Lacan's world of desire and non-relation, Jean Baudrillard's world of simulations, and Roland Barthes's world of semiotics. Beyond this list of white male celebrity theorists, though, many more writers have used theory to engage with and expose lived realities that aren't experienced and often aren't acknowledged by the majority of participants within existing political and social conditions. For Black, feminist, and Indigenous activists such as bell hooks, Audre Lorde, Angela Davis, Patricia Hill Collins, Paulo Freire, Leanne Simpson, Thomas King, and Lee Maracle (to name only a few), theory is already a playful tool of *doing*, of interruption and intervention, of spreading new vocabularies, of narrating past injustices and current ignorance, and of prompting acknowledgement and instigating political and social change. While there are increasing numbers of people in the general public and academy alike working against the continuing marginalization of voices and perspectives, such advocacy and activism are still underrepresented in the digital humanities and the humanities more generally, especially within post-secondary research, communication, and pedagogy.

Conversely, to exemplify how virtual worlds embody theories, Walt Disney's theme parks (which were created with utopian motivations and led to the ultimately unrealized "Experimental Prototype Community of Tomorrow," or EPCOT) were built to embody and facilitate a theoretical narrative of imaginative happiness.[2] However, given the complexity of virtual worlds, the parks also sustain additional, intersecting sub-narratives of capitalism, utopian nostalgia, nationalist propaganda, and homogenization. Also exemplifying theoretical embodiment, 4chan and 8chan's virtual-message-board worlds or environments, shaped by rules of anonymity and the impermanence of one's utterances, demonstrate the social consequences of the lack of accountability in a community (Saklofske 2011).

Beyond the metaphoric playfulness of this association between theories and virtual worlds, a number of potential applications emerge. If interactive, prototypical systems are akin to virtual worlds, and virtual worlds engage users with imagined and interactive mental models and theories, then this idea can be used to conceptualize and critique everything from databases and user interfaces to various forms of digital storytelling. To tease out additional potential applications for this idea in DH practice, virtual worlds are akin to Gilles Deleuze's idea of the "diagram," based on Foucault's use of the term, to describe a way to map relational power. The diagram is "a map, a cartography that is coextensive with the whole social field. It is an abstract machine" (Deleuze 1988, 30). In collaboration with Félix Guattari, Deleuze further clarifies that the "diagrammatic or abstract machine does not function to represent, even something real, but rather constructs a real that is yet to come, a new type of reality" (Deleuze and Guattari 2014, 142). But while a diagram "is a transmission or distribution of particular features," it remains unfinished and adaptable (Deleuze 1988, 73). Thus, programming, mapping, marking up—these are all diagramming practices that trace and reveal relations. Their process is forms of virtual world building and theorizing. As such, these practices *are* digital humanities.

Building a particular theory environment/spatial representation allows for and privileges particular narratives and provides space to test the limits of ideas. Interactive virtual worlds and the theories they embody can be versioned, modded, and hacked as well. And perhaps, most importantly, they are "spaces apart"—spaces removed from (but also embedded within) the everyday; arenas of playful interaction (bounded by clear rules), a "magic circle" that is distinct from the normal rules and reality of the outside world (Saklofske 2019).

However, as Edward Castronova observes (and Juul confirms), there is a permeability to these arenas, a feedback loop between the virtual and the real that inherently complicates and affects the real (Castronova 2005, 147; Juul 2011, 3). Hence, actions sanctioned by or performed within theory worlds are not completely virtual. Virtual worlds/playable theories are more than fantasy escapes—they are experiences. And experiences can and do shape the narratives that we tell ourselves and others outside of the theory space. If we imagine these theory worlds not just as ecosystems but as interactive participatory experiences (like many of our DH experiments), game-like arenas defined by theoretical "rules" which encourage particular behaviours and consequences, what kinds of play and practice are enabled within these worlds and through mechanisms of interaction, via the affordances and constraints established by the arena, procedural interactivity, and rule set(s)?

The kinds of imaginative spaces emerging from this relation between virtual worlds and theories are *speculative simulations*, which model imaginative identifications of, interpretations of, and alternatives to social and political systems. They are exploratory rather than representative or prescriptive. This expansive function of simulation and its relation to other forms of representation and the real is brilliantly illustrated by Franco Landriscina (see figure 5.3).

In Landriscina's conceptualization, *definition* is the process that one engages in when translating and transforming reality into system. A process of *representation* occurs when developing a model out of a system, *exploration* when moving from model to simulation, *revision* when iterating the model after a simulation process,

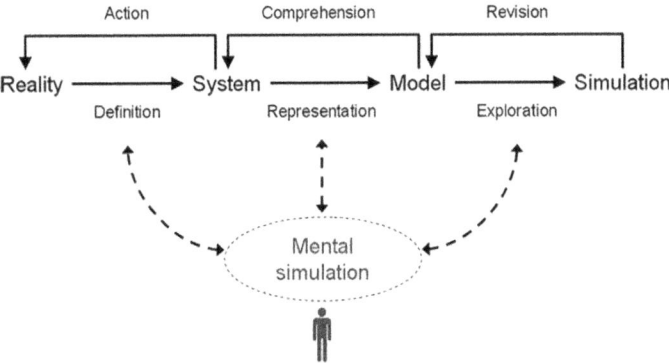

Figure 5.3. The epistemic cycle.

Source: Landriscina 2013, 204.

comprehension when illuminating a system via a revised model, and finally—and most importantly—acting on and within the real based on new understandings driven by this process of simulative transformation. The importance of this last step is highlighted by Sicart's assertion that "politics happens when play becomes political action" (2017, 73). In other words, and according to Landriscina's process, these acts of modelling and simulation enable an opportunity to extend virtual-world theory-play into political action. Regarding simulations, Landriscina also notes that "the main feature of a simulation is the reproduction of a particular aspect of an observed or possible reality. It is not, however, a static reproduction, but an active, or rather, an 'interactive' one" (2013, 12).

This interactivity, the ability for users to affect the systems and environment of the simulation and receive consequential feedback, is what distinguishes such experiences from other forms of theoretical engagement. Degrees of interactivity determine the robustness of simulations, virtual worlds, and prototypes, meaning that some simulations, like many procedural toys, reductive generalizations, or declarative games "often uncritically reflect and reproduce the mundane, so that it can be learned and assimilated" (Sicart 2017, 42). These types of simulations can and have been used to reduce systems and operations into dehumanized and dehumanizing mechanisms, reinforcing algorithmic conformity via reward, discipline, and punishment. Such restrictive virtual worlds are still theories and still political, insofar as they favour and prioritize rigid control and prescriptive interactions. However, this lack of flexibility works against the exploratory and speculative richness of the kinds of simulations Landriscina describes and the kinds of theory worlds this chapter is promoting.[3]

Sicart's emphasis on employing playful design to encourage playful interaction is important to include in the idea of theory worlds to avoid authoritarian modes of theorizing, prototyping, and world building. Play characterizes the extent of interactive possibility allowed by the parameters of a system, but Sicart importantly acknowledges that while play is usually an activity reserved for sanctioned playspaces, *playfulness* involves employing a spirit or attitude of play in environments which are not inherently zoned for play activity (2017, 21). Playfulness is thus a mode of interaction and intervention, bringing freedom and expression to the world outside play (30). Playful designs break away from designer-centric thinking and, as a result, are ambiguous, self-effacing, and in need of a user to complete them (31). Rather than imposing a context, playful designs open themselves to interpretation (31). In addition to featuring such playful design, virtual worlds and the theories they represent, like the toys that Sicart comments on, can be vehicles for play and are most effective and enabling when they embody playful opportunity for the user, when they are ambiguous, encouraging but not entirely directing play, and when the interactions they encourage involve flexibility rather than imposing authority (42). Importantly, for playable theory spaces, playful design downplays system authority, employing ambiguity to encourage but not direct play (31, 42).

Virtual worlds that embody playful design have the capacity to shift mental models, to immerse one in a playable theory environment that encourages revisionary modelling, increased comprehension and awareness, and subsequent action within and beyond the simulation. This leaves room for what N. Katherine Hayles terms "practice": an "embodied skill [that is] intimately involved with conceptualization. Conceptualization suggests new techniques to try, and practices refine and test concepts, sometimes resulting in significant changes in how concepts are formulated" (2012, 19). Braiding Sicart's and Hayles's ideas together results in the notion of *playful practice* and—as a notion essentially characteristic of virtual worlds that function as playable theories—serves as a strong foundation for designing simulations that provoke compassionate and more comprehensive understandings (which is what a DH theory or argument should strive toward) rather than just replicating the technical specifics of a system or uncritically reinforcing systemic operations and processes.

Finally, the revisionary aspects of Landriscina's cycle above also captures both aspects of Sicart's discussion of the function of toys. A toy (which—in the service of the current argument—we can also associate with a virtual world or speculative simulation) is both an expression and a thing (Sicart 2017, 36). As an expression, it encourages play through two types of spatial appropriation: intrinsic (creating a world) and extrinsic (occupying a world). The opportunity to create and occupy worlds via playful engagement with an expressive representation allows us to understand the double function of users in simulations and virtual theory worlds: users (or players) co-create and occupy that world, that imagined and interactive mental model.

In theory, all this sounds quite optimistic and positive. Is this chapter's conceptual effort just an attempt at and equivalent of utopian thinking? Is this playful world building akin to or different from Utopia's totalitarian implications? I would suggest that this effort is more akin to Ruth Levitas's "utopia as method" or IROS (imaginary reconstruction of society) method (2008, 24) She draws from H. G. Wells's idea that "utopianism is a kind of speculative sociology, an attempt to explore and predict what might be, and to expose it to judgment" (Levitas 2005). For Levitas, "the purpose of a utopian method is to bring to debate the potential structure of an alternative society" (2008, 25). Further, she asserts that, "for Miguel Abensour, glossed by Edward Thompson, the point of utopia is its disruptive function and the opening up of a space in which we can experience the possibility of being otherwise [...] a space where we can imagine ourselves desiring differently" (26). Utopia, for Levitas, is a space for the education of desire in the same way that an interactive theory world is an interventional space that encourages revisionary understandings via its performative playfulness.

Levitas's ideas of world building as speculative sociology speaks directly to the connection between virtual worlds and playable theory. But much of utopian thinking, like many non-playful game-based systems, inherently and problematically supports the idea of a universal progress toward a perceived betterment of social conditions that exclusively values certain populations while neglecting others. This flawed idea of development is singular and unidirectional, reinforcing inequitable economic models and promoting deterministic perspectives. Imaginative and interactive virtual theory worlds, inspired by a sense of playful practice, aren't necessarily chained to utopian tensions relating to betterment, forward movement, or an imagined "reconstitution" of society to some once-possessed, now-lost ideal.

More simply, this chapter's argument emerges from a desire for a diversification in the ways that theory is largely still practised and used in academia, in spite of alternative applications and opportunities such as those offered over fifty years ago by Paulo Freire in *Pedagogy of the Oppressed*. I work in a university in which theory's distribution throughout the arts and humanities is largely conceptual and cerebral. Critical theory is employed as a lens to interpretatively unpack literary and cultural artifacts; social and political theory is wielded in an almost classical philosophical sense, to engage students in classroom debates as they prepare texts (like this one) full of theory-celebrity citations and complex verbal gymnastics to prove a conceptual point and demonstrate their competency. And many theoretically aligned scholarly publications, rather than reaching broad publics, are constrained by their terminological jargon and contained in closed-access, overpriced volumes or behind exclusive and restricted subscription models. Alternative models and examples of theoretical praxis and critical engagement (such as Freire's advocacy of participatory action research and Sarah Wright's notion of "mucking in") are ironically taught and contemplated as course content rather than employed as pedagogical methods or activity (Freire 1982; Wright 2017). To fully engage with the kinds of imaginative experimentation that theoretical modelling demands, it is necessary to diversify our platforms beyond writing about writing and to think beyond anachronistic metaphoric models. However, written language still seems (whether out of habit, ease of use, or systematic efficiency) to be an ideal tool and platform for engaging with and developing theory. It is a playful symbolic tool that relies on structure but also encourages figurative flexibility, even to the point of transformatively breaking meaning, built-in knowledge, and habits of use generated by experience. But language is only one type of game, one genre of theory modelling. What about other models of research creation, of world building, that encourage alternative interactions, processes, procedures, and engagement? Theoretical modelling has the potential to critically reconstitute and transform the perspective of the inter-actant in a much more engaged and experiential way than an encounter with the printed word, and such personal transformations (via such involved activity) will encourage and facilitate transformative social interventions more broadly.

DH practice, with its multimedia and multimodal approaches, its visualiza-tion potential, and its novel braiding of qualitative and quantitative data types, has significant potential to critique the status quo, but it is also buttressed by traditional research and scholarly communication habits on the one hand and methodological uncertainty on the other. At times, the computing processes and tools that are harnessed to reveal new patterns and perspectives distance the data from the user, and engage in scales and statistics of understanding and complex-ity that negate the potential for empathy and obscure narrative particularities, which sound the heartbeat of the humanities. To resist the emergent potential in DH work of dehumanizing abstraction either through anachronistic exten-sions of "high theory" or scaled-up pattern-seeking, and in an effort to enable and encourage the possibility of applied theory via intervention, advocacy, and activism, I've been favouring the idea of a playable game space as an alterna-tive, inhabitable environment for interactive encounters with ideas, but—in the same way that we build our own arguments, perspectives, and responses out of language—after encountering the virtual worlds imagined by other thinkers, the process of responsively and responsibly building/creating/making/prototyping a playable virtual space is an equally if not more important way of encounter-ing and refining a theory/idea/argument toward an overall goal of revealing new perspectives and facilitating political action. Given such potential, the building of virtual worlds as a form of research-creation and theoretical inquiry requires values-based design.

Exploring a few already-existing examples will help to illustrate the ways that inhabitable game spaces can function as playable theories that provoke and enable playful critical reflection on the very systems that a user inhabits rather than prescribing and promoting a specific procedurality. *The Uber Game* is a free game that was released on the *Financial Times* website in October 2017 and allows players to play through a week in the life of an Uber driver (the game takes ten minutes to complete), challenging them with the question: "Can you make it in the Gig economy?"[4] This experience's condensation of time and cartoonish representation of its characters and environments implies a less-than-realistic simulation. However, while minute-to-minute realism is not the aim of the game, its ending—in which players realize that, after expenses and Uber's share of their earnings are paid out, their exhaustive efforts (where they work a 66-hour week, choosing to earn money over helping their son with his homework and going out with friends) can leave them earning less than minimum wage and unable to pay their mortgage—relies on actual interview data from Uber drivers. The simulation is simplified but features a number of realistic choices, constraints, and unanticipated challenges that can affect a driver's already-limited opportunities and impede their ambitions and goals. Even more importantly, though, as Marijam Didžgalvytė observes in her article "The Uber Game Shows the Latent Power of Political Video Games" (2018), this game's location positions it "to be played by the *Financial Times* readership—a group of people that generally praise the gig economy, and its lack of bothersome unionization." While its specific influence on *Financial Times* readers isn't possible to assess in a conclusive way, Didžgalvytė rightly points out that this particular audience is likely "not as concerned with stories about worker exploitation," and thus the purpose of the game is puzzling. Is it meant as a subversive intervention which challenges the ideas and politics of its targeted players (and likely goes against the politics of *Financial Times* subscribers) in ways that a statistics-filled article or editorial would not? Is it meant as a "safe" and accessible (simulated) critique of the gig economy's philosophy? Is it meant to humanize the experiences of people that *Financial Times* readers might not often think about in such detail? Less radically, does it simply provide a numbers-based reality check on the unrealistic claims and expectations of gig-economy promises? All these are possible intentions and outcomes, but this interactive experience's fundamental incompatibility with the economic, social, and political philosophies (ideals?) of its publisher and potential audience makes this a unique example of playable theory, an interventional humanist lab space that generates more questions than answers, provoking critical reflection and a playful re-evaluation of assumptions.

Lucas Pope's *Papers, Please* (2013) is another example of a game that gives players a seemingly simple, rule-based task that quickly becomes more challenging, complex, and emotional than anticipated (Balci 1994; Banks quoted in Landriscina 2013). The game positions its player as an immigration officer in November 1982. This officer's job is to screen people who want to cross

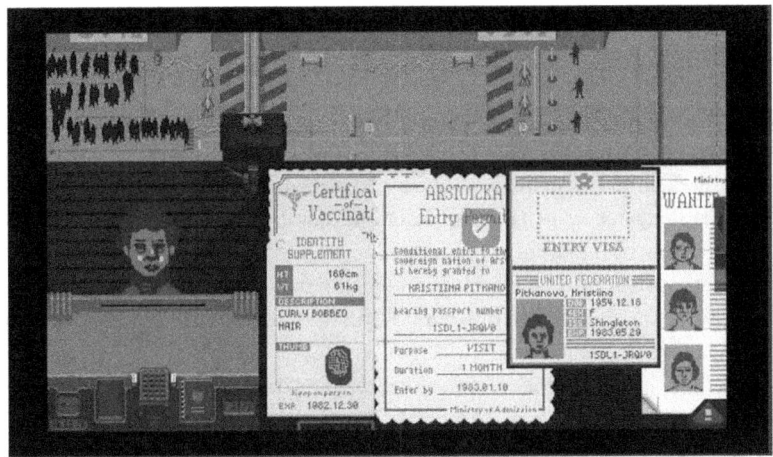

Figure 5.4. Lucas Pope's *Papers, Please.*

Source: Pope 2013.

the border into Arztotska, a fictional nation stylized to resemble the fatigued spirit and infrastructure that accompanied late-Cold War Soviet rule. As Leigh Alexander summarizes in an article that unpacks the design processes of the game, "in order to earn enough to support their struggling family, the player needs to process as many travelers as they can in one day, an objective balanced against the stated goal of only admitting those who have their increasingly-complicated documents in order" (2013; see figure 5.4). These instructions are initially simple and the tasks seem mechanical and unproblematic, but complex-ities soon surface relating to procedural inconsistencies and the narratives that accompany the people who pass by your window.

A learned efficiency competes with time constraints and quotas, establish-ing a tension between metrics and the sometimes-frustrating humanity of the people whose movement you have power over. In addition, as Alexander asserts, "subtle narrative elements dovetail with random ones—the game may ask you to look for a specific wanted individual, but the means of spotting that person are always different. Or, on a day when you're on guard for bomb threats, you may accidentally uncover a sex trafficker." Eddie Lohmeyer's scholarly review of the game acknowledges its potential as playable theory when he suggests that it "puts the player in moral predicaments that prompt a critical reflection of the social and political tensions that exist at the border of two nations" (2017, 14).

However, as Alexander observes, the game "excels as a holistic experience because of its ambiguity [...] never feel[ing] preachy, nor black-and-white [...] and the player never feels in conflict with what 'the game wants'" (2013). Like *The Uber Game*, this experience similarly functions in a less prescriptive and more situationally provocative way, raising questions about assumptions and—even while modelling a reductive overall simulation—managing to challenge players with unanticipated systematic and emotional complexities. Both examples offer instances in which players are prompted to thoughtfully and critically inhabit a simulated system / virtual world / playable theory, questioning its functions and goals alongside their own scripted motivations as they interactively contribute to an emergent experience.

During and after the creation/construction of virtual worlds and play-able theories such as those described above, or any imagined and interactive DH-based model, we need to ask: "Did we build the model right? Did we build the right model?" And—of course—what is "right?" What contexts or questions do these models serve? On the whole, this process of co-creating and occupying theory worlds is a catalyst for (or perhaps even a performance of) a lucid critique of motivations, beliefs, and principles, not to improve systematic or performative efficiencies but to activate playful critical agency.

Is it more useful to think of virtual worlds as theory and theory as virtual worlds than to relate theory to utopianism, to mental models, or to simulations? Not really—this association is not being presented as a "better" option, and this chapter is not promoting an oppositional or competitive stance. "Virtual world" is a broad categorical term that encompasses all these relational understandings while invoking a spatial, experiential "withinness"—an invitation to interactively occupy and critically navigate through the particularities of ideascapes in ways that fuse theory with praxis, that calibrate interactants toward responsive and consequential applications of, and actions resulting from, imagined and inhab-ited systems, models, and simulations.

So when I said to my colleague that the *Douche Debater* game *needs* to be made, what I was really suggesting was that Peterson is usually experi-enced through the passive viewing of YouTube videos, the reading of his words, or through overpriced debates in which he relentlessly controls the conversation through strategic subject changes and distracting rhetorical generalizations, and that this situation, along with its ethical, philosophical, and political implica-tions, needs to be theorized, explored, satirized, and critiqued within an inter-active environment. As well, participatory, interactive experiences that employ game mechanics in virtual arenas are ways and means of theoretical engagement that can extend beyond some of the rhetorical traditions relating to linguistic communication. When designed to provocatively facilitate critical lucidity, these experiences acknowledge that theories are contemplative and speculative environments, which encourage playful testing via their prescriptive/descriptive systems. Most importantly, however, virtual worlds expose theory and practice to a kind of "safe" tangibility. While Peterson's systems of debate can be simulated, his theories are still, thankfully, *virtual* worlds (for now). Who are such worlds designed to favour? How is debate constructed and constricted within such a world? What kinds of interactions are modelled? Where does the simulation break down and who/what does it exclude? How could designing a game world that embodies Peterson's ideas and techniques reveal more about his views and methods and open up possibilities for critiquing his worldview in ways that debating with or writing back to him could not?

While these strategies employ alternate forms of engaging with an idea, acknowledging that virtual worlds are theories and vice versa is not a post-critical stance. The power fantasies and idealizations at the heart of many commercial game experiences reveal the inherent potential of such interactive environments to be wielded as powerful and seductive propaganda, to condition, manipulate, or "program" users into normalizing certain systems of perception and reaction. As with theory, algorithms can be employed in sinister and prescriptive ways rather than being used as provocative calls for playful critique. Thus it is more important than ever to counter the weaponizable and manipulative potential of theories by rendering them playable, and to counter the same potential of virtual worlds by encouraging a critical systems literacy and transparency, by understanding the theoretical values that guide the design of such environments and their underlying processes and procedures. The legibility of virtual worlds depends more on a simple clarity that encourages a critical lucidity than rhetorical flourish.

To take this idea further, what's needed is more meta-critical reflection upon the notion that virtual worlds are playable theories and theories are virtual worlds. As Elyse Graham asserts regarding such metaphoric methodologies: "What meanings does this metaphor constrain, what values does it offer to overwhelm those constraints[?]" (2018, 82). While constraint is a necessary part of such combination, the greater value of this process relates to how metaphoric defamiliarizations result in a transformative understanding of each element being associated, and an interruption of the kinds of biases that emerge from habitual perceptions.

Indeed, the overall purpose of this thought experiment is to kindle discussion about the relationship between making, thinking, using, and critique in DH processes and practices that allow us to imagine otherwise. If every environment we program or design or organize in the digital humanities is constituted by and reinforces a theoretical position, and this theoretical position presupposes a particular landscape of rules, habits, and assumptions, then as both builders and users, an awareness of the interaction between the topography and topology of these theory worlds will embed a critical, consequential playfulness into our creation and occupation of them as we discover their opportunities and limits.

NOTES

1. For an extensive discussion of the various definitions associated with games, see Stenros (2017); for a more specific discussion of digital-game definitions, see Arjoranta (2019). A humorous website that randomly generates definitions of "game" from a database of phrases found throughout game-studies scholarship (and which differ upon each visit) can be found at http://gamedefinitions.com/.
2. For more about EPCOT, see the Walt Disney Company promotional film from 1966 at https://youtu.be/sLCHg9mUBag?t=630.
3. Some mechanical and procedural toys or games (such as *SimCity*) do not require us as interactants—they position us as observers and tinkerers. These are not the kinds of virtual worlds, theories, prototypes, or simulations that concern this chapter.
4. See "The Uber Game," *Financial Times*, October 2017, https://ig.ft.com/uber-game/.

REFERENCES

Alexander, Leigh. 2013. "Designing the bleak genius of *Papers, Please.*" *Gamasutra.* September 3, 2013. https://www.gamasutra.com/view/news/199383/Designing_the_bleak_genius_of_Papers_Please.php.

Amory, M. J. 2018. "Douche Debater." Hard Times, August 20, 2018. https://hard-drive.net/douche-debater-game-jordan-peterson/.

Arjoranta, Jonne. 2019. "How to Define Games and Why We Need to." *The Computer Games Journal* 8: 109–20. https://doi.org/10.1007/s40869-019-00080-6.

Balci, Osman. 1994. "Validation, Verification, and Testing Techniques Throughout the Life Cycle of a Simulation Study." *Annals of Operations Research* 53: 121–73. https://doi.org/10.1007/BF02136828.

Castronova, Edward. 2005. *Synthetic Worlds: The Business and Culture of Online Games.* Chicago, Ill.: University of Chicago Press. https://doi.org/10.7208/chicago/9780226096315.001.0001.

Didžgalvytė, Marijam. 2018. "The Uber Game Shows the Latent Power of Political Video Games" *Kotaku UK*, February 18, 2018. https://www.marijamdid.com/home/https/critical-distancecom/amber/cache/98cd2e5031f2812ee6ac51ff5a f17082?rq=uber.

Deleuze, Gilles. 1988. *Foucault*. Translated by Seán Hand. Minneapolis: University of Minnesota Press.

Deleuze, Gilles, and Félix Guattari. 2014. *A Thousand Plateaus: Capitalism and Schizophrenia*. Translated by Brian Massumi. Minneapolis: University of Minnesota Press.

Everyday Arcade. 2022a. "The Voter Suppression Trail." Accessed June 20, 2022. https://everydayarcade.com/games/the-voter-suppression-trail.

———. 2022b. "Angry Olds." Accessed June 20, 2022. https://everydayarcade.com/games/angry-olds.

Freire, Paulo. 1982. "Creating Alternative Research Methods. Learning to Do It by Doing It." In *Creating Knowledge: A Monopoly?* edited by Budd Hall, Arthur Gillette, and Rajesh Tandon, 29–37. New Delhi, India: Society for Participatory Research in Asia.

Galey, Alan, and Stan Ruecker. 2010. "How a Prototype Argues." *Literary and Linguistic Computing* 25, no. 4: 405–24. https://doi.org/10.1093/llc/fqq021.

Graham, Elyse. 2018. *The Republic of Games*. Montréal, Que., and Kingston, Ont.: McGill-Queen's University Press.

Hayles, N. Katherine. 2012. *How We Think: Digital Media and Contemporary Technogenesis*. Chicago, Ill.: The University of Chicago Press. https://doi.org/10.7208/chicago/9780226321370.001.0001.

Helms, Jason. 2019. "Play Smarter Not Harder: Developing Your Scholarly Meta." *Scholarly and Research Communication* 10, no. 3 (July). https://doi.org/10.22230/src.2019v10n3a333.

Juul, Jesper. 2011. *Half-Real: Video Games Between Real Rules and Fictional Worlds*. Cambridge, Mass.: The MIT Press.

Landriscina, Franco. 2013. *Simulation and Learning: A Model-Centered Approach*. New York: Springer. https://doi.org/10.1007/978-1-4614-1954-9_5.

Levitas, Ruth. 2005. "Imaginary Reconstitution of Society." Inaugural Lecture. University of Bristol. October 24, 2005. http://www.bristol.ac.uk/media-library/sites/spais/migrated/documents/inaugural.pdf.

———. 2008. "Being in Utopia." *Hedgehog Review* 10, no. 1 (spring): 19–30. https://hedgehogreview.com/issues/imagining-the-future/articles/being-in-utopia.

Lohmeyer, Eddie. 2017. "*Papers, Please* as Critical Making: A Review." *Press Start* 4, no. 1 (2017): 11–16.

Marino, Mark. 2020. *Critical Code Studies*. Cambridge, Mass.: The MIT Press. https://doi.org/10.7551/mitpress/12122.001.0001.

Molleindustria. 2003. "About." http://www.molleindustria.org/blog/about.

———. 2007. *Operation Pedopriest*. https://molleindustria.org/en/operation-pedopriest/.

———. 2008. *Free Culture*. https://www.molleindustria.org/en/freeculturegame/.

———. 2022. "Game Definitions." http://gamedefinitions.com/.

————. 2010. *Leaky World: A Playable Theory*. http://www.molleindustria.org/leakyworld/leakyworld.html.

————. 2019. *The McDonald's Videogame*. First version 2005. https://molleindustria.org/mcdonalds/.

Pedercini, Paolo. 2014. "Videogames and the Spirit of Capitalism" Talk given at *Indiecade East*, February 2014. https://www.molleindustria.org/blog/videogames-and-the-spirit-of-capitalism/.

Pope, Lucas. 2013. *Papers, Please*. https://papersplea.se/.

Python.org. 2022. "3.11.0 Documentation Glossary." https://docs.python.org/3/glossary.html.

Saklofske, Jon. 2011. "Inb4 404: Using 4chan.org to Challenge the Stasis Quo Illusion of Media Stability." Conference paper presented at *Media in Transition 7: Unstable Platforms: The Promise and Peril of Transition*. Cambridge, Mass.: Massachusetts Institute of Technology, http://web.mit.edu/comm-forum/legacy/mit7/papers/Saklofske%20MIT7%20Paper.pdf.

————. 2019. "Spreadable Jams: Implementing Social Scholarship through Remodeled Game Jam Paradigms." *KULA: Knowledge Creation, Dissemination, and Preservation Studies* 3, no. 1. http://doi.org/10.5334/kula.45.

Sicart, Miguel. 2017. *Play Matters*. Cambridge, Mass.: The MIT Press.

Stenros, Jaakko. 2017. "The Game Definition Game: A Review." *Games and Culture* 12, no. 6: 499–520. https://doi.org/10.1177/1555412016655679.

Wright, Sarah. 2017. "Critique as Delight, Theory as Praxis, Mucking In." *Geographical Research* 55: 338–43. https://doi.org/10.1111/1745-5871.12207.

CHAPTER 6

Reimagining Representational Codes in Data Visualization: What Contemporary Digital Humanities Might Learn from Visual Arts-based Disciplines

Julia Polyck-O'Neill

The ways visual forms of information are understood reflect the knowledge contexts in which they are situated; this generates a broad range of possibilities for interpretation. Disciplinary and epistemic variance also renders visual media vulnerable to potential misinterpretation and misunderstanding. Theorist Johanna Drucker begins her 2014 book *Graphesis: Visual Forms of Knowledge Production* with the evocative claim that, "even though our relation to experience is often (and increasingly) mediated by visual formats and images, the bias against visual forms of knowledge production is longstanding in our culture" (16). Further compounding this statement, in her explorations of the state of the emerging field of digital art history in her aptly named 2013 study "Is There a 'Digital Art History?'" Drucker provides a number of insights that suggest that the relationships between visual studies, art history, and the digital humanities have fallen out of step with those relating to other, non-visual disciplines. If, as she suggests, text has a "one-to-one relation of source to code" while, "by contrast, images do not have a natural equivalent in digital form" (2013, 8), it should come as no surprise that disciplines predicated in the visual and its interpretations might face epistemic and ontological challenges in environments constructed around the conventions of text-based media, and thus be more challenging to translate to digital methods.

On data visualization more specifically, or what she calls "the graphic expression of data," Drucker (2018) points out further inconsistencies: the "traditional work of scholarly interpretation, at the level of individual artifact or text, often seems at odds with the computational processing that produces data visualizations" (248). She argues that the visual strategies used by digital humanists—namely, the graphic forms used for data visualization—were customized to the needs of non-humanities environments ("natural sciences, social sciences, business applications," etc.), and so, among other issues, they bear "the hallmarks of positivist, quantitative, and or statistical approaches to knowledge" that render them less useful for the hermeneutic methods humanists most frequently employ (248).

Drucker's argument for the inconsistencies in translation between visual and non-visual media and representational systems is, of course, not new, and not unusual in traditional humanistic disciplinary spaces. Literary and art theorists have long debated the potential for reconciliation or translation between text and the visual, and have examined the rhetorical potential and potency of images and other visual media. In *Picture Theory* (1994), W. J. T. Mitchell, in thinking through the "theoretical" image he names the "imagetext," posits,

> if the relation of the visible and the readable is (as Foucault thought) an infinite one, that is, if "word and image" is simply the unsatisfactory name for an unstable dialectic that constantly shifts its location in representational practices, breaking both pictorial and discursive frames and undermining the assumptions that underwrite the separation of verbal and visual disciplines, then theoretical pictures may be mainly useful as de-disciplinary exercises. The working through of their formal specificity and historical functioning may leave us with nothing more than a pragmatics loosely grounded in tradition. (1994, 83)

Mitchell identifies how, in attempting to bring the two uneven frames together, disciplinary boundaries are further crystallized, focalized, and infixed. The same appears to happen when images are imported—not simply digitized, but brought, as visual information for critical analysis—to digital environments, but as Drucker elucidates, along different, if conceptually related, disciplinary and epistemological lines. And these lines are often overlooked in conventional DH environments, where visual media is often treated as second-order representation for text-based data rather than a separate field that might benefit from its own distinct frame of reference in order to enhance its applications and affordances.

One of the key aspects of art-based disciplines that differentiate them from those which deal predominantly in text or verbal media is that the way visual information is processed and interpreted is epistemologically and ontologically unique, and specific. Mitchell argues,

> the "differences" between images and language are not merely formal matters: they are, in practice, linked to things like the difference between the (speaking) self and the (seen) other; between telling and showing;

between "hearsay" and "eyewitness" testimony; between words (heard, quoted, inscribed) and objects or actions (seen, depicted, described); between sensory channels, traditions, and representation, and modes of experience. We might adopt Michel de Certeau's terminology and call the attempt to describe these differences a "heterology of representation." (1994, 5)

As such, it follows that within an arts-based framework for analysis, methods for the visualization of data might thus be interpreted differently, and with sensitivity to the inherent multiplicity of material-visual methodologies. For instance, the pluralistic exhibition/project *Object:Photo, Modern Photographs from the Thomas Walther Collection, 1909–1949*, curated by Quentin Bajac and Sarah Hermanson Meister at New York's Museum of Modern Art (MoMA) in 2015, uses conventional, primarily and recognizably quantitative, DH methods, such as the visual mapping of geographic and formal relationships with line and scatter plots, for visualizing and visually modelling information for a context that suggests chiefly qualitative experience and modes of interpretation.

In my analysis here, I briefly discuss extant definitions of and ideas framing data visualization as they circulate in generic DH contexts, and thus, in the context of representation and data, focalize digital methods for data visualization. I examine, by means of select contemporary case studies, how the application of inherently visual methods for translating, processing, analyzing, and producing data within the broad frame of visual studies and art history might have the potential to produce results that effectively transcend (and in ways subvert) the one-to-one relation of source (a location from which information is gathered) to code (the symbolic language or set of instructions used by programmers in transforming data). By considering the ways that visual art forms interact with data to produce meaning in *Object:Photo*, the Mediated Matter group's Vespers project (2015–2018), Theaster Gates's series of paintings in his 2017 exhibition *But To Be A Poor Race*, and the 2018 exhibition *Coder le monde (Coding the World)* at the Centre Pompidou, in Paris, I suggest that there are several extant models for innovative methodological collaboration between visual studies, art history, and the digital humanities, and that such creative reimaginings of the representational codes of digital visual methods enables both an intellectually productive and detailed, sensory experience of data interpretation. Catherine D'Ignazio and Lauren Klein posit in *Data Feminism* (2020) that data are contextual and pluralistic in nature, and by nature rely on invisible, embedded structural elements for analysis. The methods and models I examine connect to and draw from ideas such as those

D'Ignazio and Klein discuss in relation to data feminism as a toolkit for rethinking representation and interpretation in digital contexts.

Before I move to working definitions, it is useful to note the multiplicity of common forms and methods for visualization. Duke University's library guide, for instance, lists the following varieties of data visualization, a taxonomy based on that published in a widely cited 1996 paper by computer scientist Ben Shneiderman: 1D/linear; 2D/planar (including geospatial); 3D/volumetric; temporal; nD/multidimensional; tree/hierarchical; and network (Shneiderman 1996; Duke University Libraries 2019). These commonly used forms and methods have been adopted as the formal visual language used by DH scholars and computer scientists and employed within a wide variety of communication environments, and although they serve to effectively describe data visually, they are rarely engaged with as aesthetic information. That is, despite the formal qualities of data-visualization methods, which translate data as visual information, such information is rarely interpreted according to its poetic and aesthetic qualities; these observations also apply to the ways that visualization is used and formally defined in the language of data visualization in its contexts of practice.

To examine the specific issue of how visualization is used to represent information in DH work, I look to basic definitions, such as that published by Nikos Bikakis of the ATHENA Research Center in Greece for Springer's *Encyclopedia of Big Data Technologies* (2018):

> Data visualization is the presentation of data in a pictorial or graphical format, and a data visualization tool is the software that generates this presentation. Data visualization provides users with intuitive means to interactively explore and analyze data, enabling them to effectively identify interesting patterns, infer correlations and causalities, and supports sense-making activities. (1)

In *The Encyclopedia of Human-Computer Interaction* (2016), data-visualization specialist Stephen Few offers a simple definition in "Data Visualization for Human Perception":

> Data visualization is the graphical display of abstract information for two purposes: sense making (also called data analysis) and communication. Important stories live in our data and data visualization is a powerful means to discover and understand these stories, and then to present them

to others. The information is abstract in that it describes things that are not physical. (entry 35)

I also examined the specific definitions employed by large research institutions, such as, for instance, the Fondazione Bruno Kessler (FBK), a research non-profit public-interest entity specializing, most notably, in the development of AI. The FBK website defines data visualization as follows:

Visualization tools and techniques are crucial to the analysis of digital humanities data, especially in cases of large amounts of data. Current visualization techniques allow now a better communication of ideas and analysis results than verbal communication. Therefore, the exploration and implementation of novel visualization techniques for displaying processed material in graphical format is an important research topic, helping to mediate a message for different types of audience. (FBK, n.d.)

These definitions share several qualities that point to their contexts of use: visualization is here employed as a tool for the interpretation of data, and that the audience is familiar with the visual (pictorial, graphical) information as a specifically designed apparatus for the communication and analysis of data seems to be presumed. Bikakis's framing of data visualization as providing "users with intuitive means" and Few's acknowledgement of the "abstract" "not physical" qualities of the information and its visualizations suggests a recognition of the potential for hermeneutic ambiguity in the deployment of visualization as an analytic tool, but, for the most part, these definitions suggest that data visualization is a means for the delivery of information from user or producer to audience.

Visualization, in conventional analogue or digital contexts, remains for many standard practitioners and publics a largely misunderstood method for information processing and analysis. It is frequently used to convey quantitative and/or qualitative information—though this information is effectively here *quantified*—in an accurate and succinct manner, but often, as with much visual media, omitting steps encouraging critical engagement and, subsequently, lacking the lens or lenses necessary to intellectually (or affectively, psychologically) parse the information conveyed. Basic components like form, scale, and colour can be used to manipulate findings, and complex information is frequently reduced and oversimplified. Largely, the visualization of information is either secondary to text-based analysis, existing in dialogue with written studies, and is challenging to interpret as a stand-alone representation of information, including research findings, or it exists as a second-tier component of information processing, stemming from text-based information, including written analysis and/or statistics. See, for instance, the controversies around visualizations of the statistical information about U.S. election results, as examined by a 2016 study by the National Opinion Research Center at the University of Chicago. Historically, these results have been visually oversimplified, reducing voting results to choropleth maps, the recognizable schematics consisting of red- and blue-coloured states readily used by a variety of media sources. Not only does this visual method leave little room for narrative ambiguity, but the use of this visual rhetorical convention obfuscates important details related to the functioning of the Electoral College system, and the imbalance between the geographical size of states and their population density causes further confusion when information is communicated according to this media convention.

Borrowing evidence from the 2016 study "Digital humanities is text heavy, visualization light, and simulation poor" by Erik Malcolm Champion, a professor of media studies and the UNESCO chair of cultural heritage and visualization, aspects of literacy in the digital age require further attention in order to correspond to the rapid evolution of digital culture. Champion points out that, as UNESCO notes in its 2014 study of world literacy, while literacy is generally increasing, technology is widening the divide between those who can "read" (and understand, access, and/or manipulate this technology) and those who cannot, noting varying issues related to the accessibility of digital technology, as well as the general economic and cultural forms of gatekeeping that continues to affect the dissemination of digital research projects. However, as legal scholars Anshul Vikram Pandey et al. (2015) observe, there also seems to be a need for visualization literacy, specifically, as the public appears to be far more easily convinced by visualizations than by reading text. The implication is that, for the public, visualization literacy as a formal communicative mode is generally not as discerning.

Put into meaningful dialogue, these findings suggest that digital-visual cultural literacy is a field that requires specialized attention because it concerns not only digital literacy but particularized forms of visual literacy that impart specific effects. It is a significant challenge to call for a widespread initiative to educate scholars from outside visual-arts-based disciplines to become experts in the creation or interpretation of visual media; as with other disciplinary specializations, the acquisition of the specific knowledge base and skill sets involved in visual-arts-based disciplines requires rigorous engagements with discipline-specific epistemologies that, realistically, are not easily accessed without rigorous study and training. Instead, I suggest that scholars with expertise in these disciplines might be called on to contribute to the development of more refined forms of visual literacy that might inspire more nuanced approaches to the material and conceptual aspects of visual media in digital environments—approaches that correspond with the data at hand at a more granular level.

My exploration of these issues was partially inspired by personal experience. I attended a one-day data-visualization workshop run by Compute Canada, one of Canada's leading organizations in "lead[ing] the acceleration of research innovation by deploying state-of-the-art advanced research computing (ARC) systems, storage and software solutions" (n.d.), as part of the 2016 Digital Humanities Summer Institute at the University of Victoria, and was disappointed that there was no time dedicated to discussing the significance of issues that, as an emerging scholar in critical digital humanities with a formal background in art history and visual cultural studies, I believe to be of central importance to understanding practice. Why were issues such as the complexity of the design decisions and aesthetic arguments exercised in constructing visual components to the research not addressed? The entire introductory workshop was narrowly focused on the methodological and mechanical aspects of the building of visualization components, and operated according to the assumption that the visual rhetoric underlying the exercise and its technologies were neutral, objective, and accurate. In my follow-up research and discussions with experts in the data-visualization field, as well as with leading critical DH scholars, I learned that the reason these conversations are not often a part of the curriculum in data-visualization learning environments is because they are, in an industry context, largely considered to be secondary to the technology from the standpoint of the conventional computer-science-informed DH creator/code worker. From the almost-banal definitions and references above, we can deduce that there is a shared understanding of data visualization within the field that corresponds to my earlier generalized observations that ideas of complexity and open interpretation are frequently not at the forefront of discourse, theory, and practice, and that the concerns of specialists in the interpretation of visual media are simplified or frequently left out of the conversation.

But, as the DH field broadens exponentially, increased attention is being paid to these more "secondary" concerns at multiple levels, including from the position of specialists in visual technologies and media. As Lauren F. Klein and Matthew K. Gold explained in the introduction to *Debates in the Digital Humanities 2016*, much has changed in the years between the so-called emergence of the digital humanities as a common field in approximately 2012 and the current, accelerated state of DH scholarship. As they posit three years later, in the updated

2019 collection (the third iteration of their anthology), the progressively more nuanced critical conversations around DH scholarship have come into sharp focus, particularly around academic and media activism in the face of growing global injustice, as well as in terms of the "maturation" of the digital humanities, indicated by means of the "deepening and narrowing of scholarly niches within the field" (Gold and Klein 2019).

The boundaries and very definitions of what constitutes DH work have expanded as critical engagement and intervention have increased, creating what is effectively a pluralistic, dynamic playing field. This includes the areas of visual studies and art history, where digital humanities and digital art history were once seen as being set apart from the traditional disciplines. But one could say that this is a divide that is being increasingly bridged by the development of methods for critical making (here we might think of the field of new-media art), where the fine-tuned aspects of critical analysis within the disciplines have been merged with methodologies for making—but somehow specialist conversations about data visualization and its futures are not yet a part of this. Bringing scholars of the arts into conversation is a first step, and might begin to create shifts in the use and development of tools for the visualization of data that more accurately reflect scholarly findings. Moreover, with care and attention, such developments will have the capacity to inspire increasingly responsive interactions with visual forms of data, promoting new, nuanced forms of interpretation.

OBJECT:PHOTO AS A FLASHPOINT

One example of how visual studies, art history, and the digital humanities have come together to suggest the potentiality for bringing contemporary digital methods for mediation and sense making in a single, multiplatform, pluralistic project is the 2014–2015 project-as-exhibition *Object:Photo*, which engaged with the Thomas Walther Collection, a significant collection of photographs from the early twentieth century. I have worked on this project in previous research with media scholar Aleksandra Kaminska as a way of discussing and critically analyzing the uptake of online projects within larger arts institutions, and this project represents a moment when an influential art institution, in this case MoMA, aggressively incorporates a project using current DH methods and tools to expand their programming into digital spaces and technologies. Thus, *Object:Photo* provides a key exemplar for institutions striving to meet twenty-first-century mandates for accessibility, outreach, and the use of technology in galleries, libraries, archives, and museums (GLAM) programming—something many traditional institutions struggle with, especially as funding bodies (and arguably publics) demand the integration of this genre of project. In our research, Kaminska and I discovered that institutions largely outsourced (and continue to outsource) the digital components of projects to specialized technology workers, and that, as a result, these institutions struggle to speak to the mechanical and methodological details of projects, which, in this instance, seems to be reflected in the presentation. The maps and timelines of *Object:Photo* speak primarily to the aesthetics and functionality of a museum finding aid, providing access to relational information that is primarily positivistic in nature (facts about the artists' geographic locales and the formal qualities of the photographs from the Walther Collection), ascribing to, and upholding traditional conventions in, the history of photography. That said, *Object:Photo*, as a collective, multipart, and polyvocal project, is successful in that it is a catalyst in inspiring a wide range of visual arts and media scholars to unite, by formal academic, published, and thus archived but often disparate means in engaging with ideas of what the data representations in the various components of the exhibition represent, how they are constructed, and what is reduced or simplified.

The project's website includes four sections: visualizations, essays, topics, and a photo gallery. The visualizations are fairly straightforward, and using open-source tools, present two different forms of visualization, which use techniques for geospatial mapping and network tracking developed in the Software Studies Initiative research lab, in San Diego and New York—a robust lab and design studio run by media and visual studies (star) scholar Lev Manovich, a specialist at the forefront of the development of data visualization for visual and media studies. The published collection of 29 critical essays (2014), edited by Mitra Abbaspour, Lee Ann Daffner, and Maria Morris Hambourg, reflects on aspects and themes of the project, including the materiality of photography (such as in McCabe's "Noble Metals for the Early Modern Era") and the digital methods used in the project (Hochman and Manovich's "A View from Above: Exploratory Visualizations of the Thomas Walther Collection"), and the topics and gallery sections present as standard web features, allowing visitors to access images and information.

A 2016 College Art Association review of *Object:Photo*, authored by art historian Dana E. Byrd, presents what might be read as a characteristic response from the perspective of a specialist. In dealing with issues of digital and data literacy in its design, much interpretive nuance is obfuscated and unavailable. Byrd notes that visualization techniques in no way stand in for the complex roles of art historians, researchers, or conservators, even though these strategies seem to gesture to such ends, and suggests that the project best suits a non-specialist audience

(Byrd 2016). The very aspect of the project that makes it accessible to broader publics also has the potential to steer the interpretive possibilities, and the results of even the most benign attempts to create nonlinear pathways to analysis, such as the interactive timelines comparing materials, subjects, or styles, become staggeringly linear and flat. But, thinking now about the specialist, informed audience that Byrd's critique seems to invoke, the constellation of *Object:Photo*'s strengths and its shortcomings as an attempt to bring together so many elements with technologies that are changing the way research is conducted and information, material and immaterial, is processed and communicated, might, in effect, be the goal of the project. The project itself seems to invoke critical response, which also signals a sea change in public programming. The website specifies the following:

> As the field of digital humanities matures, we hope consortiums of like-minded researchers and institutions will seek to pool the results of analytic initiatives in an interdisciplinary fashion. Research models such as these, based on collaborative research, interpretation, and dialogue promise to bridge the allied fields of art-historical and materials research. (Abbaspour, Daffner, and Hambourg 2014)

From my perspective, looking back on this project, which was new when we first took it up as a case study for our 2015 research, it now exists as a flashpoint, signalling pathways for the convergence of arts contexts with specialized digital methods—data visualization in particular—and representing the state of DH and digital art-historical research at that time. It presents and employs representational codes, but stops short of critically engaging with them; it also highlights the discrepancies between DH conventions and the interpretive language used in arts contexts. Even for Manovich and his lab, this 2014 project served as a practical component of a longer research conversation about visual studies, data, and how that data are represented, as his ongoing work combining such topics as cultural analytics and artificial intelligence suggests. While, as it stands, *Object: Photo* shines a light on the shortcomings and discursive gaps Drucker observes in her study, contrasting digital art history to other forms of digital humanities, it also serves as a stepping stone along the path to new modes of thinking about and through potential futures in the visualization of information from the standpoint of experts in visual disciplines. As a high-profile project, it certainly brought increased attention to how visualization technologies might be integrated with the specialized interests of arts practitioners and scholars.

There are a number of other ways that visualization technologies have been brought into the art world that are less linear, which might suggest the kind of potential for a transcendence or subversion of conventional approaches to data visualization I foreground in my discussion. My analysis of the following examples suggests how artists' interactions with data and the use of digital representational strategies in art environments, or, for in at least one case, of artistic strategies in digital environments, has the potential to open new pathways for the central participation of scholars and practitioners specializing in visual and art studies in the construction of visual representational systems in DH contexts. The strange temporality of digital-material explorations, which combine ephemerality and uncertain archival practices, hinting at a kind of semi-permanence (think of the issue of techno-obsolescence), it is here that I turn to two examples of projects that are conceived according to an entirely different context for and process of making.

VESPERS: VISUAL-MATERIAL MEANING

If MoMA's curatorial team outsourced much of the digital building and expertise to external specialists in digital media and methods, digital expertise and creative maker combine in the Vespers project. The three-part serial project started in 2016 and led by designer and scholar Neri Oxman and the Mediated Matter group at the Massachusetts Institute of Technology, in collaboration with 3D printer manufacturer Stratasys, produces 3D-printed "death masks," inspired by and, in ways, recreating concepts and models of the death-mask genre.[1] The masks propose to "reveal cultural heritage and speculate about the perpetuation of life, both cultural and biological," and "express the death mask's deeper meanings and possible future use" (3D Printed Art and Design World, n.d.). Oxman and her team use data sets and algorithms to combine emerging modelling technologies with historical craft, a production process effectively combining data-visualization methods with sculpture, philosophy, archaeology, and anthropology, among other disciplines. The three parts of Vespers encode abstract-conceptual and physical-historical data in sculptural form. The death masks, made primarily from a kind of resin often used in 3D printing, often combined with various "natural" materials (microbiological and geological), take on a number of different forms, all reminiscent of the mask genre (roughly corresponding to the shape and scale of a human face), but each series responds to a different set of broad cultural-ecological queries that has to do with the ecological and philosophical uncertainty and unknowability of the future.

What is useful here, in terms of the renegotiation of the meaning and interpretive modes encoded in data visualization in the context of visual studies and art, is that Vespers asks the specialist viewer to critically engage with the data being visualized, and to consider the aesthetics and the poetics of the processes involved, and the product, the object, itself. The masks, as compelling, alien objects of beauty that occupy and manifest a hybrid category of sculpture, anthropological or cultural object, and utilitarian tool in the contemporary moment, are difficult to describe according to conventional systems of classification. Furthermore, the material, the form, and the concepts are considered in relation to one another; a relational approach to data and its visual representation encourages both an ethical and an aesthetic responsibility for how (and why) information is presented to the viewer, and asks the producer to reconsider the potential for significant (and signifying) interconnections and data exchange and representation according to a range of scales of meaning. Encountering the project requires the viewer to challenge presupposed responses to the objects as amalgamations of visual and material information, and to engage in a form of critical interpretation premised in the specifics of what is being viewed rather than broad generalizations.

THEASTER GATES: THE META-ABSTRACTION OF DECONTEXTUALIZED VISUALIZATIONS

The convergence of what might be thought of as the aesthetics and poetics of the processes and product is one aspect of what arts specialists might find lacking in conventional data-visualization practices. As such, perhaps one of the conceptual hurdles that require attention in bringing cutting-edge technologies into the GLAM context has less to do with the strategic use of digital tools and methods and more to do with their associated viewing conventions. Yes, there is something mathematical and scientific that drives the production of the visualization, but when imagining datasets and their subsequent visual representations there is room to broaden the conversation. There is room for nuance and contextual critique. The example that follows foregrounds the importance of thinking through the concept of representation: who and what are being rendered in these methods, for and by whom?

Theaster Gates's canvases in *But To Be A Poor Race*, shown at Regen Projects, Los Angeles (January 14–February 25, 2017), combine, in a quite analogue manner, painting and "statistical mapping," or data visualization. The visualized data borrow directly from the sociological findings and hand-drawn infographics (a form of analogue data visualization) of W. E. B. Du Bois, exploring information related to land ownership, education, and domestic life in twentieth-century African American households. Part of what gives these works poignancy is Gates's praxis as an activist archivist of African American heritage, as well as the deliberately ambiguous design of the images, which omit the information commonly associated with these kinds of graphs in favour of indeterminacy. For example, *Mountain Aura* (2017) is a large-scale work in latex and acrylic on an aluminum panel (182.9 × 124.5 cm) that is part of a series of related images, and takes the image from Du Bois's infographic *The Amalgamation of the White and Black Elements of the Population in the United States* (1900) and strips it of all textual and numeric elements. This erasure results in the image of an indeterminate, minimalistic, monumental form composed of four colours, organized darkest to lightest, here renamed a "mountain," with a colour scheme suggesting an "aura." Other artworks in the series are similar; they become minimalist shapes on plain backgrounds. Although these images have been decontextualized in Gates's work, they retain their connections to Du Bois's visualizations of sociological data prepared as part of his contribution to the "Exposition des Nègres D'Amérique," or "The American Negro Exhibit," a collection of graphs, charts, tables, and maps that were "generated from a mix of existing records and empirical data that had been collected at Atlanta University by [his] sociological laboratory" (Battle-Baptiste and Rusert 2018, 9). Without the inclusion of textual or numerical information, Gates's images are "unreadable," but their direct invocation of Du Bois's project and recognizable informational forms lends them conceptual and ideological weight. This tension between abstraction and signification, retraction and indexicality, is what gives them their power.

Gates's works also suggest that these increasingly ubiquitous modes of communication are often interpreted apart from the realities that form their basis, from which the data are derived. But Gates's paintings, shown alongside works based in poetry and sculpture, also enact a form of visual rhetoric, suggesting that visualized data takes on an aesthetic life apart from its communicative and translative context as a "second-tier" informational mode. Paintings such as that derived from one of Du Bois's bar graphs encode a story in their symbolic content and context, as well as their materiality.

CODER LE MONDE/CODING THE WORLD: A SURVEY OF DIGITAL MIXED METHODS

The context for the 2018 group exhibition *Coder le monde* (*Coding the World*), curated by Frédéric Migayrou, is particularly important to conversations about the specificity of viewing conventions for art versus how other informational and museological exhibitions might be viewed. The Centre Pompidou exhibition brought the history of formal methods for data representation, from early non-digital methods to complex algorithmic and AI-based methods currently used in data labs, into conversation with the institutionalized expectations of the traditional international art-museum audience, presenting data-based works as *objets d'art* and encouraging viewers to engage with information-based representational strategies within the context of art interpretation.

Earlier technology-based artworks such as Italian artist Andrea Branzi's *No-Stop City* (1967), a large-scale, minimalistic drawing related to her larger No-Stop City project (1969), featuring a series of small typewritten x's symmetrically organized on dotted grid graph paper, were shown alongside contemporary works like Mishka Henner's *Prins Maurits Army Barracks, Ede, Gelderland* (2011), a large, full-colour archival pigment print (167.6 × 149.9 cm) depicting a bird's-eye photographic image of a Google Earth landscape with sections pixelated to suggest visual censorship.

Coder le monde, which was part of the Pompidou's year-long program Mutations/Créations (2018) and occupied their Galerie 4 space, purported to offer "an introduction to the creative use of code through timelines, installations and screenings, tracing over a period of 40 years the key moments in the emergence of a digital culture that has today become a taken-for-granted part of everyday life" (Centre Pompidou 2018, para. 2). It consisted of six sections: one dedicated to the international art movement the Algorists, contemporary music, digital literatures, digital form creation in art and design, the body and code, and technologies for the visualization of code and datascapes. The resulting show presented viewers with insights into the relationship between art and digital technology, but in some ways the crowded and extensive exhibition began to replicate the disciplinary desensitization that visual elements undergo in DH environments: the information began to take precedence over the works of art. Digital and code-based artworks, like other works of art, need time and space for contemplation. In the repetitious act of looking at several interrelated digital artworks and the curatorial didactics that explain, the viewer's focus easily shifts to the text, and the visual works become secondary to their textual summaries. The neighbouring special exhibition in Galerie 3, *continuum*, featured works by Japanese sound and digital artist Ryoji Ikeda, and gave his works the kind of focused attention artworks might demand. Ikeda's artworks, such as *A [continuum]* (2018) and *code-verse* (2018), engage with ideas based in the paradigm of the digital but also ask the audience to experience the works as artworks first, as an aesthetic enterprise. In contrast to the larger group show, it served to remind that artistic viewing conventions ascribe to a different kind of code, asking viewers to adopt a contemplative and particular genre of gaze: one that approaches the visual from a critical and interpretive standpoint. In absorbing and analyzing textual information, the mind turns over to a function that loses visual-critical focus. This is not to say that visual interpretation requires a particular, modernist "white cube" atmosphere for contemplation (O'Doherty 1999, 14–15), or that visual art functions according to a kind of disciplinary "purity" (Mitchell 1994, 96); they do not. But artworks are not a second order of representation, functioning alongside text or other forms of data: they are an integral part of the whole; they function independently, according to the schematics of representation. And when data are *rendered visual* they also become an integral and independent part of a representational schema. In short, the visual requires (specific) attention.

The question of the role of the visual in data-driven environments is useful to consider. Traditionally, images and visual materials have acted as a supplement to research, and this is how most producers and viewers have been entrained to employ visualized information. But in the context of visual arts-based disciplines, the potential for visual translations of information shifts, opening up the viewing experience for new kinds of dialogue between the visual and other forms of knowledge. Such possibilities necessarily build from accessible, teachable goals, such as how one might develop the capacity to experience the aesthetic as a means to invoke intellectual and affective responses of a new order premised in critical, subjective reflection. The Association of College and Research Libraries developed the ACRL Visual Literacy Competency Standards for Higher Education in 2011, stating,

> the importance of images and visual media in contemporary culture is changing what it means to be literate in the 21st century. Today's society is highly visual, and visual imagery is no longer supplemental to other forms of information. New digital technologies have made it possible for almost anyone to create and share visual media. Yet the pervasiveness of images and visual media does not necessarily mean that individuals are able to critically view, use, and produce visual content. Individuals must develop these essential skills in order to engage capably in a visually-oriented society. Visual literacy empowers individuals to participate fully in a visual culture. (ACRL 2011, para. 1)

There are several robust international initiatives for the promotion of visual literacy as a means to gaining digital literacy, such as the English-language organization Visual Literacy Today (visualliteracytoday.org), and the International Visual Literacy Association (ivla.org), which give credence to the importance of the development of such literacies as a component of digital citizenship and competency. Reimagining the possibilities for a focus on the significance of the visual in DH environments, the creative reinterpretations of representational codes created by artists and/or interpreted by specialists from within arts-based disciplines are a compelling place to begin the conversation about what visual disciplines might impart within fields of digital study, particularly as a supplement to the integration of traditional methods for the acquisition of visual literacy skills.

In fields based in the primacy of the auratic object, where materiality and immateriality or ephemerality, broadly, are often the basis of interpretation, the material aspects of the interpretive process might need to be reconsidered and reimagined in order to open these matters up to scholars in other fields. While Drucker, in her analysis, was addressing some practical concerns about the uptake of digital methods in visual disciplines, she was also drawing attention to the specific hermeneutic, epistemological, and ontological contexts visual media and disciplinary conventions require. And while the differences between the genealogical examinations of artworks in *Coder le monde*, Gates's decontextualized data paintings in *But To Be a Poor Race*, the Mediated Matter group's death masks in Vespers, and MoMA's photographic data project in *Object:Photo* far outweigh their similarities, each project suggests concepts to be turned over in the broad use and interpretation of data visualization, indicating areas for potential exploration and creative development that call upon the particularized expertise of specialists in the disciplines of visual studies and art history. While the conclusions or solutions of the quandary to which this essay responds are yet to come, I hope that this thinking encourages a necessary critical expansion of the practice and interpretation of data visualization as a rich visual medium.

NOTE

1. The Mediated Matter group consisted of Neri Oxman, Christoph Bader, Rachel Soo Hoo Smith, Dominik Kolb, Sunanda Sharma, João Costa, and James Weaver. Other credited contributors to the Vespers project include Jeremy Flower, Kelly Egorova, Ahmed Hosny, Wendy Salmon, Tzu Chieh Tang, Noah Jakimo, Naomi Kaempfer, Boris Belocon, Gal Begun, MIT Environmental Health and Safety, Media Lab Facilities, and the Center for Bits and Atoms.

REFERENCES

3D Printed Art and Design World. n.d. "Vespers." http://web.archive.org/web /20190903064717/https://3dprintedart.stratasys.com/nerioxmanves persgallery/.

Abbaspour, Mitra, Lee Ann Daffner, and Maria Morris Hambourg. 2014. *Object:Photo.*
   Modern Photographs: The Thomas Walther Collection 1909–1949 at The Museum of
   Modern Art. New York: The Museum of Modern Art. moma.org/interactives/
   objectphoto.
ACRL (Association of College and Research Libraries). 2011. "ACRL Visual Literacy
   Competency Standards for Higher Education." http://www.ala.org/acrl/
   standards/visualliteracy.
Battle-Baptiste, Whitney, and Britt Rusert. 2018. Introduction to *W. E. B. Du Bois's Data*
   Portraits: Visualizing Black America, by W. E. B. Du Bois, edited by Whitney Battle-
   Baptiste and Britt Rusert, 7–22. Hudson, N.Y.: Princeton Architectural Press.
Bikakis, Nikos. 2018. "Big Data Visualization Tools." In *Encyclopedia of Big Data*
   Technologies, edited by Albert Zomaya, Javid Taheri and Sherif Sakr. Springer,
   Cham. https://doi.org/10.1007/978-3-319-63962-8_109-1.
Byrd, Dana E. 2016. "Review of 'Object:Photo, The Thomas Walther Collection'
   by Mitra Abbaspour, Lee Ann Daffner, and Maria Morris Hambourg." CAA
   Reviews, March 17, 2016. https://doi.org/10.3202/caa.reviews.2016.32.
Centre Pompidou. 2018. "Mutations/Créations : Coder le monde (Coding the World),
   Ryoji Ikeda, Vertigo." Troika. Accessed September 15, 2019. https://troika.uk.
   com/wp-content/uploads/2018/03/Press-kit-2018-eng_Coding-the-world.pdf.
Champion, Erik Malcolm. 2017. "Digital humanities is text heavy, visualization light,
   and simulation poor." *Digital Scholarship in the Humanities* 32, no. 1: i25–i32.
   https://doi.org/10.1093/llc/fqw053.
Compute Canada. "About." Accessed September 15, 2019. https://www.
   computecanada.ca/about/.
D'Ignazio, Catherine, and Lauren F. Klein. 2020. *Data Feminism.* Cambridge, Mass.:
   The MIT Press. https://doi.org/10.7551/mitpress/11805.001.0001.
Duke University Libraries. 2019. "Data Visualization: Visualization Types." September
   2019. https://guides.library.duke.edu/datavis/vis_types.
Drucker, Johanna. 2013. "Is There A 'Digital' Art History?" *Visual Resources: An*
   International Journal of Documentation 29, no. 1–2: 5–13. https://doi.org/10.108
   0/01973762.2013.761106.
———. 2014. *Graphesis: Visual Forms of Knowledge Production.* Cambridge, Mass.:
   Harvard University Press.
———. 2018. "Non-representational approaches to modelling interpretation in a
   graphical environment." *Digital Scholarship in the Humanities* 33, no. 2: 248–63.
   https://doi.org/10.1093/llc/fqx034.
FBK (Fondazione Bruno Kessler). n.d. "Data Visualization." https://dh.fbk.eu/data-
   visualization.
Few, Stephen. 2016. "Data Visualization for Human Perception." *The Encyclopedia*
   of Human-Computer Interaction. 2nd ed. Interaction Design Foundation.

https://www.interaction-design.org/literature/book/the-encyclopedia-of-human-computer-interaction-2nd-ed/data-visualization-for-human-perception.

Gold, Matthew K., and Lauren F. Klein, eds. 2016. *Debates in the Digital Humanities 2016*. Minneapolis: University of Minnesota Press. https://doi.org/10.5749/9781452963761.

———. 2019. *Debates in the Digital Humanities 2019*. Minneapolis: University of Minnesota Press. https://doi.org/10.5749/9781452963785.

Mitchell, W. J. T. 1994. *Picture Theory: Essays on Verbal and Visual Representation*. Chicago, Ill.: University of Chicago Press, 1994.

O'Doherty, Brian. 1999. *Inside the White Cube: The Ideology of the Gallery Space*. Berkeley, Calif.: University of California Press, 1999.

Pandey, Anshul Vikram, Katharina Rall, Margaret L. Satterthwaite, Oded Nov, and Enrico Bertini. 2015. "How Deceptive are Deceptive Visualizations? An Empirical Analysis of Common Distortion Techniques." *CHI '15: Proceedings of the 33rd Annual ACM Conference on Human Factors in Computing Systems*, 1469–78. Seoul, South Korea: Association for Computing Machinery. https://doi.org/10.1145/2702123.2702608.

Shneiderman, Ben. 1996. "The Eyes Have It: A Task by Data Type Taxonomy for Information Visualizations." In *Proceedings 1996 IEEE Symposium on Visual Languages*, 336–43. Washington, D.C.: IEEE Computer Society Press. https://doi.org/10.1109/VL.1996.545307.

CHAPTER 7

Making, Conversation: An Experiment in Public Digital Humanities

Kim Martin and Rashmeet Kaur

As one of the three founders of the Digital Humanities MakerBus, I (Kim) was busy from 2014 to 2019, creating and running a mobile makerspace and DH classroom. Ryan Hunt, Beth Compton, and I learned many things during the time we toured southern Ontario (and beyond) teaching hands-on, practical skills to kids, teens, and adults—from gamifying education and soldering jewelry to flying drones and building 3D printers. Over the years with the MakerBus, it became obvious to us that making, as with most things technological, was gendered; for example, people often walked by Beth or me when we were presenting on 3D printers to talk to Ryan or one of the other men who volunteered with us, assuming their knowledge was greater than ours (despite 3D printing being the focus of Beth's doctoral research). As the MakerBus grew in popularity and became a small business, this became all the more clear: people wanted photos of Ryan in newspaper stories, and collaborators often listed us as "Ryan Hunt and Co." on grants, despite knowing the three of us and the passion we all had for the project.

As the MakerBus travelled to nearby towns and cities, I would visit local makerspaces to see not only what was being made, but who was making it. I was genuinely bothered by the presentation of the maker movement as predominantly male. I'd ask women at makerspaces (public and academic) and tool libraries about the roles they played there, who was in charge of projects, and who used the space. It didn't take much digging to understand that "maker" wasn't a word that many women were comfortable using to describe themselves, and that the majority of makerspaces were not places where women felt comfortable. When I started my postdoc at the University of Guelph, I decided to investigate this further. Interviews with makers confirmed my suspicions: people with social and cultural identities that intersected with gender (race, ability, sexuality, etc.) (Crenshaw 1989, 1991) were less comfortable (and in some cases, unwelcome) in makerspaces that were predominantly made up of straight, white men (Legge 2016).

After approximately ten formal interviews and many casual conversations with makers, I knew that theorizing about the problems with makerspaces was only going to get me so far. I had my own lived experience to understand what these makers were feeling—but I also had the privilege of being a white, straight, cisgender woman in academia. It wasn't until a local makerspace asked me to participate in their decision-making about moving to a new location that I thought I might be able to make a difference. Together with the director and a board member of this makerspace, I applied and received funding for a community-focused research grant, which I titled the "DIYversity Project," to connect with the do-it-yourself (DIY) nature of makerspaces. The goals of my project were:

1. To conduct a multi-method feminist ethnography of female and LGTBQ+ makers, including reflective diaries, participant observation, and survey research.
2. To expand and reshape the membership of one makerspace, DIYlab, and to assess the effect of a year-long series of workshops and events for women and LGTBQ+ makers.
3. To develop a series of digital posters that outline best practices for other maker communities interested in diversification, and to showcase findings and workshop successes at an end-of-year conference.

This chapter reflects on the process of running this grant, its (few) successes and (many) failures, and recounts the discussion of what it means to "do digital humanities" in public. Rather than present this work solely from my own experiences, I have invited Rashmeet Kaur to join me.[1]

Following a brief review of the literature, we organized the body of this chapter as four conversations between the two authors, in which we reflect on the themes that emerged from the project: money, community, accessibility, and negotiating boundaries. We conclude with a set of best practices for those involved with makerspaces to consider, and we relate these considerations back to DH labs, centres, and the process of community building.

WHAT'S A MAKERSPACE?

The rise of the maker movement, merging a DIY mindset with the use of technology, is commonly attributed to the launch of *MAKE:* magazine in 2005 (Peppler and Bender 2013; Davies 2015). However, the practice of "making," in all its various forms, has existed for centuries (Burke 2014), influencing the Arts and Crafts movement of the late 1800s (Morozov 2014), the DIY movement of the 1960s and 1970s, and the hacker culture of the 1980s and 1990s (Willett 2016). In writing about the maker movement, some have gone so far as to profess that everyone is a maker (Bean, Farmer, and Kerr 2015) and it is this mindset that has inspired the formation of makerspaces all over the world (Fox, Ulgado, and Rosner 2015). These spaces are defined as "open community labs functioning as centers for peer learning and knowledge sharing in the form of workshops, presentations, and lectures" (Godfrey 2015). In short, makerspaces serve as gathering places for members to engage in knowledge-exchange practices, to learn about new technologies, and to enhance their skills (Lewis 2015).

Although makerspaces may seem like a marvellous endeavour in community building, in practice, building community can be difficult. Definitions of makerspaces often fail to capture who is and—perhaps more importantly—who isn't using these spaces and the reasons for which they are using the spaces (Voigt, Unterfrauner, and Stelzer 2017). In essence, all makerspaces are reflections of the communities that built and maintain them. So, if these spaces exclude certain people, how does this reflect on the overall practice of making? Can we truly say that everyone is a maker?

MAKING + GENDER

Anyone who has worked in the tech sector, in technology-related disciplines, in the trades, or in the sciences does not have to reflect for long to understand that these fields are deeply gendered. Maker culture has been widely critiqued for this (Bean, Farmer, and Kerr 2015; Faulkner and McClard 2014; O'Sullivan 2018; Shinnick 2019), with a recent article claiming that it is "generally considered to be about 80% male, a figure roughly in line with the demographics of the tech industry" (Whelan 2018). Whether on university campuses (Morocz et al. 2015), in grade schools (Buchholz et al. 2014), or at member-owned, public makerspaces (Riley, McNair, and Masters 2017), the lack of women in the making community is often discussed. More recently, critics of making have gone further than just addressing the gender divide; they are now also drawing attention to the fact that makers are, more often than not, white, straight, able-bodied, and middle-to-upper class (O'Sullivan 2018; Riley, McNair, and Masters 2017).

Research has been conducted on why women face difficulty accessing and using makerspaces, with Faulkner and McClard (2014) finding a lack of money, mentorship, and information being at fault, as well as a general feeling of exclusion. Women, they write, "find some makerspaces 'creepy' or unsafe, and they find cultural prejudices against women using technology" (191). One of the suggested methods of assuaging these findings is to ensure there are female role models in the maker community (Voigt, Unterfrauner, and Stelzer 2017) and to both create and uphold a community code of conduct (Fox, Ulgado, and Rosner 2015; O'Sullivan 2018). It should be noted, however, that the very belief that makerspaces should be "safe spaces" (Toupin 2013) has proven problematic. One extreme example of this occurred when a board member at a Toronto makerspace attempted to prevent abusive behaviour and ended up being trolled, harassed, and eventually banned from the space (Legge 2016).

As further evidence of harassment and discomfort, woman hackers and makers have published accounts of come-ons, innuendos, and harassment (Henry 2014), trolling on message boards (Legge 2016), and bullying, backtalk, and push back on anti-harassment policies (Wolf 2012). Despite openness being an essential part of the maker movement, concerns over safety of makerspace members continue to burden women in particular (Toupin 2013). One solution to this has been the creation of feminist spaces, where members are vetted and have to abide by strict codes of conduct (Toupin 2014). These spaces promise a safe working environment for women and LGTBQ+ makers. Feminist makerspaces have diverse, often unique requirements, with some only allowing people who identify as women (Henry 2014) and others intended specifically for women with children (Fox, Ulgado, and Rosner 2015). While these more private spaces are important and their successes (and failures) have been the focus of much of the recent literature on making, the DIYversity project was created on the idea that this cannot be the only possible "solution" for women and LGTBQ+ makers. The project was an attempt to seek change in a makerspace that already exists. After all, how can a makerspace call itself a "community workshop" when less than half the community feels welcome within its walls?

DH + MAKING

There is little that separates the concept of makerspaces from that of DH centres. Indeed, when I started interviewing makers, I did so partially at DH centres where making (or physical computing) was involved. Digital humanities, for all its definitions, is deeply connected to creating—whether it is writing code, building a database, or making a video game—and to sharing openness and care (Nowviskie 2015; Sample 2016). Junctures of digital humanities and making usually occur at sites of sharing. The Global Outlook::Digital Humanities minimal computing group, for example, was spawned after a conference in Cuba brought to light the trouble of deeply complex and design-heavy DH websites, which made viewing and interacting with these sites impossible for those without strong Internet access. This group has since reflected on digital and physical making (GO::DH, n.d.), which has led to workshops that investigate the use of microcontrollers (Arduinos) and microprocessors (Raspberry Pi) to problem solve. In 2017, the *Making Things and Drawing Boundaries* collection in the Debates in the Digital Humanities series helped to cement the connection between making and digital humanities as a shared "tendency toward speculation or unlearning rather than proving or 'wrangling' things with technologies" (Sayers 2017).

In a recent talk, I argued that digital humanities was also about another type of building—building community. This is where I see the true connection between makerspaces and DH centres/labs—the bringing together of like-minded individuals with a common purpose, and usually, but not always, with an eye to the larger public good. However, there are specific things about DH centres on university campuses that cause them to operate differently than a public makerspace, and one of the most important is funding. Where DH centres in Canada might be funded institutionally or with grants, public makerspaces usually run on a co-op model: each member pays a monthly fee for access to the space, tools, materials, etc. Ownership is shared by the collective (or at least, it is imagined to be) and is not led by a single researcher or a small group. While this difference might seem insignificant, it has direct impacts on what happens in each space, and as will be detailed below, moving from one space (a grant-funded DH centre) to another (a public, co-operative makerspace) can be complicated. Building community is never easy, but my own negotiation of these two types of spaces demonstrated more challenges for public makerspaces than were apparent at the start of the grant.

CONVERSATION #1: MONEY, MONEY, MONEY

During the process of writing the grant, DIYlab members came forward to participate in various ways: offering to run workshops, to journal their thoughts, and even to co-lead the project. In every conversation, however, it was obvious that no one thought we'd be successful. DIYlab had previously—and unsuccessfully—applied for many community grants, and members shared their doubts openly. Others were focused on getting their new location organized and simply did not see gender disparity as a concern, much less one where potential funding should be focused. Things changed when the grant was successful. Here we reflect on funding expectations, who benefitted from the grant, and how money impacts work at DIYlab.

KIM MARTIN

When I applied for the Social Sciences and Humanities Research Council of Canada (SSHRC) Partnership Engagement Grant, I was hoping to engage members at DIYlab to run a series of workshops that inspired new members to join the space. The funding was primarily intended to support women in leading and attending these workshops, an intervention that was intended as a direct response to the discussions I'd had with makers throughout southern Ontario and to concerns in the wider literature. I was even able to get funds to repay child-minders, which was very unusual for funding agencies to approve. I wanted to pay women to run grant-supported, low-cost workshops for other women, but I recognized that this might not always be possible because of the makeup of DIYlab's membership.

RASHMEET KAUR

I think Jenn² was the only woman we were able to find from DIYlab's membership who was willing to run a workshop for our project. And, at first, I remember her not wanting to be paid for the incredible amount of time she put into her woodworking workshops—all that planning, organizing, and purchasing of materials. I was consistently amazed by her passion for teaching workshops: for one of her workshops, she made a custom jig for participants to use on the table saw, which must have taken so much patience and skill to create. Even Andrea severely undervalued her time for the lovely succulents workshops she organized for us. She managed to make DIYlab's stuffy atmosphere bright with her string lights and her care for detail!

KM

I agree! It is well documented that women undervalue their work. I also knew from previous experience that artists and creatives, especially non-professionals, drastically undervalue their own work, but I didn't realize this would also be the case for the work they put into running workshops. At the beginning, it took me sitting down with them and reminding them how hard they were working, how skilled they were, and how much of their time they were giving to convince them to take the money I had worked so hard to procure.

RK

Yeah, I think I went through something similar. This project definitely taught me how to properly value the time that I put into work and elsewhere. I am grateful that the SSHRC funding was able to pay for my development as an undergraduate research assistant. I learned so many skills that I would've otherwise never had the opportunity to foster. For example, before this project I would've never even imagined being around, much less using, tools such as soldering irons and laser cutters. Now, I can proudly say that I am no longer scared of power tools and have begun to incorporate them into how I approach mechanical problems. These skills definitely opened up avenues I didn't know existed. Also, I think it was even more empowering when I was able to pass these same skills on to the participants at our events. It felt amazing to see the awe and sparks of fascination on participants' faces as they watched the laser cutter perform the actions that they coded in my laser-cutter-basics workshop. They took home more than a tangible product at the end of the evening—they took home the knowledge that these skills belonged to them too!

KM

We've both learned so much from the project, and watching you grow as an artist and a scholar over the past two years has been really exciting for me. Working with a public non-profit, however, it has been difficult for DIYlab members to understand how academic grants work and that their primary function is to train and educate students through paid positions such as yours. I think that people at DIYlab expected me to just throw money at the new space or purchase new tools, and a couple of them were definitely unimpressed when I told them that everything purchased with the grant needed to be for one of our workshops or open nights. This is why we never ended up purchasing a 3D printer for the space—no one volunteered to run a workshop on 3D design or printing, and I couldn't have justified the expense. Do you remember what happened to the woodworking supplies that we purchased at the start of the grant? They were all taken by someone and never replaced. Having to lock all of our supplies away or take them with us after each class really hurt—this lack of trust made building a community very difficult.

CONVERSATION #2: COMMUNITY BUILDING

Building a more cohesive, diverse community was the intended focus of this grant. The call for action came not only from my previous experience and research but also directly from the DIYlab community and two of its board members. The new location was meant to reinvigorate the community: there was a beautiful woodshop installed, space for a community classroom, and small offices to encourage local start-ups to join DIYlab. When we started, it seemed like everything was in good order for our initiative, which was launching a series of workshops and open nights specifically for women and LGTBQ+ folks. Things started off well, with a successful open house and engaged participation in our first couple of workshops. However, we overlooked a few things.

RK

I think one of the major problems, which we realized midway through the DIYversity project, was that we weren't retaining workshop and open-house attendees as DIYlab members. In a sense, we were building our own temporary community at DIYlab—a community that only existed within the three-hour evenings we had set up to make, craft, and play—but failed to make these interactions more permanent. As far as I can tell, none of the participants visited DIYlab outside of our events and they definitely did not invest in memberships. I remember thinking it was bizarre that we had the same recurring workshop participants at so many of our workshops, but they didn't want to be involved at DIYlab after their projects were finished. These were the participants that had set up email alerts to notify them when we posted new events, so they could be the first to sign up. These were also the participants who were eager to know when we would be hosting our next event, what new skills they would learn, and often offered suggestions for workshops we could organize next. I thought that these would be the participants that would purchase memberships to explore new tools and techniques to expand their skills beyond the ones we were offering—participants who could see themselves at DIYlab beyond our workshops and open houses. Reflecting on this now, I understand why these participants could not see themselves as a part of DIYlab, or at least not on their own. DIYlab failed to make their community presence known at our workshops. Our cluster of participants and workshop leaders existed in their own protected bubble on Thursday nights. We were a community that we made for ourselves, and this community left with us as we walked out the door at the end of each evening. The participants did not have the chance to have meaningful interactions with DIYlab's community and this made it difficult for them to conceive of DIYlab as a space where they could grow.

KM

Yes, this takes me right back to what I've always thought about makerspaces—you can put as much technology and as many tools as you like in them, but they'll never thrive as spaces without people there to do the work of community building. And community building is hard work. From what I've heard about DIYlab in its earlier years, there were many people that came together to make the space work: to join in on community events, to problem solve, and to purchase items collectively that would help attract more members. But things change, people get busy, start families, get new jobs, etc. The thing is, I thought that the jump to a new location was a good time to try and restart, to invite the community anew and reignite the passion that many of those original members had for making and creativity. But three hours a week does not a community make, and we didn't get the buy-in from the existing members that I had anticipated.

RK

Exactly! Remember when I was naïve enough to think that we could count on the DIYlab members to help put together a community barbecue across the road in the park? I had set up a spreadsheet with things we'd need and empty spaces for people to sign up and contribute. Sadly, the only things folks seemed interested in contributing were excuses and reasons why we couldn't hold such an event.

KM

It would have been a great form of community outreach. When a new community moves into a neighbourhood, the existing community wants their curiosity satisfied. They want to know what's going on in the new space and how they might work together, and these opportunities weren't offered. DIYlab didn't extend that necessary reach out into the community at all.

RK

If I could go back and do things differently, at the beginning of each workshop I would have had a DIYlab member introduce the space and advertise the current membership fees and services available. This could have been a really easy way for DIYlab's community to get involved with our workshops, open houses, and attempts to build a larger community. Also, this would've cut out on the terrible amounts of confusion we had when we had the occasional participant expressing interest in becoming a member at DIYlab but could never find the right person to give them a proper tour and more information.

KM

Yes, we definitely could have organized that better. My main regret is not getting to know many of the DIYlab folks on a one-to-one basis. By the time we were through running most of the workshops, it became clear that there was more of an "us and them" dynamic. A new board of directors had been formed, and we were getting direct feedback that what we were doing wasn't working. I remember how odd I found it when we were asked not to do another workshop on terrarium building, despite it being our most successful, because people didn't feel it reflected what DIYlab had to offer. We managed to shift back to planters and combine plants with woodworking but knowing we could have filled the space with another terrarium workshop was frustrating. Andrea was one of the workshop leaders who I could have seen being quite instrumental at DIYlab, had a connection been formed, but instead we were all left feeling like what we were trying to do didn't belong.

RK

I'm not sure how much of a difference it would've really made trying to "hang out" at DIYlab. I used the space fairly often for personal projects as I discovered that I could make all these cool projects with the laser cutter. When I was there on my own, people either went straight past me, as if I didn't exist, or nodded their heads in acknowledgement as they passed by. The few conversations that I did have while waiting for my projects to finish were always initiated by me, usually just asking people what projects they were working on. It was almost as if folks didn't want to hold conversations, they were there for business and that didn't include small talk. It's really difficult to build a community when people don't want to talk.

KM

Interestingly, one place that discussions did happen was on the DIYlab Slack workspace, but again, this was only a subsection of the community. Others relied on the listserv, which I didn't even realize existed until well into the grant. Slack was where I first realized that many people were starting to rely on the shared tools at DIYlab to create items in order to earn a living. When a machine was broken or a piece needed replacing, it was urgent because it was holding up people's ability to make money, and this is a much more delicate situation than a space meant solely for pastimes and creative ventures. This became even clearer as COVID-19 closed the makerspace down and you could see people beginning to panic. As DIYlab goes through a staged reopening, it's been the people who use the space for "essential manufacturing" that have been allowed entry first, showing that individual necessity and money have priority over community. If I had realized this when I first met people at DIYlab I doubt very much I would have proceeded with the grant but looking back at the interviews I conducted at the start, those members really believed in the community aspect of DIYlab. It's difficult to remember that the grant having failed to draw this community together is not a personal failing, but rather a situation in which too many needs and desires conflicted.

CONVERSATION #3: ACCESSIBILITY

Perhaps the worst unresolved situation from our time at DIYlab was the lack of an accessible entrance to the makerspace. We knew moving to the new location meant that there would be two floors with no elevator between them but didn't realize until tools were moved over that the woodshop would be upstairs (though, of course, something had to be). This caused problems for several regular members who used woodworking tools in the original shop but were unable to access them in the new location, but it seemingly was not enough of a concern for anyone to find a solution.

RK

It's still exhausting to think about the (still nonexistent) ramp at DIYlab. Since the beginning of the DIYversity project when I realized how problematically inaccessible DIYlab is, I've been pushing—and politely screaming—to have a ramp built for the front entrance. This is the one issue that personally committed me to becoming my most annoying self. I would slip the building of a ramp into quite literally every conversation I had with DIYlab members but was never met with anything constructive. The only person I chose not to bring this up with was the one female board member because I knew that if I kept asking her, she would feel the need to work on this project on her own. It was an awful experience having to tell participants that DIYlab was physically inaccessible. What felt worse was when I had to physically maneuver a participant using a wheelchair through the front entrance that was not flush with the ground, through the hallway, and up over a considerable step just to get into the classroom. Out of all the promises that failed to be kept, this is the one that upsets me the most. DIYlab is a makerspace. It has all the required tools. We were willing to pay for materials. Why does this space not have a ramp?

KM

Because it never directly affected them, I think? This was something that I saw over and over again at DIYlab—people would only change things that would benefit themselves in some way. It's part of the reason that the community didn't foster, you can't build a community when everyone is just out for themselves. Fairly early on in the project, I was approached by George, a man with a physical disability who had made frequent use of the woodworking machinery at the old location, and found that he was unable to carry his materials and projects with him up and down the stairs at the new location. He wanted to write a grant to get a small elevator for DIYlab, and took this up with the director, who admitted to being too overwhelmed at the changes that would be required for the woodshop to be accessible, but very much encouraged George to move forward with an application to the federal government's accessibility fund. George, one other interested member, and I got together to discuss the grant, and I became aware of the lack of empathy in the DIYlab community through conversations I had about the application. In other makerspaces I've visited, working to make the space usable by everyone has been a community-building effort in and of itself, but conversations I had about the accessibility grant showed that people were more concerned that they wouldn't be able to get to the tools they used while this construction was happening than they were that someone else couldn't access them at all. In the end, the government grant was not available the year we met, and no one, me included, followed up on this. I never saw George at the makerspace after our grant conversations and thought of him every time we helped our participant in and out of the classroom in her wheelchair. When we held woodworking classes, this same participant knew she couldn't attend, and that was something that really upset me. It was also something we didn't talk about enough.

RK

Yeah, I also noticed and was consistently frustrated by the DIYlab community's attitude of focusing on personal interests rather than collective interests. At the Future of Making Unconference in December 2019, I found out from Trevor and Scott that there was a DIYlab member who had to bring their own ramp every time they wanted to access this space. I was beyond shocked and at this point probably a little exhausted.

KM

Ugh. I wasn't even aware of yet another member having to navigate accessibility. It's very problematic. I mean, it would be problematic anywhere, but in a space with a full woodworking shop and the materials and skills to make a ramp easily at hand, it's more than that. It speaks directly to the priorities of the members and the Board of Directors. That said, I could have pushed harder, ensuring the materials were in place for the ramp, and even built the thing myself.

CONVERSATION #4: NEGOTIATING BOUNDARIES

As with most academic projects, the DIYversity project was one of many things that both of us were juggling, and we had to navigate how to balance the workshops and our desire to build up a community with several other commitments. A lack of time made much of what we wanted to do impossible, but it was a combination of this, disinterested responses from several DIYlab members, and, finally, a global pandemic that led to the decision to wrap up the failing DIYversity project.

KM

When I think back about all the "coulda, woulda, shouldas" of this project, and the ways in which we failed to build or diversify a maker community, I believe I did the best I could with the time that I had. It was very easy to feel guilty for not pouring my heart into the DIYversity project the same way I had with the MakerBus, but as we continued to run workshops and hold events for women and LGTBQ+ makers, it felt more and more like we were working for very little reward. Rashmeet, I could see you becoming frustrated with the situation at DIYlab and I didn't want you to feel burdened by the project!

RK

You're right, I definitely did pour my heart into the DIYversity project, and that is probably why it felt so terrible not seeing much change. Some weeks I also felt a bit stretched thin. Although most open nights and workshops at DIYlab served as a creative and social outlet, some evenings added to my worries and stress. One specific open night comes to mind. I was responsible for hosting the participants on my own this evening because I think you had other business to attend to—though I imagine things would have gone a lot differently if you had been there. At this point, I felt comfortable being at DIYlab on my own, but I'm glad I brought a friend along with me for this event. My friend and I were prepping all the materials to make these rad comic book coasters, as we were expecting quite a few participants. We waited, but no one showed. Then Dan, a DIYlab member and his child decided to join us. After perusing the selection of comics, Dan began commenting about the "greatness" of guns, scanty clothing on women, and the violence in comics. Another DIYlab member, sitting in a corner of the classroom, did not verbally agree or disagree, but his silence and occasional nods were definitely a part of the problem. This commentary continued despite blatant attempts from both my friend and I suggesting these issues were problematic and this was not a conversation we were comfortable having. To try and turn the conversation to something more constructive, I asked Dan about building a ramp for DIYlab. He briefly talked about not having the necessary wood and I offered to ask Kim about funding the costs before he returned to making the space feel uncomfortable.

KM

I only heard about this evening as we came together to reflect and write this piece, and it's so upsetting to me. I had wondered why I didn't see Dan anymore, as he often brought his child to our open nights, but assumed he was busy like so many others. The fact that he has said things that were inappropriate on one of the few open nights I wasn't on site makes me furious. It also deeply upsets me that I put you in a position in which you were made to feel uncomfortable. I'm truly sorry for that. Despite having gone through research ethics board training and having very similar experiences at other spaces myself, I thought that DIYlab seemed safe, and I was wrong.

RK

Kim, this was not your fault. Honestly, I thought that DIYlab was pretty safe too—I would never have agreed to host an event on my own if I felt at all uncomfortable. This was an open night meant to invite women and LGBTQ+ folks into a space that they already didn't feel welcome in. Yet, this experience left me and my friend both feeling unsafe and unwelcome. It was somehow worse— humiliating—that despite all the readings and work I had been doing for the DIYversity project, I was experiencing the exact same uneasiness in the very same space that I was trying to make more inclusive and welcoming. It was further upsetting when Dan's partner, who was passing through DIYlab for a different project, decided to join us. It was after her arrival that Dan stopped speaking. It made me feel inadequate—for some reason it was okay for this member to disregard my friend's and my feedback.

KM

But not in front of his partner, or me, as someone closer to Dan in age. It's unacceptable. He knew exactly what we were trying to do at DIYlab and this seems like a pretty direct response to that; seeing what he could get away with and enacting this on those who he likely felt had the least chance of retaliating. The fact that his child was present just makes it worse.

RK

My friend and I did speak up, but weren't heard, which I feel reflects on how the broader DIYlab community chooses to approach most problems. People are ignored, or their problems are just not visible enough for other people to care or even notice. At the Future of Making Unconference, I remember how taken aback Trevor and Scott, two DIYlab members accompanying us, were at all the things they were only just learning about. It was almost as if they didn't believe that DIYlab had all these problems despite the fact that we were in their space trying to solve them. It took being at a conference where the issues and solutions we'd been advocating for over the past year were being spoken about by others for the issues to finally begin to hold significance for them. I'm glad we got there but am still upset that people didn't take us or our work seriously. All community building revolves around having constructive communication and healthy negotiations, but that becomes difficult when folks fail to try and view things from other perspectives.

KM

Yes, and just as we came close to having a mutual understanding between a couple of the board members and our own goals, the pandemic hit. We were all thrown in different directions, and neither of us have set foot in DIYlab in over a year. The distance we've been forced to take from the project, however, has really allowed us both to reflect and recognize what we learned.

CONCLUSION

Perhaps naming this experience the DIYversity project meant that it was bound to fail from the get-go. Aiming for gender equality was one thing but, as we have discussed in this chapter, any attempt to reach women at intersectional identity categories was negligible. Guelph is 84 percent white, and, besides Rashmeet, we only saw three people of colour come to DIYlab, and two of those visitors only attended our workshops. Although we met two of the project's three original goals, we did not succeed in building the community we'd set out to. That said, we did have some great experiences. We met a wonderful network of women makers in Guelph and through southern Ontario. We picked up new skills and came to understand our own positionality in new ways that we will apply to future projects. We also defined the following set of best practices based on these experiences, which we think apply to both public makerspaces and DH labs on university campuses.

BEST PRACTICES

HAVE CLEAR AND PURPOSEFUL LANGUAGE AROUND CONDUCT, COMBINED WITH ACCESSIBLE, REGULAR CONVERSATIONS ABOUT PROBLEMATIC SITUATIONS THAT OCCUR AND HOW TO ADDRESS THEM

In talking to DIYlab members at meetings and on Slack, it was apparent that there had never been an opportunity for many of them to talk openly about gender diversity, or about any type of diversity. Some people confided in us about their own experiences privately and wished us well. Some said they were excited to see what changes the project would bring about, but this wish for change was never shared among a group or in a meeting. Reflecting on this situation led us to suggest this first practice: provide time and space for people to come together in conversation about community, change, and conduct. Having a code of conduct is great, but someone needs to enforce it, and conversations around conduct must be had if people are going to work toward mutual understanding and respect.

BE FLEXIBLE: ADAPT TO PROVIDE OPPORTUNITIES

Jenn might be the project's best success story, and really all we did was offer her a space to showcase her talents. When we first met Jenn, we knew she wanted more women in the space; she recognized that there was an issue but wasn't sure what her role in our project could be. Early in our time at DIYlab, Jenn ran a workshop for women on woodworking. It was so well organized. Despite this, she was nervous, unsure, and kept checking in with us during the event to see how things were going. We had only planned to fund each person to run a single workshop, but we saw Jenn's confidence skyrocket as a result of her workshop, and we were thrilled with how well the group connected with her as instructor. We shifted our plans and Jenn organized two more workshops for us and has gone on from there to run other workshops and sell her wares (signage, furniture, etc.) both locally and online.

LOOK BEYOND YOUR WALLS: INVITE COMMUNITY IN AND WORK WITH (NOT FOR) OTHERS THROUGH BROAD COMMUNITY INITIATIVES

One of the most important takeaways from this project has been understanding what it means to build a community and hold it close. Although we struggled to build our vision of community at DIYlab, we felt embraced by all the wonderful people in other, related communities we are fortunate to be part of—especially the THINC Lab at the University of Guelph and the surrounding DH community. Over the course of our project, we participated in collaborations that brought together the various maker groups in Guelph and brought more community through DIYlab's door. It was through these collaborations that we built a strong network of both personal and professional connections. For instance, our 3D self-portrait scavenger hunt brought together five different local maker communities, including DIYlab; this event sparked conversations on future collaborative work. Knowing that we are not alone, and that we could contribute in meaningful ways to work being done by others, gave us the energy to continue with our project.

INVITE PLAYFUL COLLABORATION AND EXPERIMENTATION

One of the events that stood out was a soldering open house we ran quite early on. Not only were the tickets sold out for this event, but DIYlab members also showed up! We had put out a call to borrow extra soldering irons on the Slack channel and, to our surprise, some members dropped off their personal soldering irons and others showed up to help. It was wonderful to see the participants asking questions and enjoying themselves, and to see the DIYlab members engage with participants' curiosity and learn things too. This is what we wanted, not just for one event, but for every single event. So, as another practice, we propose creating events and spaces that foster exploration in collaborative settings.

RECOGNIZE THAT IT'S OKAY TO FAIL

"Learning through failure" was a slogan that the MakerBus used when teaching kids DIY projects and is one that is deeply connected to maker education in general. Recognizing that failure can be beneficial has also become a recent theme in DH literature (Dombrowski 2019; Graham 2019). Reflecting on the DIYversity project for this chapter and learning to accept our failures has been difficult, but we hope the best practices that have emerged from our project and the conversations we share above are useful not just for ourselves but for anyone looking to build community.

NOTES

1. Rashmeet joined as an undergrad research assistant on the DIYversity project in the summer of 2019 and she has been invaluable in so many ways: she's led workshops and open houses at the makerspace, and she has her own lived experience in and around this space from which to draw.
2. Pseudonyms have been used in place of the names of makers at DIYlab.

REFERENCES

Bean, Vanessa, Nicole M. Farmer, and Barbara A. Kerr. 2015. "An Exploration of Women's Engagement in Makerspaces." *Gifted and Talented International* 30, no. 1–2: 61–67. https://doi.org/10.1080/15332276.2015.1137456.

Buchholz, Beth, Kate Shively, Kylie Peppler, and Karen Wohlwend. 2014. "Hands On, Hands Off: Gendered Access in Crafting and Electronics Practices." *Mind, Culture, and Activity* 21, no. 4: 37–41. https://doi.org/10.1080/10749039.2014.939762.

Burke, John J. 2014. *Makerspaces: A Practical Guide for Librarians.* Lanham, Md.: Rowman & Littlefield. https://doi.org/10.3163/1536-5050.103.4.016.

Crenshaw, Kimberlé. 1989. "Demarginalizing the Intersection of Race and Sex: A Black Feminist Critique of Antidiscrimination Doctrine, Feminist Theory and Antiracist Politics." *University of Chicago Legal Forum*: Vol. 1989, Article 8.

———. 1991. "Mapping the Margins: Intersectionality, Identity Politics, and Violence against Women of Color." *Stanford Law Review* 43, no. 6: 1241–99.

Davies, Michelle. 2015. "A Brief History of Makerspaces." *Curiosity Commons* (blog). https://curiositycommons.wordpress.com/a-brief-history-of-makerspaces/

Dombrowski, Quinn. 2019. "Towards a Taxonomy of Failure." *Quinn Dombrowski* (blog). http://quinndombrowski.com/?q=blog/2019/01/30/towards-taxonomy-failure.

Faulkner, Susan, and Anne McClard. 2014. "Making Change: Can Ethnographic Research about Women Makers Change the Future of Computing?" *Ethnographic Praxis in Industry Conference Proceedings* 1: 187–98. https://doi.org/10.1111/1559-8918.01026.

Fox, Sarah, Rachel Rose Ulgado, and Daniela K. Rosner. 2015. "Hacking Culture, Not Devices: Access and Recognition in Feminist Hackerspaces." In *CSCW '15: Proceedings of the 18th ACM Conference on Computer Supported Cooperative Work & Social Computing*, 56–68. Vancouver, B.C.: ACM.

Godfrey, Beth. 2015. "Making Gender: Technologists and Crafters in Online Makerspaces." Master's thesis, Georgia Institute of Technology.

GO::DH (Global Outlook::Digital Humanities). n.d. "Thought Pieces." Minimal Computing: A Working Group of GO::DH. Accessed March 23, 2021. https://go-dh.github.io/mincomp/thoughts/.

Graham, Shawn. 2019. *Failing Gloriously and Other Essays*. Grand Forks: The Digital Press at the University of North Dakota. https://doi.org/10.31356/dpb015.

Henry, Liz. 2014. "The Rise of Feminist Hackerspaces and How to Make Your Own." Model View Culture. https://modelviewculture.com/pieces/the-rise-of-feminist-hackerspaces-and-how-to-make-your-own.

Legge, Melissa. 2016. "#makergate." Talk given at the Sherman Centre for Digital Scholarship, October 27, Hamilton, Ont.: McMaster University.

Lewis, J. 2015. "Barriers to Women's Involvement in Hackspaces and Makerspaces." Access Space. https://access-space.org/portfolio/barriers-to-womens-involvement-in-hackspaces-and-makerspaces/.

Morocz, Ricardo, Bryan D. Levy, Craig R. Forest, Robert L. Nagel, Wendy C. Newstetter, Kimberly G. Talley, and Julie S. Linsey. 2015. "University Maker Spaces: Discovery, Optimization and Measurement of Impacts." 122nd ASEE Annual Conference & Exhibition. https://smartech.gatech.edu/handle/1853/53812.

Morozov, Evgenvy. 2014. "Making It: Pick up a Spot Welder and Join the Revolution." *New Yorker*, January 13, 2014.

Nowviskie, Bethany. 2015. "On Capacity and Care." *Bethany Nowviskie* (blog). http://nowviskie.org/2015/on-capacity-and-care/.

O'Sullivan, Em. 2018. "Excellence in the Maker Movement." *Journal of Peer Production* 12. http://peerproduction.net/issues/issue-12-makerspaces-and-institutions/.

Peppler, Kylie, and Sophia Bender. 2013. "Maker Movement Spreads Innovation One Project at a Time." *Phi Delta Kappan* 95, no. 3: 22–27. https://doi.org/10.1177/003172171309500306.

Riley, Donna M, Lisa McNair, and Sheldon Masters. 2017. "MAKER: An Ethnography of Maker and Hacker Spaces Achieving Diverse Participation." 124th ASEE

Annual Conference & Exposition. *American Society for Engineering Education.* http://hdl.handle.net/10919/82443.

Sample, Mark. 2016. "The Digital Humanities Is Not about Building, It's about Sharing." In *Defining Digital Humanities*, edited by Melissa Terras, Julianne Nyhan, and Edward Vanhoutte. London, U.K.: Ashgate. https://doi.org/10.4324/9781315576251.

Sayers, Jentery. 2017. "Introduction: 'I Don't Know All the Circuitry.'" In *Making Things and Drawing Boundaries*, edited by Jentery Sayers. Minneapolis: University of Minnesota Press. https://dhdebates.gc.cuny.edu/projects/making-things-and-drawing-boundaries.

Shinnick, Stacey. 2019. "Stories in the Making: A Phenomenological Study of Persistent Women Techmakers in Co-Ed Community Makerspaces." *Theses and Dissertations.* 1060. https://digitalcommons.pepperdine.edu/etd/1060.

Toupin, Sophie. 2013. "Feminist Hackerspaces as Safer Spaces?" *DPI: Feminist Journal of Art and Digital Culture* 27. http://dpi.studioxx.org/en/feminist-hackerspaces-safer-spaces.

Voigt, Christian, Elisabeth Unterfrauner, and Roland Stelzer. 2017. "Diversity in FabLabs: Culture, Role Models and the Gendering of Making." In *Internet Science.* 4th International Conference, INSCI 2017. *Lecture Notes in Computer Science* 10673. Cham, Switzerland: Springer. https://doi.org/10.1007/978-3-319-70284-1_5.

Whelan, Tara. 2018. "We Are Not All Makers: The Paradox of Plurality in the Maker Movement." In *Proceedings of the 2018 ACM Conference Companion Publication on Designing Interactive Systems*, 75–80. New York: ACM. https://doi.org/10.1145/3197391.3205415.

Willett, Rebekah. 2016. "Making, Makers, and Makerspaces: A Discourse Analysis of Professional Journal Articles and Blog Posts about Makerspaces in Public Libraries." *Library Quarterly* 86, no. 3: 313–29. https://doi.org/10.1086/686676.

Wolf, Asher. 2012. "Dear Hacker Community—We Need to Talk." *Cyber Wanderlust* (blog). 2012. https://cyberwanderlustblog.wordpress.com/2012/12/29/dear-hacker-community-we-need-to-talk-by-ashe/.

Part 2

Digital Poetics

Canadian Poetry and the Computational Concordance: Sandra Djwa and the Early History of Canadian Humanities Computing

Sarah Roger, Paul Barrett, Kiera Obbard, and Sandra Djwa

In the 1960s, Sandra Djwa was a graduate student at the University of British Columbia, where she was working on a dissertation on Canadian poetry. Canadian literature was then an emerging field of study, and Djwa's project was timely in at least two respects: she was writing her dissertation in the shadow of Carl F. Klinck's recently published *Literary History of Canada*, and the Computing Centre at UBC was willing to accept literary projects. Djwa was interested in responding to Northrop Frye's "Conclusion" to Klinck's literary history, particularly Frye's identification of a "cruel north" thesis—the manner in which Canadian literature and poetry responded to an unforgiving and indifferent environment. When preparing her comments on nineteenth-century Canadian poets, Djwa found that Frye's conclusion did not support much that she was reading for her dissertation and set out to challenge it.

Djwa developed computational concordances supported by the UBC Computing Centre. These concordances—the first instance of DH research in Canadian literature—employed digital methods to map poetic diction and themes in Canadian poetry. Djwa's DH work aimed to demonstrate continuity of poetic theme and diction from the earliest Canadian poets to modern writers and to demonstrate that those continuities were often somewhat different than Frye's conception of a cruel north. She explains her project thus:

> Between 1966 and 1968, the published books of seven poets, Isabella Valancy Crawford, Sir Charles G. D. Roberts, Archibald Lampman, Duncan Campbell Scott, E. J. Pratt, Earle Birney and Margaret Avison were keypunched. Between 1968 and 1970, seven other poets, Charles Mair, Charles Sangster, Bliss Carman, A. J. M. Smith, A. M. Klein, Irving Layton and P. K. Page were added.
>
> The procedure followed was the same in all cases. Each poet's published books in chronological order were key-punched on computer cards

at the rate of one typographical line per computer card [see figure 8.1]. The [300,000] computer cards containing the poet's canon were then fed into an IBM 7044 computer for printout. Following [manual] proofreading and necessary corrections, the computer drew up a word frequency count. This is an alphabetical index listing every word that a poet uses and indicating its frequency of appearance [see figure 8.2]. On the basis of the critic's understanding of a poet's work, and taking into consideration both the frequency of occurrence of particular words and the apparent collocations or associations of clusters of words, a selected list of words under the heading of thematic categories was then drawn up by hand. (Djwa 1970, 44)

The scale of Djwa's achievement should not be underestimated. This early digital project involved conceptualizing new methods of interpreting and understanding a collection of texts, working with new modes of data storage in the form of punch cards and tape reels, manually editing and correcting punch cards,

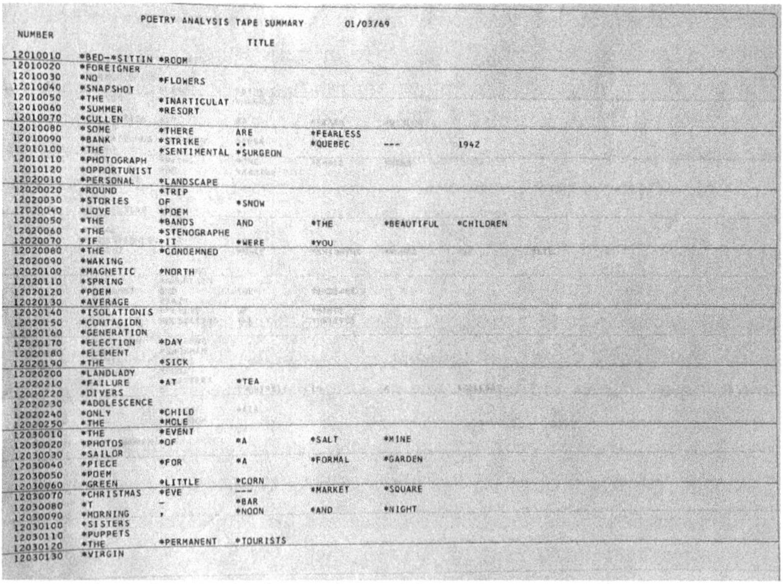

Figure 8.1. Poetry analysis tape summary.

Source: Personal archive of Sandra Djwa (photo by Kiera Obbard).

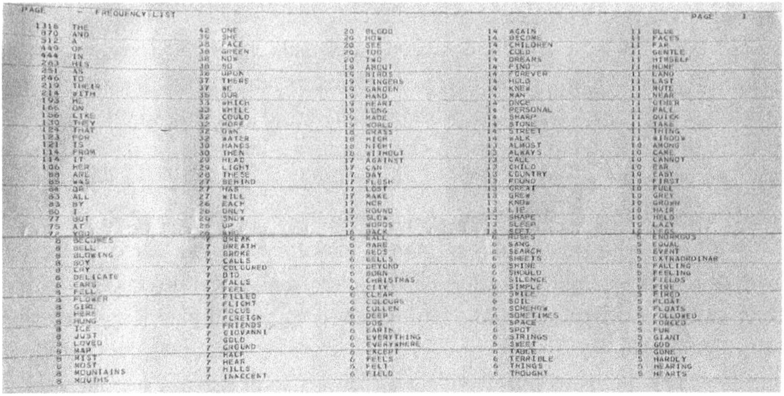

Figure 8.2. Word frequency list.

Source: Personal archive of Sandra Djwa (photo by Kiera Obbard).

and arranging word counts into "apparent collocations" and "clusters of words." This required a substantial amount of labour, planning, and project management on the part of a graduate student. Furthermore, these are not singularly technical concerns but, in contending with the relationship between the computational concordance and her own interpretation of the texts, Djwa was grappling with what Alan Liu calls the "meaning problem" in digital scholarship. Liu writes that "it is not clear epistemologically, cognitively, or socially how human beings can take a signal discovered by a machine and develop an interpretation leading to a humanly understandable concept" (Liu 2013, 414). Djwa's use of word counts and "clusters of words" as evidence for literary interpretations requires separating digital signal from noise and motivating the evidence as part of an argument about the meaning of the broad corpus of Canadian poetry.

Despite the novelty of Djwa's approach as well as the impressiveness of her technical achievement, her project was not met with enthusiasm. After completing her dissertation and starting her career at Simon Fraser University, Djwa found little support for her computational work: "I received no academic credit at all for it, and the university was not very willing to take on the responsibility of the files." Djwa attributes this to the "lack of knowledge in terms of my colleagues" who "continued to affirm that this was just some foolish supplementary activity."

Indeed, despite her efforts, Djwa's computational concordance did not find a publisher. Her sole publication on the work, "Canadian Poetry and the

Computer" (1970; reprinted in this collection as chapter 9) has only been cited twice in the fifty years since it has been published, although "The Directory of Scholars Active" (1970), published annually in *Computers and the Humanities* listed it as one of the forty-odd initial projects in literature in North America. Despite becoming one of the foremost experts in Canadian literature (and the editor of Klinck's literary autobiography), Djwa's concordance remains largely unknown in both Canadian literature and digital humanities.

We begin with this historical contextualization to frame our conversation with Sandra Djwa. Over the course of two years, Sarah Roger, Kiera Obbard, and Paul Barrett corresponded with Djwa, received paper and digital copies of her concordance, and reviewed her work. This exchange concluded with the following interview that goes into detail about Djwa's early digital experiments, the implications for Canadian poetry and literature, and how she reflects on the concordance today.

PAUL BARRETT

To set the stage for the concordance, can you take us back to how you ended up working on a PhD in Canadian literature at UBC? What was the path that brought you there?

SANDRA DJWA

Well, I was a mature student when I went to UBC. I had worked in a church and I had done social work, and I discovered Canadian literature at UBC. Up until then, it was Newfoundland literature [that I knew about] because I had come through Memorial University in the 1950s. In 1956 I was taught by George Story, who edited the *Dictionary of Newfoundland English*. In my view, Story inspired the Newfoundland literary renaissance of the 1970s.

So, when I came to UBC, I had backup from Memorial. I knew E. J. Pratt, I'd written verse in imitation of Pratt when I was in high school. At UBC, my professor, Roy Daniells, was [Head of English] and one of the pioneers of Canadian literature. He instituted a course at UBC in Canadian literature, and then a graduate seminar course, and then a series of courses. Up to that time, Canadian literature was usually taught at the end of a course in American literature. That's how it was taught at the University of Toronto. This was the system that Margaret Atwood went through. So, Daniells had instituted Canadian literature at UBC and I began to realize, "We have a Canadian tradition!" And I became a literary nationalist, which I remained all my life, I think.

I decided to do my honours essay at UBC on E. J. Pratt's "The Titanic," which led me to make all sorts of discoveries about sources, and why he'd written it the way he did. Most importantly, I discovered the UBC archives with its wonderful collection of Canadiana. And later, in the '70s, then teaching at SFU, I wrote a small book on E. J. Pratt called *The Evolutionary Vision*, trying to show that he was both a Victorian and a modernist.

When I came to do my PhD, Daniells was part of my committee; Donald Stephens, my thesis supervisor, was another member, as was Bill [W. H.] New, then newly appointed. I had done well at the bachelor's level and received the gold medal in English and honorary mention for first place in the Faculty of Education, so the English Department decided that if I met their standards for an initial year, I could go directly to the PhD, which I did. I decided that I would write my dissertation on Canadian poetry. I wanted to show that there was a continuity from some of the older poets—in the Confederation group especially—to some of the moderns. Of course, the moderns didn't think this at all!

In the meantime, the first version of Klinck's *Literary History* had come out in '64 and it included Frye's "Conclusion" about a cruel nature and a "garrison mentality." Well, I had read William Carlos Williams's *In the American Grain*, and I knew that D. H. Lawrence had also spoken of the North American "garrison" and "palisade" in similar terms. So, I thought to myself, how can I test Frye's conclusions? Because the Canadian poems that I read didn't seem to be talking about garrisons and palisades at all—at least the majority did not, there were exceptions, of course. But if you took Pratt's poems and looked backwards, especially from the perspective of *Brébeuf and his Brethren* through to Archibald Lampman's "At the Long Sault" you could conceivably see the tradition in that fashion. So, I thought: "Why not look at poems in terms of frequency, or distribution of metaphors, which eventually brings you down to the word level?"

It was not until much later that I learned that Frye, then a graduate student at the University of Toronto, was indebted to Pratt and fond of him. Around 1935–1936, Pratt brought [Frye] in as his assistant on the *Canadian Poetry* magazine. Frye later told me in an interview that Pratt had taught him "when a poem was worth money." Moreover, he attributed much of his understanding of poetry to Pratt's practical tutoring.

But at UBC in 1966, I thought I'll try to do a kind of concordance. The English Department was a little surprised at this. But the head of the Computing Centre at that time was Alvin Fowler, a pragmatic engineer. I sat in his office and explained to him that I wanted to do an inventory, a concordance of sorts, of specific Canadian poets. He thought the project possible, was encouraging, and assigned a programmer, John Coulthard, to work on the project. Both were very supportive. The real problem, of course, was inputting the poetry. How many thousand computer cards did the UBC Library say we had? An enormous number.

PB

I think it's 300,000 according to one of your publications. But I've always wondered—that can't be right.

SD

It probably was right. Remember, it's line by line.

PB

So, one line of poetry equals one card?

SD

That's right.

SD

Enter Daniells—Daniells being one of my thesis supervisors, and a product of the University of Toronto and Pratt, whom he also admired. Daniels took me to lunch at the Faculty Club when I was still an undergraduate to quiz me about Pratt. We became friends. Daniells's good friend was Walter Koerner. From the Koerner Foundation, Daniells obtained funds to help me and another graduate student, Lillian Rodman, to input this incredible amount of poetry. For two summers we holed up in an old Quonset hut at UBC and typed to get the poetry into the computer. The curious thing was at the same time I was writing my dissertation, and I wasn't paying too much attention because we didn't have the results. But I wrote the dissertation and, at the end of it, we did have the results, and so I submitted [the concordance] along with my dissertation. And that's how it came about.

PB

So. it sounds like the concordance was kind of running in parallel—you were chipping away at it as you were working on the dissertation.

SD

That's more or less the case. I did my bachelor's in '64 and between '66 and '68 [the poetry was input, and] I graduated in '68. It might have influenced something at the very end, but basically, I would have had my oral in April of '68 and went to SFU in September of '68. I didn't have much time to really apply [the concordance]. Then, of course, when I got to SFU, I discovered that many of my colleagues were new Canadians who didn't believe that a Canadian literature existed. They had been taught British and American literature but no Canadian literature.

PB

Did you have any background in computers before any of this? Had you worked with any sorts of machines or was this completely novel to you at the time?

SD

None whatsoever. But I had a sense, you know, this is a new technology, it obviously has really useful implications, and in fact a number of scholars in Canadian literature immediately recognized this. Early in 1970, I had a letter from Klinck, so he understood at once that it had possibilities. And I also had a letter, curiously enough, from P. K. Page, who wanted to know what metaphors predominated in her print out—and of course, I told her.

SARAH ROGER

Was she surprised when you told her what you'd found?

SD

Well, she didn't really say whether she was surprised but she was particularly interested in knowing whether or not the word "mute" appeared in her vocabulary. And after I thought about it for a bit—you know, after I had written her biography—I thought, well there was a period in P. K.'s literary output when there was a hiatus and she was mute. But then I wondered did it have larger ramifications, was it associated with gender perhaps? But in those early days, I didn't have these perceptions.

PB

Can you tell us how the concordance was received at the defence? If it was raised at all or was the focus primarily on the dissertation itself?

SD

Well, the very curious thing is that at my oral there were far more people from the Department of Computer Science than there were from English. They had come to cheer me on because it had been, from their point of view, a successful project.

SR

So, the people in Computer Science were largely supportive of what you were doing?

SD

Very supportive.

SR

Did you encounter any tension with people feeling like the computer resource was being "wasted" on a humanities project?

SD

No, no, not at that point. No. It was seen as a new and legitimate way of using [computer resources].

PB

In the area of digital humanities, which is probably what your project would be described as today, they have something called the meaning problem, which is, in general terms, how do we take a computer signal and turn that into some phenomenon we can describe in cultural or humanistic terms? In other words, when does a certain cluster of words turn into something you can interpret or you can make an argument out of? Did you encounter anything like that where a word count seemed meaningful in that way?

SD

Well in some cases they did seem meaningful—you know, in the ones that I wrote about in the article that was published in *Canadian Literature*. In other cases, I was uncertain about how they would fit in with the whole structure. I didn't really have a chance to go on with [my analysis].

It was an awkward situation; I was moving between two universities. UBC had been supportive and continued to be supportive for about ten years after I left. But I remember in the summer of 1973, I was sitting in a meeting of the university tenure committee [at SFU]—I was being turned down basically, and it was a question of whether I would get promoted. It was clear that my computer concordances were regarded as not significant or important by members of my department. So, it was a complete washout. I got no academic credit at all for it, and the university was not particularly willing to take on the responsibility of the files, which UBC was willing to turn over. I think I was simply too early. Initially there was a lack of computer knowledge, but in four or five more years all of my department had computers and would have learned there was such a thing as a computer concordance, and it had some academic respectability. But I received no back credit for my work, and my colleagues continued to think this was some useless supplementary activity.

PB

How was Canadian literature seen at SFU and did that factor in? Did people say, "Oh, perhaps you can do a concordance but at least do it on writers that matter"?

SD

Well, at that point a number of my colleagues did not really think that a Canadian literature existed, which was a very large hurdle to overcome. The problem was that if you were educated in England, you certainly studied American literature or you had the possibility of doing so. If you studied English literature in America, you studied English literature, and the moderns. So, you would need to have on your assessing committee someone who had been raised in Canada, and who had some notion of Robert, Lampman, Carman, Scott—and who, in addition, were willing to assert that a Canadian literature existed.

The irony is that in my published articles I was teaching those poets and writers who are now recognized as major international writers: Margaret Atwood, Leonard Cohen, Margaret Laurence, Alice Munro, and so on. But I was preaching to the unconverted for a long time. The Symons report [on Canadian studies,

To *Know Ourselves*] came out around 1975 or 1976, and made people feel a little uncomfortable, but it didn't greatly change the academic climate. It was not until poets like Atwood and Cohen began to be accepted in the United States as major contemporary figures that there was some recognition that there must be something in Canadian [literature].

PB

Did you have any discussions with Klinck or Frye about the cruel-north thesis and the garrison mentality?

SD

Oh, Klinck didn't think it existed. I didn't dare put the question to Frye. I did a number of interviews with Klinck because I was helping him get the story of the *Literary History of Canada* on paper—nobody had told the story of the *Literary History*, and I had urged him to put it down on paper. And he said, "Well, I'm not going to do that now," and I said, "But you could dictate it." So, we cobbled up a letter to the Social Sciences and Humanities Research Council of Canada. He signed it and sent it off. The director of the SSHRC gave him twice as much as he'd asked for, and then Klinck asked one of his former students, Wilma Graber, to help with the tape recording. At the end I put together the essays together with an introduction. It was published as *Giving Canada a Literary History*.

PB

Your thesis on its own doesn't singularly stand as a kind of refutation of the cruel north thesis, but it is more in conversation with the concordance. I'm wondering if there's some rough connection there.

SD

I wasn't applying the concordance, really, but I think I know how Klinck felt about the conclusion. Klinck had believed that Frye would read all of the essays submitted for the whole book and come to some conclusion. But what Klinck said—if I can paraphrase what I can remember from what he said—is that it was Norrie [Frye] and it was wonderful but it "wasn't literary history." But, you know, Klinck had got the idea of a literary history for Canada after a hearing particularly fine literary talk by Frye. Klinck admired Frye for his scholarship and because he had put his weight behind Klinck to secure funding to get *The Literary History of Canada* out. Frye was really a very helpful person. He was helpful to me because he admired Pratt, and I was trying to write an introduction to Pratt's poems. Also, he advised me on the Klinck book, suggesting I leave much of Klinck's language as it was, rather than attempt to do a more formal editing.

PB

You attempted to get the concordance published, is that right?

SD

I did. I sent out a few letters asking if publishers would be interested in it, but, in the meantime, I wanted to get promoted with a decent salary, and it was pretty clear I wasn't going to get anywhere with the concordances. So, I put my energy into writing and publishing other essays and attempting to get a national association [the Association for Canadian and Québec Literatures, formed in 1975] going.

SR

At the time, did that feel gendered to you?

SD

Oh yes! Gender was one of the major problems for women faculty in the period from the sixties to the turn of the century. To some degree, it is still a problem.

PB

Did you experience any gender-based resistance like that as you put together the concordance?

SD

Well, there wasn't any in the Computing Centre. They all thought, "Oh boy, isn't this interesting. Look what we can do with the computer." We—Lillian and I—were working away faithfully with our punch cards, producing them. I can't imagine now how we did it. At the Quonset hut at 9:00 a.m., or so, and clickety-click until noon, and then come back after lunch and start again. I didn't spend the whole summer at it, but quite a bit of it.

SR

Were you running the cards in batches? Did you punch all of the cards at once before you went to process them?

SD

I think we decided on getting it all through before we tried to do a total run, and it must have gone over two summers. I first did the poets of Confederation because I thought to myself, well, if Frye is mostly looking at Pratt, who bulks so large in the moderns, we'd better take a look at the early verse, which is bound to be derivative and romantic and the rest of it, and see how this really applies.

SR

You didn't do some sort of test case?

SD

No, no. You're not dealing with a scientist here. Well, I didn't quite hive it off to [the Computer Science Department]. We sometimes talked a bit about the way we were doing it, but most of it passed me by. I was content, Alvin knew what he was doing, John knew what he was doing. My job was simply to produce the punch cards and let them at it.

SR

From a purely practical point of view, the physical punch cards and the space in which you did the work, the university was happy to just provide that to you?

SD

Yes. The English Department housed us in one of the Quonset huts. I really can't tell you even where we worked, but I know we had the same place and we went there for two summers. Daniells, once or twice, came in to see what was happening. He was gleefully anticipating the results, and Klinck was also later most interested because, of course, he wanted to know, too.

SR

And they had no objections? They said, "here's a box of 300,000 punch cards"? It can't have been cheap—the computing time or the physical resources, the electricity.

SD

There was no thought of the cost, or the difficulties—[or they] were never directed to my attention. I did know that they would sometimes do a run when a poet was finished in the wee hours of the morning after they'd done their regular work, and I just assumed, oh well, they've got the thing going and they're just filling up extra time with [my work].

SR

That's really interesting because so much of the work we do now, there are careful calculations about the infrastructure costs. You want to use a computer? You need to account for exactly how much of this computer you're going to use, and for how much time and how much space. To think that they gave you all those punch cards and just let you go for it is really incredible.

SD

They did. But it did raise its head after 1974. I recall that when I was being interviewed about my computer work by the SFU Vice President, he raised questions about the infrastructure. He pointed out that SFU was not willing to pay for computer work on it and the next time I applied to the Canada Council I should ask [for funding] because the Canada Council was now willing to pay [for] infrastructure. But of course, I never got to that point.

PB

I've never held a punch card in my life. What do you do with a punch card?

SD

You put them into a special typing machine that punches the card. I seem to remember there were certain instructions about how you entered a title as opposed to how you entered a sentence, and you had to be very careful to get in the punctuation. And the end product, the so-called concordance, listed the sentences with a space between so you could see what was there. You had the problem of having to go back and look at the context, but [the concordance] was pretty direct: it just gave the one sentence with the word embedded. There were two documents—there was firstly a document that gave a priority word list. I had that document first, so I could look at the list and sometimes that would tell you what a possible cluster could be. You know, if "dreams" comes enormously high on the vocabulary of a certain poet, you're going to look for cognate words and things of that sort. Very little time was spent by me on the actual collocation and requesting of cluster material because it came so close to the end of my dissertation, which had to be handed in if I was going to get out of UBC and get a job. And when I got to SFU, it was not regarded as appropriate scholarship.

PB

So, in many respects, the concordance was produced but there wasn't the opportunity to go back and work with it and try to say something with it.

SD

Yes, that's exactly it. And then, when I wrote it up, I did so at the request of George Woodcock, the editor of *Canadian Literature*, who was quite interested in what [the concordance] would prove. I tried to remember, well what did I do—I did this, I did that, and such and such. Reading it now is quite a new experience for me because I've forgotten much of it.

PB

When you say you wrote it up, is that for the article "Canadian Poetry and the Computer"?

SD

Yes.

PB

The priority word list—can you explain what that is? Is that a list of the twenty most-common words?

SD

It was just a numerical list saying, for example, that in reference to Pratt's poems here are all of the words that he used. Pratt, incidentally, had the highest list of different words of any of the poets. If you looked down the list, you might find a great number of words that are variants on "water." In addition, there might be a large number of variants on "fear," "dread," "horror," etc. You might then start exploring to see if the two sets of terms have some connection in context.

PB

We have specific questions about clusters [...] I'll read them out and you tell me if any of this rings a bell. So, for example, we were looking at the frequency list and you had words like "mouth" or "throat," which have a kind of lower frequency, but are included in the thematic cluster under "body," which also includes words like "flesh," "hand," "head," and "heart." Do you have any recollection of how you made these decisions? Did that seem more subjective, or like an interpretive moment?

SD

Yes. I can't tell you on what [exact] basis I made that judgment. It may have been on the recollection of a specific poem. It may have been just simply: these are parts of the body that if the poet is emphasizing this then one might inquire further if this is a part of a description.

PB

Perhaps there's some way in which you're moving between the concordance and the poems and going back and forth to say here's how they align?

SD

Yes, yes. And none of it is very solid because, you know, this was a sort of gambit. In some ways, [the concordance] fulfilled its function because there were not the kinds of words that you would have anticipated to find if you were working with the notion that it was the cruel north that predominated in Canadian poetry. So, I could—and I undoubtedly did have that information before I drew my conclusions to my dissertation. Whether I say it or not in the dissertation is another thing.

PB

You do say, in your conclusion, "The isolation of cold North hypothesis is an oversimplification when discussing the development of Canadian poetry."

SD

Yes, well that's interesting. I think it was.

PB

You know, one of the knocks against this sort [computational] of work even today, and I'm sure you heard it when you were doing your concordance—in fact, there was an article published almost with this very title—is "just read the texts." Stanley Fish has notoriously spent the last thirty to forty years, whether it was stylometry in the 1970s, computational literary studies in the 1990s, and digital humanities now, basically saying, "Just read the texts!" And my argument against that is we never just read the texts. There's always an interpretive framework, whatever that happens to be, there is no such thing as just reading the texts. If I were to put his question to you, why not just read the texts? Did you need all those giant mainframes? When I talk about your work in class, I always show a picture of the mainframes—so I hope it's as big as I imagine.

SD

It would take up almost a whole room.

PB

Okay, so I'm not lying to my students. I've never seen an IBM 7044. So, why not read the texts? What does the computer really do?

SD

I think [the computational work] reminds you of your blind side. There is so much that we read [...] well, first of all, everything we read is conditioned by our own experience and our ability to enter into the world of the poet. But we have so many fortresses. There are so many subjects which we do not pick up, we do not see. And one of the things that computer concordance does is to alert you to the fact that there is something out there that you haven't really seen. And so, it can push you toward further recognition.

PB

I think that's totally right. One of my last questions is your selection of poets. I'm curious about how you selected your group of poets. You've got some Quebecers in there; obviously Pratt is a Newfoundlander. These are major figures—did that factor in? Did you have to exclude some people? I mean, you're working just in the English tradition?

SD

Well, I couldn't use the Québec tradition because one of the things we discovered when we started what we called the Association for Canadian and Québec Literatures was that Québec literature has quite a different tradition than English Canadian [literature] simply because it turned to the poets of France, whereas English Canadians turned to the poets of England. And I would have been thoroughly incompetent—even supposing I could find the texts.

PB

So, to your mind, you chose the poets who were best representative of an English Canadian tradition, is that a fair assessment?

SD

Yes, but I should have included F. R. Scott. One of the amusing things was that I didn't include Scott, whose biography I eventually wrote. I had intended to, but we ran out of time and money. We had done Duncan Campbell Scott who was quite significant in terms of the earlier poetry, and I really regretted that I hadn't done F. R. Scott, but I didn't fully realize this until I started looking at the list of poets—I assumed somehow that I had included him and I hadn't.

PB

That's the trouble when you have two writers with the same last name.

SD

Three writers. Frederick George Scott, Duncan Campbell Scott, and F. R. Scott.

PB

I think at some point we'd like to try and recreate some aspect of the concordance, so maybe we can do it as you did it and then include the missing Scott—unless, of course, it proves Frye's thesis somehow, then we'll exclude him.

SD

I think that would be unlikely [...] it's a different view of nature!

PB

So, it seems to me there's this interesting tension between the kind of stated goals of your project—to trace this continuity of the Canadian poetic tradition—which seems to be a kind of humanist project in a general sense, and the methods by which you set out to achieve them, which is not humanist in one sense of thinking about it. And this is somewhat suggested in the conclusion.

SD

In one sense, you're going into the camps of the philistines. If you want to prove a thesis, one way of going about it is to go from the "fact" perspective. And you can't argue against numbers. So, yes, it was a humanist project and, yes, it was an odd means through which to be seeking a corroboration, but I found it an interesting experience.

PB

You presented some of your results at a Learned Society conference in Winnipeg in the early 1970s—do you recall what the reaction was to the paper? I don't know if that's when you met Page and she asked about it, or if those are unrelated?

SD

No, I invited P. K. to come and speak to my poetry class at SFU in April 1970.

SD

You know, at the Learned Societies when I gave my paper on "The Computer and Canadian Poetry," the audience didn't say very much. They sat there and listened. I had a note from Robin Matthews [poet, academic, and political activist] when it was first published. "Brilliant," he said, referring to the article. What he was talking about, I think, was the joy that there was some factual way to show that there was a transmission from one generation of Canadian poets to another. Robin and I were sometimes loosely associated in the subsequent years because we were both ardent Canadianists—I was known as a strong nationalist. And I was. It's curious for a Newfoundlander—I was a strong Newfoundlander too, and I saw nothing wrong with Canadian nationalism. In fact, I became part of the Canadian tradition through Confederation. So, in terms of response at the Learneds, I think they were just a little stunned. They didn't quite know what questions to ask me. And I was just as happy they didn't. I don't remember any fuss or bother; it was just clapping. We were all very polite and Canadian.

REFERENCES

"Directory of Scholars Active." *Computers and the Humanities* 5, no. 2 (1970): 84–128. https://doi.org/10.1007/BF02402288.

Djwa, Sandra. 1970. "Canadian Poetry and the Computer." In "The Frontiers of Literature," special issue, *Canadian Literature* 46 (autumn): 43–54.

Liu, Alan. 2013. "The Meaning of the Digital Humanities." *Publications of the Modern Language Association of America* 128, no. 2: 409–23. https://doi.org/10.1632/pmla.2013.128.2.409.

CHAPTER 9

Canadian Poetry and the Computer[1]

Sandra Djwa

W hen reading through the works of the English Canadian poets of the 1880s, the critical reader is sometimes taken a little aback by the continued repetition of certain words and phrases such as "dream," "sleep," "vision," "trance," "spell," "secret," "mysterious," "unknown," or, if we prefer, there is "mystic spell," "charmed vision," "visionary moment," and "inappellable secret."

This insistence, at the diction level, on variations of the dream experience borders on the ludicrous and we are soon tempted to blue-pencil whole passages in Carman as examples of romantic excess, and to suggest that Roberts and D. C. Scott might have done well to edit their styles a little. Yet, is this approach ultimately helpful? Is an appreciation of the poets of the 1880s related to a stylistic norm which stresses neatness and economy, or does their very excess at the diction level point toward some fundamental understanding of the nature of things—a world view, a myth or a cosmology?

It is possible to dismiss this whole cluster of diction as simply vague transcendental aspiration, the Canadian backwash of Victorian romanticism. And there is no doubt that there is a certain amount of this involved; historically speaking, Canadian poetry has always been derivative. However, granted this fact, and granted that the common terms of diction are also very probably inherited, a more helpful approach might be the question of whether or not our poets did something unique with their particular inheritance. Did they construct a particular myth or cosmology from the common terms of romantic diction; and, if so, was there any continuance of myth or diction from the poets of the 1880s to those of the 1920s?

In Roberts's case, the reader soon becomes aware that he consistently uses the word "dream" and that it most often collocates with "sleep," "vision," "spirit," and "mystic." To determine whether or not these constant references to "sleep" and "dream" are simply the common coin of romantic diction as in, say, Keats's "Sleep and Poetry," or whether they are associated in a structure unique to Roberts's poetry, it would be necessary to classify each occurrence of the word "dream" together with its most commonly collocated words; this would include categories such as the common night dream, the impossible wish, the day dream, and the waking vision, that moment which Wordsworth describes in "Tintern Abbey" when the poet is "laid asleep in body, and becomes a living soul" and is so enabled to "see into the life of things."

For Roberts's poetry, the purpose of the classification would be to determine whether he adopts any of these particular aspects of the dream consistently and whether or not each occurrence reinforces a particular myth of the poet's experience in nature. Further, because we already know from Roy Daniells's fine study of the 1880s poets in the *Literary History of Canada* that "dream" is also a very strong metaphor in Lampman's work, it might be worthwhile to attempt to determine if there is a complex associated with this word which passes from Roberts into the poetry of Lampman, Carman, and Scott. But, the amount of listing and cross-referencing in a project of this scope would be quite prohibitive for any one person, and it is at this point that the computer comes into its own as a useful listing device.

Between 1966 and 1968, the published books of seven poets, Isabella Valancy Crawford, Sir Charles G. D. Roberts, Archibald Lampman, Duncan Campbell Scott, E. J. Pratt, Earle Birney, and Margaret Avison, were key-punched. Between 1968 and 1970, seven other poets, Charles Mair, Charles Sangster, Bliss Carman, A. J. M. Smith, A. M. Klein, Irving Layton, and P. K. Page, were added.[2]

The procedure followed was the same in all cases. Each poet's published books in chronological order were key-punched on computer cards at the rate of one typographical line per computer card. The computer cards containing the poet's canon were then fed into an IBM 7044 computer for printout. Following proof reading and necessary corrections, the computer then drew up a word frequency count. This is an alphabetical index listing every word that a poet uses and indicating its frequency of appearance. On the basis of the critic's understanding of a poet's work, and taking into consideration both the frequency of occurrence of particular words and the apparent collocations or associations of clusters of words, a selected list of words under the heading of thematic categories was then drawn up by hand. This listing under headings was key-punched as a thematic index and the computer then printed out concordances to the selected words from its memory bank.

After the works of Roberts had been key-punched and a word-index produced, it soon became apparent that except for function words and grammatical symbols, "dream" and words associated with it did indeed form the largest category of diction in Roberts's canon, as it also did in the works of Lampman and Scott. "Dream," "sleep," "vision" and its variants occur 217 times in Roberts, 368 times in Lampman and 221 times in D. C. Scott. In each case, it has the highest frequency of any thematic word occurring (an average occurrence would be from five to fifteen times) and indicates that for each poet the cluster of words associated with "dream" has primary significance. Further, by their continued appearance with a recognized structure of value delineated by a particular diction cluster, it was found that certain words such as Crawford's "love," Roberts's "dream," Klein's "little" and Margaret Avison's "sun" come to take on metaphoric significance. This is not to suggest that these elements of diction are always used as active metaphors. Yet, most often, the key terms emerge in context as a metaphor representing a larger myth.

In Roberts's work, the "dream" emerges primarily as a description of the poet's aspiration towards "the Spirit of Beauty" beyond nature. As this metaphor is explored through the thematic concordance, it can be documented that it becomes associated with a whole mythic structure in which Roberts expresses life as a "dream" emerging from the great "sleep" of Eternity, which is, in turn, a "dream" of God. Through the human "dream," man is put in touch with this eternal world. Referring to the dream experience, Roberts has two sets of terminology which he uses interchangeably; one set is connected with Darwinian evolution while the other is primarily Christian in nature.

This process is quite explicit in a poem such as "Origins" where the germ of life emerges from Time: "Out of the dreams that heap / the hollow land of sleep"; it then develops by evolutionary processes, only to return to its divine maker, God. Similarly, in his poem "The Marvellous Work," Roberts praises the evolutionary God whose "Eternal Cause":

> Is graven in granite-moulding aeons' gloom;
> Is told in stony record of the roar
> Of long Silurian storms, and tempests huge
> Scourging the circuit of Devonian seas...
>
> Athwart the death-still years of glacial sleep!
> Down the stupendous sequence, age on age,...
>
> In the obscure and formless dawn of life,
> In gradual march from simple to complex,
> From lower to higher forms, and last to Man.

In effect, Roberts has taken over the general aspects of the Wordsworthian-Keatsean transcendental dream, associating it with poetic comfort. However, he changes a few of the essential terms of the dream experience to accommodate some of the problems raised by the Darwinian hypothesis. But, if the primary function of the dream metaphor is to alleviate pain, Roberts's choice was particularly unfortunate as it carries along with it its own built-in negation—that of the nightmare. So, although Roberts's poetic decorum precludes evil as a subject, whenever evil or death intrude into his poetry almost despite the poet, they do so, as does the nightmare, through the dream. The blinding of Orion, the capture of Launcelot, and the sick soul of the poem "One Night" all emerge from the dreaming state.

Archibald Lampman adopts Roberts's dream metaphor and with it much of his poetic myth including the "sleep" of time, the "dream" of human life and the possible evolutionary progress of the human soul. However, Lampman's concept of the poet is that of the passive observer who, standing a little apart from himself and from nature, is empowered to see into the nature of things. In this formulation, the unconscious creatures from the world of nature, such as the frogs and cicadas, become poetic emissaries from the world of dream which underlies the universe. This relationship is quite explicit in the poem "The Frogs."

In effect, the peace and comfort of the eternal dream, unconsciously known by the frogs, is passed on to the poet who lays himself open to this experience. But if the voice of the frogs can bring assurance of the eternal plan, the "dream" which underlies existence, there are other voices which remind Lampman of the fear and sorrow which are also a part of human life. The voice which comes out of the darkness, "the crying in the night" of Lampman's much anthologized "Midnight" would seem to be part of a larger sequence of poems dealing with the nightmare aspects of existence often specifically associated with the loss of the comforting "dream" as in the poem "The Loons."

The "dream" in D. C. Scott's work is first associated with "rest," "death" and "magic." In poems such as "The November Pansy," "The Height of Land" and "Lines in Memory of Edmund Morris" the transcendental attempt to reach a "mystic world, a world of dreams and passion / that each aspiring thing creates" is unsuccessful and the "secret" beyond nature remains "unutterable," a "something" that "comes by flashes / ...—a spell / golden and inappellable." When the transcendental dream does succeed, as it does in a series of "magic" or fantasy poems, it results in death for the mortal concerned, as in the poems "The Piper of Arll," "By the Willow Spring," "Avis," and "Amanda."

In Scott's early work the death theme is associated with the dream and with rest; in his later work it becomes associated with a dying world. In the poem "The November Pansy," he suggests that a "seed" of life might be dropped from the dying world to re-kindle life elsewhere. This linking of human death with the suggestion that the earth is growing old is dominant in Scott's later work and it seems to mark the end of a cycle in which Roberts's evolutionary "germ" of life has burst up into fruition and is now decaying.

In consideration of this analysis, it would appear that a critical re-evaluation of the work of Sir Charles G. D. Roberts is necessary to point out that Roberts did establish a poetic myth with his inherited romantic diction, that the function of this myth was to reconcile the Darwinian germ of life with the Christian world spirit, and that this myth was adapted with some slight variations by Roberts's major successors, Lampman and D. C. Scott. Further, it would appear that the early work of E. J. Pratt, supposedly a sport in the Canadian stream, might have developed in response to the poetry of Roberts.

Pratt's ode, *The Iron Door* (1927), provides a good transition from the 1880s to the 1920s because it is a poem which has its roots in the earlier group, yet, in development, it rejects the transcendental dream. The whole visionary experience of the poem is specifically contained within a human "dream," undercut by contrast with the reality of "terrestrial day." But, if Pratt rejects the dilute romantic aspirations of the earlier "dream" poetry, he does so by turning to the law of tooth and claw which he finds explicit in Roberts's tales of the wild and some of the later poetry. "The Great Feud," for example, has its genesis in the first two chapters of Roberts's book, *In the Morning of Time*, which was first published in 1919, just as Pratt was beginning to write. In chapter 1 is the setting for "The Great Feud"—the red clay estuary complete with giant lizards, the prototypes for Tyrannosaurus Rex, and bloody internecine battle. Here too are members of an evolving man-like species associated with the rudiments of reason—prototypes for the ape mother and her brood.

Similarly, a prototype of the battle between cachalot and kraken in Pratt's poem "The Cachalot" (1926) is to be found in a tale entitled "The Terror of the Sea Caves" from Roberts's book, *The Haunters of the Silences* (1907).[3] It is substantially Roberts's concept of the national epic and the military sea poem (cf. "The Shannon and the Chesapeake") which recurs in Pratt's later poem, "The Roosevelt and the Antinoe." In addition, the whole iceberg section from *The Titanic* (1935), including suggestions of the berg's eventual disintegration as a part of a natural cycle, can be shown to have a strong relationship with Roberts's poem "The Iceberg," first published in the *University of Toronto Quarterly* in 1931.

Pratt's progress would appear to be contained within the framework of the older Darwinism established by Roberts and Lampman. The difference between Pratt and his predecessors (and in parallel development to the later poetry of Scott) is that he continually uses the earlier pre-formulated world view to suggest its opposite. "The Great Feud" is a dominantly atavistic structure emerging from the evolutionary Darwinism of one of Roberts's later romances. A second difference between Roberts and Pratt is that the latter shifts the focus from external to internal nature as he explains in *Newfoundland Verse*: "the fight / with nature growing simpler every hour, / her ways being known." Man using the full resources of his courage, reason and self-sacrifice can resist the primal forces of the sea; however, when the primitive forces of external nature are internalized within man, "these blinded routes" are almost without cure: "the taint is in the blood." So that where Roberts searches external nature for the "secret" of "beauty" or "life," Pratt turns inward in an attempt to find the existential "why" of human behaviour.

It is at this point in clarifying the details of a poet's myth, that the computer can be of considerable help to conventional scholarship. One of the great surprises of the Pratt word-index was that the encompassing metaphor appeared to be that of "blood" rather than the expected "sea" or "water," although, of course, "sea" is a larger category than blood. Yet, as each reference to the word "blood" was followed through the thematic concordance, it began to appear that Pratt internalized the tides of the sea within the veins of man, as is explicit in the lyric "Newfoundland":

> Here the tides flow,
> And here they ebb;
> Not with that dull, unsinewed tread of waters
> Held under bonds to move
> Around the unpeopled shores —
> Moon-driven through a timeless circuit
>
> Of invasion and retreat;
> But with a lusty stroke of life
> Pounding at stubborn gates,
> That they might run
> Within the sluices of men's hearts...
>
> *Red is the sea-kelp on the beach*
> *Red as the heart's blood,*

This is a natural metaphor for a Newfoundlander, but, more importantly, it is also a natural metaphor in terms of Pratt's modified Darwinism. Man, evolving from the sea, still carries part of the sea within him. In Pratt's myth, the blood stream becomes an evolutionary battleground where the forces of instinct (associated with cold-blooded creatures) and those of higher reason (associated with warm blood) are continually at war. As Pratt writes in "Under the Lens":

> Along the arterial highways,
> Through the cross-roads and trails of the veins
> They are ever on the move —
> Incarnate strife,
> Reflecting in victory, deadlock and defeat,
> The outer campaigns of the world,
> But without tactics, without strategy.
>
> Creatures of primal force,
> With saurian impact
> And virus of the hamadryads,
> The microbes war with leucocytes...
>
> Once it was flood and drought, lightning and storm and earthquake,
> Those hoary executors of the will of God,
> That planned the monuments for human faith.
>
> Now, rather, it is these silent and invisible ministers,
> Teasing the ear of Providence
> And levelling out the hollows of His hands,
> That pose the queries for His moral government.

As is suggested by these examples, Pratt internalizes both good and evil and associates them with a physiological metaphor of the bloodstream. In Pratt's published books of poetry, "blood" and its variants (appearing 265 times) are primary nouns and have the same significance in Pratt's poetic myth as does that of the "dream" in Roberts's world view. Clustered about the metaphor of blood is a series of related nouns: "vein," "artery," "love," "hate," "instinct," and "reason." As might be expected, "red" (with its variants of "crimson" and "scarlet") is the dominant colour.

In Pratt's work, the blood line not only determines the pedigree of the creature, but it also establishes its physiological possibilities for good or bad. It is this aspect of the blood metaphor, suggesting the Biblical "sins of the fathers," which is evoked by the woman representing universal humanity in *The Iron Door* when she asks why "blood" and "time" should always bring forth a "Cain." Cyrus, on the other hand, in *The Fable of the Goats*, evolves a sport "leucocyte" in his Aryan bloodstream which enables him to make peace with the Semite goat and so save universal humanity through moral evolution. Consequently, as has been expressed in the critical formulation of John Sutherland, Northrop Frye and Desmond Pacey, Pratt's poetry moves from "stone to steel" or between the ethical norms of "the temple and the cave"; what has not been noted, however, is that it does so along the metaphor of the bloodstream.

In this connection, it is important to see that for Pratt the whole process of life from microscopic spore to man constitutes the evolutionary process. In his structure, Christianity is the evolved pinnacle of human conduct, and when man falls away from this ideal, he can only fall into atavism:

> But what made *our* feet miss the road that brought
> The world to such a golden trove,
> In our so brief a span?
> How may we grasp again the hand that wrought
> Such light, such fragrance, and such love,
> O star! O rose! O Son of Man?

Because of this, an understanding of the relationship between Roberts's book, *In the Morning of Time*, and Pratt's poem, "The Great Feud," is important to an understanding of Pratt's work. Roberts was writing of man's evolutionary progress at the very time when Pratt, a pacifist, sick at heart at the carnage of World War I, was coming to the conclusion that man was not progressing but retrogressing to his animalistic past. *In the Morning of Time* provided a structure of immense ferocity embodied in animal form which perfectly expressed Pratt's feelings regarding the bloody, brutal and unreasoned precipitation of World War I. Then, too, Robert's stress on "reason" and evolutionary "progress" indicated to Pratt the precise lines of argument with which he must disagree. "The Great Feud," with its perversion of reason and the moral law, its bloody internecine combat and the concluding implications of cyclic recurrence, is Pratt's atavistic answer to Robert's evolutionary progress.

Pratt's substitution of an atavistic myth for Roberts's evolutionary Darwinism is, perhaps, the key to much of Pratt's work. This explains Pratt's fascination with the giant creature, the survival of the fittest, and the emphasis on the power of the superior creature, be it man or machine. And because Pratt is also holding in suspension Wilhelm Wundt's mechanistic physiology which stresses the un-reasoned mechanical response, that which links the animal, fallen man, and the machine is precisely this mechanical instinctive response. When man or his representative (such as *The Titanic*) falls from reason to instinct, there is a magnificent rush of unbridled power. And it is this response to the removal of reason which fascinates Pratt.

Further, as Pratt accepts that aspect of popular Darwinism which suggests that inheritance is carried along the bloodline, these evolutionary or atavistic struggles are always carried on in that arena. "The Witches' Brew," Pratt's farcical version of Milton's *Paradise Lost,* establishes an underwater Eden where the fall from cold-blooded to purely human (warm-blooded) sinning is accomplished through an alcoholic apple in the bloodstream. "The Great Feud" is again about the fall from instinct to reason and the return to brute force through demagoguery and a "yeasty" ferment in the blood. *The Titanic* also invokes a fall from steel to stone, and Brébeuf, associated with hubris, falls from Christianity to demonism or black magic. Similarly, the characteristic technique of Pratt's shorter poems is the flashback to the primal past, as in the reversion to the wordless hate of "Silences" or to the void before the earth began in "The Ground Swell."

As this documentation would indicate, there was, in fact, a very close relationship between the major poet of the 1880s, Sir Charles G. D. Roberts, and the major poet of the 1920s, E. J. Pratt—a bend in the stream of Canadian poetry rather than the sharp break suggested by present critical comment. Further, it is possible that A. J. M. Smith, D. C. Scott and A. M. Klein, although busy carving out new provinces for poetry, were also fully aware of the work done by their predecessors and contemporaries.

In this transmission, D. C. Scott would appear to be significant. One of the surprises of the A. J. M. Smith concordance was a substantial "death," "love," "beauty," and "dream" complex not unlike the formulation of D. C. Scott's concordance. This is not to imply that D. C. Scott's sometimes metaphysical *Beauty and Life* (1921) was to Smith as Roberts's work was to Pratt, for Smith's whole canon is much more profoundly influenced by Eliot's fertility myth. Yet, there are significant parallels with the older poets in Smith's work. In this connection we might compare D. C. Scott's "Variations on a Seventeenth Theme"—a series of modulations on death using the primrose, Eliot-fashion, as an organizing metaphor—with Smith's habitual practice. Then, too, Smith's poetic technique of metamorphosis, often related to successive shadings of reality, would seem to be quite close to Scott's poetic (cf. Scott's "The Tree and the Birds," and Smith's poem "The Fountain"). Similarly, there are continual parallels with D. C. Scott's concept of the timeless geological North ("Lines in Memory of Edmund Morris") in F. R. Scott's work, as well as a strong emphasis on the evolutionary concerns of Roberts and Pratt.

If it is Pratt's concern that man is in danger of reverting to his animalistic past, in the poetry of Abraham Moses Klein, man very often *is* an animal, and a predatory animal at that, as in this description of Hitler:

> Fed thus with native quarry, flesh and gore
> He licked his whiskers, crouched, then stalked for more.

Hitler is also specifically identified with an atavistic fall and the concept of inherited evil: "Judge not the man for his face/ out of Neanderthal! / ... the evil of the race/ informs that skull!" "Animal" is Klein's largest category of diction, recurring some 400 times with "blood" also a substantial category, occurring seventy-six times. In Klein's work, "blood" is most often associated with the spilled blood of the small and innocent creature. Klein's poetry appears to suggest two worlds: one is the world of the "Black Forest" ethic where the good little man is pursued by the ravening beast; the other is the reconciling art world of Biblic wood and fairy tale where the small boy of "Bestiary," hunting at his leisure, can stalk the "beast, Nebuchadnezzar."

It is one of the ironies of the development of Canadian poetry that E. J. Pratt and A. M. Klein, both fundamentally kind and compassionate men, should, by virtue of their differing historical and religious perspectives, have been fundamentally influenced by diametrically opposed aspects of the same myth or world view. Pratt, strongly influenced by the Darwinistic superior creature, is fascinated by the spectacle of immense strength and power, the giant whale, the enormous iceberg, the largest ship the world has ever known; Klein, who has been made tragically aware of the immense danger of unbridled power during the Nazi era, holds as exemplar the good little man, the homoculus, the dwarf.

This would imply that Roberts's evolutionary Darwinism has become atavism in the works of E. J. Pratt and that the whole concept funnels into the Aryan myth, where it is picked up by A. M. Klein in the late thirties. In a real sense, Canadian poetry has been a direct response to a world view or *weltanschauung*, and if it may be hypothesized that an appreciation of the Puritan mythos is essential for an understanding of the poetry of the United States, it might be equally hypothesized that for Canadian poetry, coming as it does 300 years later, an understanding of the ramifications of popular Darwinism is essential.

But, although Canadian poetry has developed in response to the prevailing popular philosophies and literary influences (even Pratt has a few poems suggesting Eliot's fertility myth structure), it does not seem possible to argue that literary climate alone can explain the links of connection between our poets. Current interest and mere chance do not seem adequate explanations for the fact that Roberts and Pratt choose to write of the struggles of cachalot and kraken; that Lampman and Smith invoke machine hells in corresponding accents; that D. C. Scott, F. R. Scott and Earle Birney turn to the North land as the new Eden; that Pratt and Klein both write ironic litanies of progress noting that man has turned to the beasts of the field for his instruction; that Klein's little hunter seeks out the enemy "spirochete" in Pratt's "whispering jungle of the blood"; that Birney uses the following terms: "Andromeda" (1), "apotheosis" (1), "architrave" (1), "Armagadding" [Armageddon] (1), "Betelgeuse" (1), "cordite" (1), "hiero glyphed" (1), "narwhal's" (1), "pleiades" (1), "saurian" (1), "saurians" (2), "trilobites" (1), "tyrannosaur" (2), usually once, and with implications of Pratt's schemata; or further, that when insisting on man's need to accept responsibility for his own evil, Birney equates man's potential savagery with the iceberg of Pratt's *Titanic*, suggesting "the iceberg is elective."

These persistent linkings suggest that we need to re-evaluate one of the major issues of the 1940s—the question of the continuity of Canadian poetry. The term "continuity" has an unfamiliar ring in this context. In most critical texts we stress not continuity, but the division of the Canadian stream into four unrelated groups: those of the pre-1850s, the 1880s, the 1920s and the post-1940s. If such a continuity does exist, how may it be indicated? Northrop Frye, reviewing A. J. M. Smith's *The Book of Canadian Poetry* in 1943, states that he senses a "unity of tone" in Smith's selections. In his later essay, "The Narrative Tradition of English Canadian Poetry" and his "Conclusion" to the recent *Literary History of Canada* (1967) this has been expanded to suggest a unity of tone achieved by a dominant thematic pattern—one of the cruel North characterized by a forbidding nature and a "garrison mentality." However, if we are to accept John Sutherland's angry dismissal of Smith, Frye, and the Canadian tradition in his preface to *Other Canadians* in 1947 or the tacit editorializing of his lineal descendants, Louis Dudek and Michael Gnarowski in their recent anthology of criticism, *The Making of Modern Poetry in Canada* (1967), there was not only no continuity in Canadian poetry prior to 1940, there was no Canadian poetry worthy of consideration prior to 1940.

Disregarding the question of poetic worth, I think this assertion can be disputed on the basis that there simply has not been enough work done in the area to be able to make so final a statement. I am inclined to agree with the later Sutherland, writing in *Northern Review*, when he suggests, somewhat elliptically, that it might not be a bad idea if the Canadian poet were not unaware of his place in the tradition of Canadian poetry. The necessity for this is obvious, and I would think that it would apply equally to Canadian criticism, too. Without an understanding of our own development, we cut off our poetic roots: without Roberts, Pratt is not entirely explored; without Pratt, we negate aspects of Klein and Birney; without D. C. Scott and Lampman, aspects of Smith's poetic are incomplete. Similarly, Layton's insistence on the image of man as a "dis-eased animal," Cohen's "Lines from My Grandfather's Journal," Avison's preoccupation with the technical terms of space, Page's "dream" metaphors, and Atwood's *The Journal of Susanna Moodie* do not emerge from a cultural vacuum, but are intimately related to the development of writing in Canada.

NOTES

1. Reprinted with permission from *The Frontiers of Literature*, a special issue of *Canadian Literature* 46 (autumn 1970): 43–54.
2. The research from 1966 to 1968 was supported financially by the President's Grant Fund of the University of British Columbia and the Koerner Foundation, and that from 1968 to 1970 by the Canada Council and the President's Research Fund of Simon Fraser University.
3. In the case of Pratt's poem "The Cachalot," there is also very likely an intermediary text, Frank Bullen's *The Cruise of The Cachalot; Round the World After Sperm Whales*, 1898. From notes contributed by both Pratt and Roberts to an anthology of sea poems for school children entitled *Verses of the Sea* (1930), it would appear that both poets were familiar with Bullen's work. As Bullen's work came after *Moby-Dick* and does share some similarities with it, this supports Pratt's contention that he did not read *Moby-Dick* until after the completion of "The Cachalot."

"saga uv th relees uv huuman spirit from compuewterr funckshuns": Space Conquest, IBM, and the Anti-digital Anxiety of Early Canadian Digital Poetics (1960–1968)

Gregory Betts

With some halting precedents, the digital age begins in Canada in 1949 with the onset of the construction of the first bona fide computer, the University of Toronto Electronic Computer Mark I (UTEC for short) (Bateman 2016). Built largely from scratch with wire and solder by a group of eight academics (one director, three professors, and four graduate students—with guidance from Alan Turing in Cambridge), the UTEC used an integrated series of International Business Machines (IBM) mechanical calculators and vacuum tubes as its core processor. Before it was completed and fully usable, however, events transpired across the Atlantic that shifted the relevance of this first computer significantly. In a moment of political austerity, the U.K. Atomic Energy Authority lost funding to support construction of the world's second manufactured commercial computer, a much better and more powerful machine. The University of Toronto snapped it up for $300,000, with funding from the National Research Council and the Defence Research Board (the precursor to Defence Research and Development Canada), with key support from the Atomic Energy of Canada Ltd. (Williams 1994, 10). Work on UTEC halted almost immediately, and the computer was stripped for parts. The new machine was named "Ferut" (pronounced "ferret") as a combination of its U.K.-based manufacturer, Ferranti, and University of Toronto (Hume 1994, 13). This computer is responsible for the earliest known recording of computer-generated music, playing a medley based upon "God Save the King" as made by the BBC toward the end of 1951, with the programming being done by Christopher Strachey. It moved to Canada on the first ship to travel down the newly constructed St. Lawrence Seaway and was launched in the summer of 1952. Poignantly, Ferut was introduced to the world with a word-game contest that the machine lost. Though the computer was bested in a competition of the *Toronto Daily Star*'s Tangle Comics (a game where the object is to untangle the letters and spell out as many of the

Figure 10.1. Tangle Comics $10,000 tiebreaker.

Source: "Tangle Comics," *Toronto Daily Star*, June 13, 1952, 14.

names of the characters from the newspaper's comic pages as possible; see
figure 10.1) by Mrs. Richard Pearson of Wallaceburg, reporter James Y. Nichol
concludes that "Ferut will help to keep Canada strong and up to date in both
military and economic spheres" (J. Nichol 1952).

This combination of military nationalism, atomic-energy interests, IBM, and
visualized word play presents a compelling nexus point to mark the onset to the
computer age in Canada. It was, indeed, the exact combination of these forces
that divided the Canadian literary avant-garde in the next decade, prompting a
debate about the meaning and implications of computers and their relationship
to art. Publisher, poet, and all-around champion of the avant-garde bill bissett,
in British Columbia, was immediately suspicious of machines that relied upon
closed-loop linguistic logics. He was already battling the lingering tentacles of
Aristotelian categorical thinking in the literary scene in Vancouver as part of a
broader political stand against Western codices of individualism, imperialism,
and capitalism. He and various members of his *Blewointment* cohort (a group
loosely defined by their participation in bissett's Blew Ointment Press publishing house, *Blewointment* magazine,[1] and the general intermedia downtown scene)
turned against things that divided the body, the art object, and the world. They
explored things like hand-written poems and illustrations, handmade books—
really, anything bearing the irreducible imprint of the body. These were quickly
understood as an antipodal mode of art production from the digital. One of his
cohorts who did not follow him in this organicist turn, however, was Vancouver-
expat, Toronto-based concrete poet and publisher bpNichol. For Nichol, the
computer was but another tool—alongside genre, grammar, page space, even

the alphabet itself—that could be conscripted into the "language revolution," an international movement that, Eric Schmaltz has argued, pulled Canadian poets into a widespread radical opening of poetry (2018, 9–10). Nichol would eventually liken himself a "kid of the book machine" and, working with Steve McCaffery, make a concerted effort in thinking through (and subverting) the mechanical apparatus of language and literature (McCaffery and Nichol 1992). Nichol's experiments with digital poetry began in 1968 and culminated in his iconic suite of digital kinetic poems called First Screening in 1984, a date that coincides with the publication of SwiftCurrent, the world's first online literary magazine, which featured Nichol and some of bissett's West Coast peers.

As Nichol was by no means a conscript into the cold calculation of IBM's economic pursuits (which included supplying the Nazi Party with machinery; see Black 2001), and bissett (a cinephile and typewriter artist) by no means a technophobe, the divide was less an essential philosophical or political rift between them and more of a disagreement about methodology and aesthetics. For Nichol, the mechanics of language could be integrated into the poetics of text production in a way that exposed their trappings and helped to reanimate poetry. He was, as Richard Cavell has argued, attempting to become a postmodern medieval poet (2002, 149) by realigning the spoken aural text with the spatial mechanics of typewriters and other machines. This involved a constant cycle of creative destruction, and his texts are accordingly filled with words shattering into letters, reforming into earnest articulations of love and myth, before spiralling off into bad puns, imaginary frames, comic-book interruptions, or merely semiotic noise (oxymoron intended). As every element of text formation is interrogated, the computer or any digital device did not interfere with his process. Eric Schmaltz, in chapter 11, points out how Nichol made use of even hidden comments in the coding of the digital poem as a space for creative intervention. For bissett, however, such a theoretically infused approach leaned too heavily on abstraction, on detachment from the material, such that the work, however ostensibly resistant, became complicit in oppressive systems. For him, poetry held the potential to actualize presence, rawness, and direct vitality. The computer, in contrast, with its integration into the military and economic machine of society, became a symbol of a violent detachment that by hyperbolic extrapolation permitted violence on the scale of a Vietnam or Nagasaki. In this chapter, I look at how this difference manifested in a limited number of key texts of early Canadian computer poetry and to highlight the anxiety felt by Canada's avantgarde about the intersection of computers and poetry. To be clear, allies throughout it all, both bissett and Nichol could agree that the interface is never neutral, that the government and army were not to be trusted, and that some chance for hope remained in a grim world.

Although both would end up in Toronto, bissett and Nichol began their literary careers in Vancouver, which by the mid-1960s had become a genuine hub for avant-garde, cross-genre experimentation. As the intermedia community matured, Vancouver art increasingly privileged direct experience, engagement, and romantic knowledge in Wyndham Lewis's sense of being "popular, sensational, and 'cosmically confused'" (1927, 35). Lewis's discussion of the lingering romanticism in modern art from which that definition emerges in *The Enemy* (1927) is particularly useful in highlighting the mistrust of positive science and technology that became a common concern of the group, even as they embraced a utopian, McLuhanesque potential of new media. The imperialist war in Vietnam, the rise of environmentalism in Vancouver (Greenpeace was founded in the city in 1971), psychedelics, and the emergence of communal lifestyles all contributed to a deep sense of mistrust in Western notions of technology-led progress. McLuhan's theories of the increasing "tribalization" of citizens of the electric age, the denizens of a global village, also offer some insight into why science and rationalism were no longer deemed sufficient. Electricity was reprogramming the sense ratios away from the detachment of the visual system in favour of the more immersive auditory system (McLuhan 1964). The shift presaged a decline in rationalism and individualism as people experienced themselves within environments rather than detached from them. In this way, the electric age privileges multiplicity over singularity, communal engagement versus individual detachment. McLuhan, who now functions as the prophet of the digital age, was then read as a prophet of the age of Aquarius dawning in the city. The downtown artists (defined in contrast to Vancouver's more Aristotelian university poets) became emphatic followers and celebrated his theories of emerging multi-consciousness and his deployment of an anti-scholarly mosaical prose form. McLuhan lectured to the downtown community at the New Design Gallery on Pender Street in 1959, and later at the 1965 Festival of Contemporary Arts. The latter event was subtitled "The Medium Is the Message" in tribute to his rising influence (Turner, n.d.). The iconic phrase had, in fact, first been uttered at an event in Vancouver in 1958.

bissett, for one, worked to incorporate McLuhan's ideas into his writing. In the preface to *We Sleep Inside Each Other All* (1966), he explains the collection as a direct response to McLuhan's sense of epochal change:

Marshall McLuhan sz we are poisd between th typographic individualist trip the indus trial revolution & th electronic age we have been in for sum

time, between a unique dis tance and alienation privacy well now iullbe
in th study for th rest of th night with my nose in a boo k & th corporate
image tribally we are a part of out extensions do reach now have been
reach thruout all time th historical jazz consumd in th greater fire of mo
view t v & lo ve. (1966)

McLuhanism, acid, jazz improvisation, anti-imperialism, anti-individualism,
mysticism, and the emergence of new publishing technologies (the mimeograph
and the typewriter especially) all facilitated the sudden prominence of collage
experimentation in the city. Indeed, Vancouver was arguably the most fertile
ground in the country for technological experimentation with new media, especially in the fields of visual art and literature. *Blewointment*, bissett's publishing
venue and literary magazine, was one of the epicentres of that endeavour. His use
of the standardized spacing of typeset letters to create puns, expose hidden word
paragrams, and create poignant neologisms in that passage above, for instance,
in isolating the "trial" in industrial or uncovering "view" between movies and
TV, introduces a playful but serious approach to language as interface, even site
of potential revelation hidden in the material of its letters. There is a sensual,
embodied experience to such expression operating in spite of the typewriter's
standardization.

bissett's work as a poet, as a collage artist, and even as a publisher brings out
the sensual, embodied experience of text—language as a mouthy material, meant
at once for the eyes and ears and tongue (my copies of his early books still smell
of a wood-burning stove and cigarettes from the room in which they were assembled). He writes with his whole body, resisting the division of the senses. In his
delightfully, revealingly titled *Rush: What Fuckan Theory* ([1972] 2012), bissett's
most developed articulation of his embodied poetics, he explicitly describes the
reliance on categorical approaches to literature as an extension of Western
imperialist logic; it is time, he writes, to go "byond the ol war / lord purposive
aristotul categorees" (89). Ezra Pound, who helped to inspire bissett's rivals,
would not have disagreed with this critique, for he too described Aristotle dis-
missively as the "Master of those that cut apart, dissect and divide. Competent
precursor of the card-index" ([1952] 1970, 343). bissett's criticism is not just
directed at evaluative modes of writing or academic interpretations, for he also
connects Aristotle to the breath line and to the sequential orientation of projective verse: "i do not accept any significant correlation between one instant and the
next '   ' but a poem can be tension of rejecting accepting this [at the] same time
since no events 'really' occur at the same time tension achievd" ([1972] 2012, 44).

The point he is making is subtle and slightly unhinged; by rejecting plot and the sequencing of events, he rejects all categories of human knowledge for being arbitrary and systemic rather than natural, inevitable, and *catalogueable*. He does acknowledge the appeal of truth and meaning and its use value in art, but these are tensions, not absolutes or laws. A new kind of art, predicated on incoherence rather than coherence, emerges: Zarmsby Potsmurth writes in *Blewointment* 1.2 to assert that "collage...the arrange-/ meant that is pleasurable / because there is no / arrangemeant not a pictorial / corresspondence to a / supposed absolute" (1963; spelling as in original). This model of a human universe is not structured by observable laws, nor by a sense of timeless order. Peter Culley observes in a 1993 paper that "a vast reassessment of accepted formulations of the real that was occupying the entire culture. It was not categories that were in question, but the whole notion of 'category' itself. In the maelstrom of the sixties, what was held on to mattered less than enthusiastically surrendering to the flow" (193). Postmodernism emerges in this categorical renunciation, in this interrogation of the medium and the interface as discursive elements contributing to the production of meaning.

Blewointment, with its radical openness and investment in eclectic disharmony, is emblematic of the transition between late modernism and postmodernism. In direct contrast to typical modernist editorial exclusivity, bissett published anything that was sent to him, prompting bpNichol's witticism that the magazine was "more interested in the news than in preserving great literature" (quoted in Reid 2002, 23). In sharp distinction to the typical aesthetic of modernist magazines, it was aesthetically irrational, unlimited, individual, and each issue unique and irreducible (everything a computer is not, in other words). The magazine also disrupted other publishing habits: some issues of the magazine excised author names from texts, while others included complex collages of various people's writing without easy attribution. (I published an index of *Blewointment* magazine, wrestling with its various resistance manoeuvres to data processing and management.) Spelling and grammar were never standardized, let alone any other aspect of typesetting or book design. bissett often stapled or pasted individually handcrafted textual objects of varying sizes inside and outside his magazine, or even hand painted on the covers. If the book is a machine, *Blewointment* represents a pointed rejection of the industrial, assembly-line aesthetics of its machination.

Blewointment's radical openness—what Tim Carlson calls its "pure provocation" (2002, 43)—anticipates Roland Barthes's 1974 attention to the "function" of the text within the "infinite play of the world" (1974, 3, 5). *Blewointment* was a concerted attempt to reimagine writerly communities outside of oppressive and hierarchical ideologies, to reconceive the function of the author, the publisher, and the reading community at large as coterminous. Barthes posits that the "goal of literary work (of literature as work) is to make the reader no longer a consumer, but a producer of the text" (4). In like manner, bissett sought to open up the book machine, energize and radicalize readers, and create communities open to possibilities beyond transnational capitalist imperialism. Thus, instead of celebrating the margin of the page as the site of avant-garde conquest, bissett proposed the eradication of centre and margin, countering that "yu dont need the margin" because "yu are already here" ([1972] 2012, 46).

In his 2012 investigation of the interface, and the limits of representation, contemporary media theorist Alexander Galloway marks a poignant distinction between artists invested in "a coherent, closed, abstract aesthetic world"—pre-existing or fashioned by them—versus "the disorientation of shattered coherence" that "makes no attempt to hide the interface" and "turns the whole hoary system into a silly joke" (2012, 39). This distinction reminds me of a line from a 2013 letter written to me from Talonbooks founding editor Jim Brown on concrete and visual poetry: "In the sixties it was a fun part of writing poetry. Nobody took it seriously. We all had fun with it in one way or another" (2013). Despite his dismissal, this space of not taking it seriously (which he contrasts to the serious writing of Olson, Duncan, Bowering, and so on) is actually a gateway to a disturbing and systemic alienation from language and expression. Nichol, when asked of this kind of sentiment, retorted: "Many people don't realize that play is serious.... Games demand work and involvement" (1974, 133). Galloway writes that the melding together of form and content "aims to remove all traces of the medium" as a barrier to representation (2012, 46). In contrast, dirty, messy, sprawling deliberately incoherent works—such as we see in bissett's concrete poems, and throughout *Blewointment*—objectify "the necessary trauma of all thresholds" and stage "the upheaval of social forms" (Galloway 2012, 46). Making the page visible as interface, as technology shatters sayability: the ink spreads like a dark silence (like nothingness), the joyful autonomy of language overwhelmed by the materiality of its machinery.

Judith Copithorne's poem "Famine" from the second issue of *Blewointment* in December 1963 articulates this consciousness of inarticulation:

> So much moves by in cold and emptiness
> The silence of only oneself.
> And the voice in ones head far back

as in a cathedral or a machine
far away.
One is often frightened because—last
time—or the time before remembered
when there was a famine
What happened then
Never will the famine go
For leaving is a time.
There is no leaving
Coming is a misleading word
perhaps. [...]
Hear you all hear you tell us listen
Receive all receive I am trying to
listen to the famine.
The page flips You were flipping
the page when we came.
We want you to listen. Listen.

The poem achieves a liminal ground between a debilitating fear of inexpressibility and recognition of the dire need to both hear it and speak through it. This famine is a prophecy of the collapse of meaning and the recoil of language. The flipped page (of a flippant text) distracts from a harrowing truth surrounding it. In the same issue, bissett writes: "We are not the same as we were inside THE BODY, or as we were coming to it or taking our departure. We have become outside remembrance and forgettings, its illusions and skills, outside time" (bissett 1963). Such harrowing alienation belies the preternatural confidence of empire's drive to overwrite, maximize the world.

Though it was a popular trope or cliché for writers at the time, especially for writers linked to the Beat movement, the downtown poets in Vancouver shunned the romance of the typewriter as a technology endowed with special representational power or as a force for autonomy for the materialist and ideological contest from which such an interface emerges. The political implications of the material world imply a disturbing complicity with a system bissett was actively fighting. As he wrote to Diane di Prima in 1971: "yr struggle is to bring down th pentagon—ours is to keep it out. its 1812 all ovr again & its 1850's red river & we th long hairs the metis again—peopul here who nevr done time for innocence talk uv hopeful change thru the existing govt—I don't know" (bissett 1971c). It follows from this nascent cynicism that, as a poet informed by the messy body, bissett's exploration of the use of the typewriter was antithetical to the mechanical precision celebrated by the likes of Charles Olson. bissett produced grimy and ambiguous texts wherein letters crash upon other letters, overlap, distort, obscure, and prevent reading. bissett's typewriter art functions as a libidinal eruption of repressed drives: highly sexual, openly revolutionary, and logically disturbing. Ink spills over the words, even the letters, darkening communication by its black fog. There is a spirituality at play in this destruction, but it begins with a slag heap of culture, with the recognition, as Copithorne notes, of famine.

More generally, the downtown poets were ambivalent about technology, aware of their function in transnational hegemonies, but drawn to the euphoric potential in mixing media and experimenting with machines. Schmaltz, in his chapter, draws attention to the pun of digital referring to both the manual and the computer. At the least, this creates an opening for radical experimentation, for a simultaneous self-effacing mystical-cum-spiritual organicism that decries machinery for a nascent environmentalist equilibrium that yet allows them to glom onto new technologies (video, tape recorder, radio, film, mimeograph, photocopier, and so on) and repurpose those machines to what were held to be subversive ends. Gerry Gilbert's 1974 poem "Thought for Penny" highlights the fact that this ambivalent position on technology includes writing and language itself:

> the pen hits hardpan right on top of the paper page here
>
> you'd naturally expect the pen to sink deep into the paper
> like the paper was strange albino moss saturated with invisible waters
>
> which would dissolve all inks so that all writing meets
> & all pages are perfectly clear

The erasure of text and the reconnection of paper technology with the ancestral forest is not a nihilistic vision, but one entwined with what Galloway might describe as "radical alterity, the inhuman. ... We shall call it simply truth" (2012, 50). Gilbert, in writing this poem, violates the *histoire*, the illusion of reality making in literature, for *discours*, the self-conscious artifice. His breach of the silence is allegory of this rupture. Roy Kiyooka—whom George Bowering has called "the first Vancouver postmodern poet" (1993, 113)—similarly breaks *histoire* by speaking to the importance of silence: finding nothing in art, literature, and even conversation is connected to overcoming artifice and the hollowness of

Western culture; "what we come to in our solitude is the recovery of our / single-ness defined, by s i l e n c e . // to recover our solitude the immeasurable silence that lies / at the heart of things is to define the future of Art" (Kiyooka 1975, 66). The process of recovering from the alienation of internationalist structures, the progenitor of so many technologies, begins with recognizing technology's complicity in the global violence of transnational capitalism. Poetry is never completely detached from the systemic structures of transnational global conflict. Thus, Kiyooka continues, "any definition of a masterwork (whatever moves you) must / answer to The Silence it commands.... [T]he BOMB continues to grab the headlines. So also the Space / Cadets, Vietnam & Gerda Munsinger... no one pays for silence." A new art, or a new role for art, emerges in careful defiance of the noise of the contemporary moment, its mass media and spectacle, which are entwined with war and global conflict. Smaro Kamboureli notes that his book *Transcanada Letters* evokes an "unimaginable community" undoing the constitution of locus through an excess of incommensurable details (2007, x). The book interface, like the letter and the post office, becomes (hopefully) a machine of subterfuge.

So, while cameras and printing presses, radios and televisions, could be recruited into the revolution of sensibility, the service these were all pressed into was, ideally, the unravelling of transglobal, capitalist consciousness. It was not an obvious extension of the work, though; as Pierre Coupey admits, "the war in vietnam is also a result of the paint flowing from my fingers" (1968). These artists were extraordinarily aware of their proximity, of their art's proximity, to the levers of powers that shaped the world, thus the repetition of the importance of finding nothing, which functions something like a mantra for the avant-garde of the period. Similarly, bissett's 1971 book IBM *(saga uv th relees uv huuman spirit from compuewterr funckshuns)* draws attention to the commonality in the dual implications of the acronym for International Business Machines and intercontinental ballistic missiles: a deadly paragram. The book attempts to rehumanize the tarnished letters of the name by spinning each letter into a series of visual poems, puns, and drawings. Computers, and both IBMs by extension, are rejected for "food poisoning," "grenade[s]," and support of "na / palm." Every page is handwritten in sloppy, hasty letters with similarly loose and gestural visual ornamentations. Upon reaching the end of this alphabet-quest, a Tibetan prayer wheel is invoked, Zeus is re-enthroned, food is passed, and "th computer bites the dust." One page of binary o's transmutes into a Buddhist chant of "om"s.

Copithorne, that same year, expressed a similar sentiment in her richly illustrated *Misstree's Pillow Book* (1971) when she writes, "IBM / YES NO / ZABSURD

[...] THE CURTAIN GOES UP IN THE COLONIAL MAGIC THEATRE." There is no escaping the (absurd) theatre of colonial violence. This mistrust of technology can be connected to the proliferation of handwritten poems and hand-drawn visual poems, especially among women, that sought to re-inscribe the material presence of the body into the text. This gestural basis for Vancouver's visual poetry helps to distinguish the work emerging there from the international concrete-poetry movement that harvested its aesthetic, however parodically, from corporate advertising models. It wasn't just external technologies, however. While many writers sought an organic basis for the alphabet, such as McCaffery and Nichol's interest in Alfred Kallir, who posited the origins of the alphabet in the sexual anatomy of humans, bissett recognized language as technology, culturally embedded and complicit in contemporary ideological antagonisms. In one of the most authentic and unique political gestures in the history of Canadian literature, upon recognizing this complicity, bissett proceeded to alter the orthography of all his writing, refashioning its very DNA—including in poems, prose, essays, and correspondence—to a free and flexible phonetic spelling. Random, individualized spelling is perfectly anathema to the standardization required for digital platforms and programmed languages.

In contrast, Nichol's work in digital poetry began at the same time that bissett rejected programmatic models of language for complicity in wider social systems. In 1969, for instance, Nichol used his grOnk Press to publish a computer-assisted poem by the Vancouver (and UBC) poet Earle Birney. Called "Space Conquest: Computer Poem," the poem took advantage of Birney's residency at University of Waterloo and their recent opening of a mathematics and computer-science building. The full colophon of the poem reads:

#Created at the University of Waterloo, Ont., February '68. 12 Lines chosen from 1066 5-Syllable Lines Supplied by a Computer Programmed to a Random Order of the Words Composing Meredith's "Lucifer in Starlight" and MacLeish's "End of the World."
Printed on an IBM/360 Computer For Inclusion in Gronk 2 Series 4. (Birney 1968)

The poetic act, in this case, involves orchestrating the IBM computer's programming, selecting the lines that will be sampled, and then curating the output of the permutation by selecting the most aesthetically pleasing lines. Dean Irvine notes that Birney had been planning this kind of computer-human interaction since at least 1965, and he believed it could introduce a new kind of meta-textual poetic production (Irvine 2015). Rather than introduce new models of authorship, however, Birney imagined the author function reasserting itself in the selection process by clipping out "passages with sufficient unity of theme and image and enough provocative overtones," and thereby maintaining aesthetic coherence (quoted in Irvine 2015). The text's twelve found lines assembled together present a poem about willful blindness, cancelled darkness, and starless skies. If one were so inclined, one might read such images as responsive to an older model of author function despite post-authorial challenges to the text via new technologies.

For Nichol, who was invested in scribal models of texts that recovered the intersection of word to ear, the typewriter and, indeed, the computer presented a radical standardization of page space that exposed a weakness in (or, perhaps, portal out of) the interface. Working from McLuhan's attention to the long legacy of Gutenberg's typographic standardization, and how it might have contributed to establishing Western feelings of detached individualism, Nichol was increasingly attuned to the role of the book in creating or permitting modes of thinking that reconnected the individual to their ground (McLuhan 1962). In an essay on the "book as machine," Nichol, with his collaborator Steve McCaffery, extended this figure/ground bias to the visual properties in language. Echoing their commitment to the language revolution, in admittedly a more muted, less revolutionary form, they propose "a new way of perceiving in which the visuality becomes, not the end product of an interior psychological process, but rather the beginning of a whole *new* method of perception" (1992, 62). By reversing the spatial implications of type, for instance, reversing the figure/ground bias of the page space, Nichol believed he might uncover a medium of relative freedom suggestive of different, perhaps ancient ideologies. Not only would this teach "users" to acknowledge the book as machine, but the negative, unseen page space starts to vibrate with meaning and intensity—serving a wide array of functions for poetic deployment, from frame, light, embodied silence, and so on. Nichol sought to use this kind of awakened visuality to disrupt the sensorial interface of language, the primary interface of authorship, and move his readers to expanded, materialist possibilities of expression. Indeed, his early concrete poems or "ideopoems" aspire to the jarring effect of seeing and exposing the hidden elements that comprise the medium of literature (and comics, etc.). This was, of course, the era of the "Earthrise" photo by astronaut William Anders in December 1968 from the Apollo 8 mission that showed humans their home for the first time. Poignantly, the Apollo missions were coordinated by the same computer used to help make Birney's poem at Waterloo, hence the complex pun of its title.

Nichol's first home computer (not an IBM) was the Apple IIe, released in January 1983. He immediately integrated the machine into his writing process, even creating digital poetic work. He released his suite of 12 digital poems the next year on a floppy disk under the title *First Screening: Computer Poems* ([1984] 2007). These texts follow a similar pattern as his concrete poems by either animating or interfering with the literal meanings of the words while using oscillations between figure and ground. The word "wave," for instance, is repeated with coded spacing that makes the word flicker and seem to wash on the screen against the more stable word "rock" (without the spacing) in the poem "Island." The word "hoe" grows its own fertile "hoe rizon" between the sun and field in "Reverie." Ironically, as Jim Andrews et al. (2017) highlight in their online edition of these poems, such cutting-edge animations of the page appear in a much less stable interface than the page it purports to supplement: the software becomes obsolete and inaccessible with astonishing rapidity. Furthermore, I think it is worth pointing out that the model of authorship informing these works is largely consistent with pre-established norms—with the obvious exception of the need of programming the machines in order to write the poems. They expand the domain of the author without disturbing the category of authorship. Similarly, Frank Davey and Fred Wah acknowledge the resistances they faced in trying to use the world's first online literary magazine, *SwiftCurrent*, to transform distribution models of text as well as the command function of authorship. Alas, libraries wanted hard copies and the pool of contributors who might be transformed limited to those "who can establish their seriousness as a writer and who owns a computer" (1986, 8).

In contrast, Nichol's early conceptual, page-based poems offer a stark reimagining of text production and author function and an exceptional expansion of the technology of the book. Consider his celebrated poem "The Complete Works," a pale-blue pamphlet that he self-published in 1968. The text is a reproduction of his typewriter's QWERTY keyboard that cleverly takes advantage of the asterisk key to sneak in a footnote: "any possible permutation of all listed elements." It is a simple joke that yet manages to convey a Wittgensteinian sense of language as a finite system and expression as a game within that closed system. Like Birney's computer poem, it is a unique combination of found and authored work, but goes beyond simple text production to expose, as Borges did with his "The Library of Babel" story, the mathematical basis of alphabet-based writing systems. It is witty, true, and theoretically infused—no wonder filmmaker Justin Stephenson chose it as the centrepiece of his documentary on Nichol's poetry, *The Complete Works* (2017). It is also, perhaps even more importantly, a (mostly) plagiarized text. In the issue of the avant-garde literary journal *Lines* before Nichol's own poetry began appearing in the magazine, editor Aram Saroyan published his own poem "The Collected Works" (1965), which is identical to Nichol's but for the footnote. Thus, Nichol's poem's challenge to the boundaries of authorship, moving from writer to creative thinker, also includes a dialogical acknowledgement and response to a previous text; creativity being limited to making explicit something implicit in the original. In this model, an author gives individual expression to a collective process of text formation, working responsively toward a shared idea of the systemic nature of the act. The author is transformed from the manifestation of a body's self-expression to a depersonalized function within a larger system. A further productive tension in the work emerges from the contrast between the sharply delineated machinic type inside the pamphlet versus the more casual, misaligned handwritten cover text that announces the title and authorship. I detect in this division a similar turn to the body of the author as in bissett's rejection of computer poetry. Rosi Braidotti talks in *The Posthuman* (2013) about how the moral and political "discomfort" of witnessing the awesome power of military technologies upsets the fundamental category of the human (9). If technology is so obviously, so grotesquely linked to a post-human violence, is it any wonder that Nichol and bissett encode a contrapuntal humanism into their texts? It is worth noting that this contrapuntal humanism, as Braidotti insists, is not incompatible with other post-humanist aspirations but is, in fact, a definitive belief of the "post-structuralist generation" (23).

There is another allusion, intentional or not, in Nichol's minimalist text to the tradition of post-authorship, avant-garde "permutation" poems. Before Nichol, bissett, and Birney's computer poems (but coincident with McLuhan's interest in electric and digital media), the Canadian expatriate poet Brion Gysin was experimenting and inventing a variety of new literary forms to strategically disempower the author. Though he developed the technique from his friend Tristan Tzara, Gysin was the great pioneer of cut-up methods that taught William S. Burroughs, Kathy Acker, David Bowie, and many others to give up authorial control and open themselves up to the potential of found writing. He also championed a method of permutation writing that examined the full implications of a stock phrase (such as "Junk is no good baby" or "Rub out the word") by listing every possible permutation of all elements in the sentence. Some of the resultant phrases make sense, some are incomprehensible, but, regardless, Gysin documents them all and, in doing so, shifts authorial control over to a mathematically exhaustive list. These efforts were accelerated by Gysin's collaboration with mathematician Ian

Somerville, who programmed a Honeywell computer at Cambridge to automate the permutations and print them out in massive lists. Christopher Funkhouser historically positions Gysin's experiments as the second attempt to harness computers for the creation of poetry, after Theo Lutz's stochastic poems from 1959 (Funkhouser 2017, xix). The digital text was instantly recognized by Gysin as important to reinvigorating the base categories of author and poetry:

> The permutated poems set the words spinning off on their own; echoing out as the words of a potent phrase are permutated into an expanding ripple of meanings which they did not seem to be capable of when they were struck into that phrase.
>
> The poets are supposed to liberate the words—not to chain them in phrases. Who told poets they were supposed to think? Poets are meant to sing and to make words sing. Poets have no words "of their own." Writers don't own their words. Since when do words belong to anybody. "Your very own words," indeed! And who are you? (Gysin 1978)

For Gysin, then, as for bissett and Nichol, the machine was interesting only in as much as it dismantled the legacy of Western imperialist violence, specifically inasmuch as that violence lay nested in Western legacies of individualism and authorship. In 2019, David Pocknee documented 43 permutation poems by Gysin between 1958 and 1982, and describes them as "a revolutionary and technologically progressive piece of art built with cutting-edge technology...that had not even existed a few years earlier" (2019). Although Arthur C. Clarke imaged permutation poems in his 1953 work "The Nine Billion Names of God," the first algorithms for computer-based permutations were only produced in 1956.

These early experiments with computer-based poetry highlight a key ambivalence to the power of the machine to overcome the commercial and military forces that had steered them into existence in the first place. With some hesitation, they were recruited into the task of expanding the experience of textuality, but never valorized as an exclusive or unencumbered tool for literatures of the future (after the language revolution). Gysin collaged his computer poems into paintings and chanted the permutations as if they were strange Delphic prayers. Nichol used computers to animate the tension between figure and ground, as he had done with his early ideopomes, but didn't challenge the categories of author or text as surely as he had done with his conceptual poems. bissett, meanwhile, dismissed the early computers altogether and worried that the integration of machines into our daily lives—including television and radio—would contribute to an ongoing "drifting into war." Indeed, in his 1971 collage masterpiece *Drifting Into War*, bissett offers a scream about the connections between the machinery of the electric age and

> chek out th radio nd TV wave scene in th country then see how yu feel abt abulvision yankee dog lovr a cours yu may think that art shud nevr deal with anything real or imagind that its supposd to be bullshit courz that then cud be yr problem but dont wondr why when th life bcums death eh nucklear baby

This text (poem? manifesto? screed?) at the bottom of a collage of comic books and letters, is followed by a concrete poem of the word "data," a parodic sneer at those who worship and "adore" "great" data despite it being a mere repetition of a limited number of keyboard strokes (see figure 10.2). Though the text creates a post-human grid, the various errors in spacing affected by imprecise typewriters and human imperfection subtly reinscribe the artist's agency. Such early Canadian ambivalence to the arrival of the digital age, even by those who experimented with computer poems, highlights the anxieties of the Cold War, the lingering horrors of the Second World War, and the ongoing hunger to dismantle the power dynamics of empire by those poets yet caught within its nets.

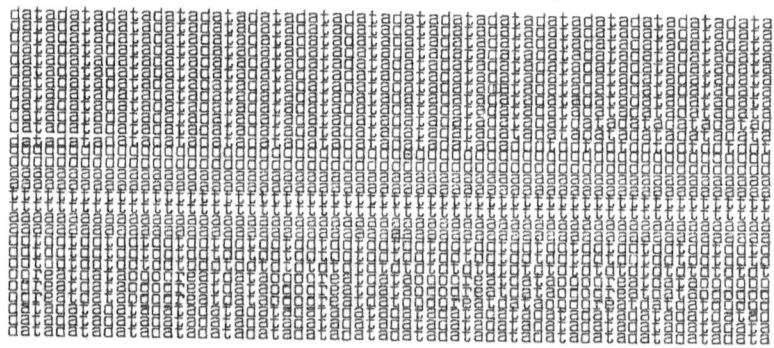

Figure 10.2. Page detail from *Drifting Into War*.

Source: bissett 1971a.

NOTE

1.   Editor's note: Chapters 10 and 11 refer to bissett's magazine and press using the spellings employed by bissett at the time of each reference. Variations such as blewointment, Blewointment, and Blew Ointment are all correct.

REFERENCES

Andrews, Jim, Geof Huth, Dan Waber, Lionel Kearns, and Marko Niemi. 2017. Introduction to *First Screening: Computer Poems by bpNichol*. March 2017. http://vispo.com/bp/introduction.htm.

Barthes, Roland. 1974. *S/Z*. New York: Hill and Wang.

Bateman, Chris. 2016. "The Story Behind the First Computer in Canada." *Spacing*, November 12, 2016. https://spacing.ca/toronto/2016/11/12/first-computer-canada/.

bissett, bill. 1963. "we are haunted…" *Blewointment* 1, no. 2 (December).

———. 1966. *We Sleep Inside Each Other All*. Toronto, Ont.: Ganglia Press.

———. 1966. "what im doing in these pomes." Draft of *We Sleep Inside Each Other All*. Number 8 of 10 special editions. Lakehead University Special Collections, Thunder Bay, Ont.

———. 1969. "Space Conquest: Computer Poem." *Pnomes, Jukollages, and Other Stunzas*. Toronto, Ont.: Ganglia Press.

———. 1971a. *Drifting Into War*. Vancouver, B.C.: Talonbooks.

———. 1971b. *IBM (saga uv th relees uv huuman spirit from compuewterr funckshuns)*. Vancouver, B.C.: blew ointment press.

———. 1971c. Letter to Diane di Prima. bill bissett Fonds. York University Special Collections, Toronto, Ont.

———. (1972) 2012. *Rush: What Fuckan Theory: A Study uv Langwage*. Vancouver, B.C.: Blew Ointment Press Reprint, Toronto, Ont.: BookThug.

Black, Edwin. 2001. *IBM and the Holocaust: The Strategic Alliance Between Nazi Germany and America's Most Powerful Corporation*. New York: Crown Publishers.

Bowering, George. 1993. "Vancouver as Postmodern Poetry." *Colby Quarterly* 29, no. 2 (June): 102–118.

Braidotti, Rosi. 2013. *The Posthuman*. Cambridge, Mass.: Polity Press.

Brown, Jim. 2013. Email message to author, January 13, 2013.

Carlson, Tim. 2002. "bill bissett." In *bill bissett: Essays on His Works*, 13–49. Toronto, Ont.: Guernica.

Cavell, Richard. 2002. *McLuhan in Space: A Cultural Geography*. Toronto, Ont.: University of Toronto Press.

Copithorne, Judith. 1963. "Famine." *Blewointment* 1, no. 2 (December).

———. 1971. *Miss Tree's Pillow Book*. Vancouver, B.C.: Intermedia Returning Press.

Coupey, Pierre. 1968. "An Allegory of Love." In *Poets Market*. Vancouver, B.C.: Talonbooks.

Culley, Peter. 1993. "Because I Am Always Talking: Reading Vancouver into the Western Front." In *Whispered Art History: Twenty Years at the Western Front*, edited by Keith Wallace, 189–97. Vancouver, B.C.: Arsenal Pulp Press.

Davey, Frank, and Fred Wah. 1986. Introduction to *The SwiftCurrent Anthology*, 7–9. Toronto, Ont.: Coach House Books.

Funkhouser, Christopher. 2007. *Prehistoric Digital Poetry: An Archeology of Forms, 1959–1995*. Tuscaloosa: University of Alabama Press.

Galloway, Alexander. 2012. *The Interface Effect*. Cambridge, Mass.: Polity Press.

Gilbert, Gerry. 1974. "Thought for Penny." In *Journal to the East*. Vancouver, B.C.: Blew Ointment Press.

Gysin, Brion. 1978. "Cut-Ups Self-Explained." In *The Third Mind* by William S. Burroughs and Brion Gysin, 34–37. New York: Viking Press.

Hume, J. N. Patterson. 1994. "Development of Systems Software for the Ferut Computer at the University of Toronto, 1952 to 1955." *IEEE Annals of the History of Computing* 16, no. 2: 13–19.

Irvine, Dean. 2015. "MISSION CONTROL: An Operator's Manual for Compulibratories." *Amodern* 4 (March). http://amodern.net/article/mission-control/.

Kamboureli, Smaro. 2007. Preface to *Trans.Can.Lit: Resituating the Study of Canadian Literature*, vii–xv. Waterloo, Ont.: Wilfred Laurier University Press.

Kiyooka, Roy. 1975. "miscellaneous 4/66." In *transcanada letters*. Vancouver, B.C.: Talonbooks.

Lewis, Wyndham. 1927. *The Enemy: A Review of Art and Literature*, 1: 35.

McCaffery, Steve, and bpNichol. 1992. *Rational Geomancy: The Kids of the BookMachine: The Collected Research Reports of the Toronto Research Group, 1973–82*. Vancouver, B.C.: Talonbooks.

McLuhan, Marshall. 1962. *The Gutenberg Galaxy: The Making of Typographic Man*. Toronto, Ont.: University of Toronto Press.

———. 1964. *Understanding Media: The Extensions of Man*. New York: Signet Books.

Nichol, bp. 1968. "The Complete Works." Toronto, Ont.: Ganglia Press.

———. 1974. "Interview with *Nicette Jukelevics*, 26 June 1974." In *A Bibliography of Canadian Concrete, Visual and Sound Poetry 1965–1972*, 127–34. Master's thesis, Sir George William University.

———. (1984) 2007. *First Screening: Computer Poems*. http://vispo.com/bp/index.htm.

Nichol, James Y. 1952. "$250,000 Electronic Brain Too Slow for Star Puzzle." *Toronto Daily Star*, August 9, 1952, 4.

Pocknee, David. 2019. "The Permutation Poems of Brion Gysin." http://davidpocknee.ricercata.org/gysin.

Potsmurth, Zarmsby. 1963. "on Collage being All." *Blewointment* 1, no. 2 (December).

Pound, Ezra. (1952) 1970. *Guide to Kulcha*. New York: New Directions.

Reid, Jamie. 2002. "The Pome Wuz a Store end Is th Storee: th Erlee Daze uv Blewointment." In *bill bissett: Essays on His Works*, edited by Linda Rogers, 13–32. Toronto, Ont.: Guernica Editions.

Saroyan, Aram. 1965. "The Collected Works." *Lines* 5: 11.

Schmaltz, Eric. 2018. "The Language Revolution: BorderBlur Poetics in Canada, 1963–1988." PhD diss., York University.

Stephenson, Justin, dir. 2017. *The Complete Works*. 42 min.

Turner, Michael. n.d. "Expanded Literary Practices." Vancouver Art in the Sixties. http://expandedliterarypractices.vancouverartinthesixties.com/.

Williams, Michael R. 1994. "UTEC and Ferut: The University of Toronto's Computation Centre." In *IEEE Annals of the History of Computing* 16: 4–12.

CHAPTER 11

The Digits in the Digital: Bodies in the Machines of Canadian Concrete Poetry

Eric Schmaltz

In the December 7, 1972 issue of *Rolling Stone* magazine, American writer Stewart Brand writes about the emergence of accessible, personal computing and its consequent formation of subcultures in research labs and corporate offices in North America in the late 1960s and early 1970s. "Reliably, at any nighttime moment (i.e., non-business hours) in North America," he writes, "hundreds of computer technicians are effectively out of their bodies, locked in life-or-death space combat computer-projected onto cathode ray tube display screens, for hours at a time, ruining their eyes, numbing their fingers in frenzied mashing of control buttons, joyously slaying their friends and wasting their employer's valuable computer time" (Brand 1972, 50). Curiously, Brand's article for *Rolling Stone* describes these new computer enthusiasts as being "out of body" but follows that comment by hyperbolically detailing the supposed effects of the popular game *Spacewar!* on the players' bodies. In doing so, Brand effectively captures the emergent enthusiasm for personal computing while also inadvertently distilling a recurring debate focused on computing's relationship to corporeality that would capture the imaginations of artists, poets, and scholars in proceeding decades (see Hayles 1999, 2002; Kelly 1994) and remain relevant in the popular discourse around computer gaming today.

The complex relationship between digital media and corporeality has been well considered in the late twentieth and early twenty-first century, often with divergent perspectives on this relationship. Brand's comments, for example, identify one end of the spectrum of debate that characterizes digital computing as a medium that appears more cerebrally demanding than it is physically. Despite Brand's comments, however, I suggest that the body has always been in the digital machine. Etymologically, the word "digital" comes from the Latin word *digitalis*, meaning "measuring a finger's breadth" or in post-Latin, meaning "relating to the finger."[1] Only later, circa 1978, would the word enter popular usage as it is commonly known today within the discourse of computing technology (of course, stemming from its other early usage to describe numeric digits). So, while some writers, like Brand, posit early computing technology as part of a markedly "out of body" experience facilitated by cathode-ray-tube technologies, the history of the word highlights the long-standing entanglement of bodies (metonymically represented by hands) with these machines. As documented by Paul Ceruzzi and Lori Emerson, early personal computing was a hobby for dexterous persons since it was guided by a do-it-yourself (DIY) open ideology that invited users to tinker with and customize their machines (see Ceruzzi 2003; Emerson 2014). Poet and critic Johanna Drucker makes a similar observation when she suggests that "computational media" are "overwhelmingly material— requiring rather large amounts of hardware to perform what was formerly done in rather minimal means (paper and pencil)" (2009, 134). The involvement of the user, in terms of construction, design, and labour for these machines, she suggests, is generally much greater than manual writing technologies.

Media theorist and artist Anna Munster is among the critics who have carefully considered the entanglement of digital media, literature, and the body, notably in her *Materializing New Media: Embodiment in Information Aesthetics*. In that text she argues against the binary opposition that situates the body and computer against one another: "[N]ew media technologies are held to be responsible for privileging consciousness over embodiment in virtual environments or favoring the machine over the human in the design of computer interfaces" (2006, 10). Munster argues that we need to do away with these assumptions about the relationship between the human body and computer to instead "treat the matter of humans and the materiality of technologies as open-ended propositions that are continually in the process of being made and unmade" (13). Such propositions catalyze "the transformation of a human capacity through the rearrangement of aspects of aesthetic of sensory life. This transformation takes place through the differential hybridizing of body and technology" (18). For Munster, computers, as information-transmitting devices, have dynamically reconfigured the relationship between body, writing, and machine, suggesting that new kinds of computational media pose new kinds of human experiential capacities, rather than overtake or efface them.

Building on the work of critics such as Munster, Emerson's *Reading Writing Interfaces: From the Digital to the Bookbound* engages the various ways that "human-to-hardware devices" have reconfigured processes of reading and writing on physical and cognitive planes of human experience (2014, x). In particular, she traces the trajectory of computing technologies as they transition from open, customizable devices into "user-friendly" devices which limit DIY involvement, thereby displacing user control from the device. At the time of this writing, smartphones—especially Apple's iPhones—are premier examples of computing's lack of customizability. The user's inability to modify their device is a point of contention for Emerson because it has significant ideological connotations: a lack of modifiability suggests a reduction of a user's agency. The lack of modifiability of these devices starkly contrasts the capacities for customization and user control of earlier computers, which could be tinkered with and adjusted to the user's desired preference and functionality. And it is this DIY philosophy of early computing technology of the 1980s that offered new possibilities for poetic composition.

In particular, Emerson's chapter "From the Philosophy of the Open to the Ideology of the User-Friendly" looks toward the work of poets bpNichol, Geof Huth, and Paul Zelevansky, and suggests that poems composed on early personal computers during the 1980s—especially the moderately customizable Apple IIe—mark a transformation for poetry from writing to coding. Coding, for a poet like Nichol, was a trial-and-error process of tinkering with BASIC. Tracing Nichol's shift from writing to coding, Emerson describes Nichol's pathway to learning to code as a laborious process since it typically would require many steps to execute the precise end results Nichol might have envisioned with the commands he created (Emerson 2014, 65). As such, access to the Apple IIe marks a transformative moment for poets who, by working through the conventions and protocols of BASIC, must reconceive their relationship to language, labour, and the poem. As I will demonstrate later in this chapter, the demands of coding languages are cerebral but also extremely physical in the amount of labour, bodies, and time required by the writer. Emerson's work highlights the bodily relationship between writing and digital machine in two ways. First, she suggests that "the digital computer has an entirely different effect on the body than that of a reading/writing machine such as the typewriter [...] the absent presence of the body" (67). Second, she traces the ways computers demand a new kind of handiwork from poets that is distinct from previous analogue typewritten and handwritten modes since code is syntactically and semantically different from standardized writing in English (63).

This chapter takes up the problems posed by the body, machine, and writing in poetic engagements with computers and digital media, focusing predominantly on work by Canadian poets bill bissett and the aforementioned Nichol. It follows Gregory Betts's astute analysis of the collision between digital computers and Canada's literary avant-garde, but I am particularly interested in the relationship between the body and computer in Canadian avant-garde poetry, especially in thinking along a continuum that begins with the analogue media-based works of Canadian poets in the 1960s and its transformation into digital-born works in the 1980s. In particular, I examine bissett's IBM *(saga uv th relees uv huuman spirit from compuewterr funckshuns*) and Nichol's *First Screening*, each of which presents a divergent view on the role of the computer in the poetic context. Such assessments demonstrate for us momentary conceptions of the relationship between poetry, body, and emergent digital media during a period when this relationship's expanse and limits had not been fully explored or understood. I argue that, while both Nichol and bissett have seemingly opposing viewpoints on the poem's relationship to the computer, they both privilege the body as a core component of poetic work. In other words, while they discretely conceive of the emergent digital media/body relationship, they both see the body as being an integral aspect of poesis. Extending these points further, I conclude with brief, tangential thoughts on how these works indicate specific ideations for how digital media and poetry might relate to notions of social belonging.

ANALOGUE AND DIGITAL MEDIA: CONCRETE POETRY AND EMBODIMENT

Nichol and bissett, whose writings began to appear in the 1960s, recognizably developed their poetic practice within a worldview that was being defined by Canadian theorist and media critic Marshall McLuhan, and his works such as *The Gutenberg Galaxy: The Making of Typographic Man* (1962) and *Understanding Media: The Extensions of Man* (1964). Among McLuhan's many theories that captured the imaginations of artists, poets, and the public, one of his core theses—media dramatically alters the conditions and experiences of physical and psychical human life—is useful to this context. "[A]fter more than a century of electric technology," he maintains in *Understanding Media*, "we have extended our central nervous system itself in a global embrace," and suggests further that media is an "extension, whether of skin, hand, or foot, [that] affects the whole psychic and social complex" (1964, 19). According to McLuhan, new media compelled emergent collective formations between audience and media users; thus, it offers new possibilities for a person's body to exceed its physical and psychical limits. For example, the radio enabled persons to broadcast distant voices into thousands of homes, far from the original source of the speaker's mouth. Relatedly, the typewriter engendered new possibilities for the placement of language upon the page. Driven by uniform type and a linear, monospace grid, the typewriter gave way to new typographic possibilities for poets to mark the breath on the page, for example, as it is announced in Charles Olson's 1950 "Projective Verse" essay-manifesto. Emerson confirms McLuhan's influence over poets like Nichol and bissett in *Reading Writing Interfaces*.

McLuhan's thinking strikingly appears as a prefatory note in bissett's first book of poetry *We Sleep Inside Each Other All*, published by Nichol through Ganglia Press. bissett writes,

> Marshall McLuhan sz we are poisd between th typographic individualist trip th industrial revolution & the electronic age we have been in for sum time, between a unique dis tance and alienation privacy well now iullbe in th study for th rest of th night with my nose in a boo k & th corporate image tribally we are a part of out extensions do reach now have been reach thruout all time th historical jazz consumd in th greater fire of mo vies t v & lo ve. (1966)

As a preface that frames the reader's experience of the book, bissett locates his writing at the theoretical vanguard, articulating, through McLuhan, an awareness of the shifting nature of the mid-twentieth century from the industrial age toward the electric age (or, more commonly known now as the electronic or information age). It is not entirely clear if bissett is positioning *We Sleep Inside Each Other All* as a response to McLuhan's theorization of the age, for there is a sneering quality to his quip "iullbe in th study for th rest of th night with my nose in a boo k" (bissett 1966). bissett's tone may be unclear and he does have a general distrust for academic modes of thought and writing; however, a survey of bissett's poetry from 1966 onward suggests that his writing is conversant with many of the cultural conditions identified in McLuhan's writing.

Nichol most evidently engaged McLuhan's thinking in 1982 in an essay that was unpublished until 1989 (featured in a special issue of *Journal of Canadian Poetry*). The text was originally intended to appear in a book on McLuhan, presumably edited by Fred Flahiff and Wilfred Watson, but was never published (Nichol 2002, 480). Nichol draws a direct connection between his work and McLuhan's writing style by way of the pun, a literary device beloved by both. "No one punned more seriously than McLuhan," writes Nichol, and suggests that McLuhan's punning "is not trying to fix 'a' or 'the' reality—he wants to open realities" (299). This inclination toward openings is one that McLuhan and Nichol share, especially openings that media-conscious modes of writing can produce. While McLuhan examined media and the changes they make to human life, Nichol's work similarly explores how media could expand the possibilities for poesis. Nichol often experimented with various methods and technologies, including the typewriter, comic strip, drawing, and, as is the topic of this chapter, the Apple IIe personal computer.

bissett's and Nichol's interest in McLuhan provides a baseline for thinking about their poetry's relationship to media and writing technologies—both analogue and digital. Nichol and bissett are known for pioneering what is referred to as "dirty concrete" poetry, "a deliberate attempt to move away from the clean lines and graphically neutral appearance" (Emerson 2014, 99). This clean appearance is aptly demonstrated by previously published concrete poetries by poets and artists such as Eugen Gomringer, Décio Pignatari, Haroldo de Campos, Augusto de Campos and others who became active in the 1950s. Like bissett and Nichol, these poets worked with language as a visual medium and explored possibilities afforded by new technologies and techniques from graphic design, and were influenced by the discourses of science and cybernetics (see Hilder 2016). The resulting poetries are rich in content while often privileging a minimalist aesthetic, grounded in grid-like structures and legible text. Gomringer's well-known visual poems such as "Wind" and "Ping-Pong" are demonstrative of this aesthetic. While Nichol and bissett were not necessarily reacting against this preceding wave of poets, their work is remarkably distinct. They at times swerved away from "cleanliness" in the poem and instead sought to create works that looked untidy, typically characterized by a denial of semantic and syntactic content, textual overlay, and illegible text. In doing so, these poets misuse their media to hack "the page, the book, and the typewriter in order to renew them, to turn them from transparent carriers of meaning to objects meaningful in themselves" (Emerson 2014, 126). These poetries, however, are also opportunities for registering the body onto the page in assemblage with the writing device of choice.

Sharon Nelson describes the various ways that bissett's poetry foregrounds the body, especially his own body, in writing. Nelson focuses on the breath as a motif: "The body serves as a metaphor for the human totality [...]. Breath symbolizes the integrity of spirit and substance and also references human relations and connection" (Nelson 1997, 47). And further: "For bissett," she writes, "the body is the locus from which social and cultural commentary flow and a symbol by which the personal is extended into the historical and political" (46). A significant addition to this discussion would be from the purview of a media-specific analysis to highlight the writing device—especially the typewriter—as a conduit for breath and body. By misusing the typewriter, as bissett does in poems such as *Ready for Framing*, for example, he attempts to capture the impulses of the body. For this work, as in others, bissett does not use the typewriter to type letters for correspondence; rather, he extends the typewriter as a visual medium to create typewritten portraits—many of which notably portray entangled bodies and genitalia. bissett redefines the rigid, monospace grid of the typewriter, which is meant to guide the placement of letters on the page, as a way of artistically depicting the human form. With that being said, the typewriter has its expressive limits. bissett's *Ready for Framing* contains numerous typewritten abstract artworks and portraits of nude bodies with dangling genitalia (see figure 11.1); however, even those bodies appear stiff, box-like, and somewhat robotic, indicative, perhaps, of the inadequacies of the machine as a means of representing the body on the page.

Digital technologies, which were becoming more accessible and more prevalent in the late twentieth century, posed new challenges for creating dirty concrete poetry and for locating the body within the poem. Following the work of McLuhan, Jean Baudrillard considers these changes in *The Ecstasy of Communication*, a work that turns away from his influential theories of simulation and simulacra to engage the increasing dominance of digital technologies and their relationship to the body. For Baudrillard, the rise of digital technologies was indicative of radical shifts in human life and the role of the body within these shifts. With the rise of the personal computer, Baudrillard, like Brand, saw a withdrawal of the physical body from everyday life: it reveals that "the sexual and social horizons of others has disappeared, and whose mental horizon has been reduced to the manipulation of his images and screens" (Baudrillard 1987, 43). Baudrillard warns that "the increasing cerebral capacities of machines would normally lead to a technological purification of the body" (37). In these comments, I find a parallel between Baudrillard's concern for a seemingly unnatural purification of the body and the work of poets like bissett and Nichol who actively sought to present the body in their media-based, "dirty" poetics. For Baudrillard, digital technologies present a world wherein the body with all its messiness is less a part of the world and, while this thinking is now over thirty years old, it effectively captures anxiety around digital technology and its increasing erasure of the human body in everyday life.

These issues, and their implications for dirty concrete poetry, are explained by Darren Wershler, who, in an ICQ chat with Brian Kim Stefans, confirms the challenges concrete poetry hold for creating a dirty art form: "I'm not sure 'dirty' is meaningful at all when you can control everything on a pixel-by-pixel level.... Scan in the nastiest, grungiest piece of Xerox art, and its immediately transformed into this other thing, because of the new (digital) context" (Stefans and Wershler 2003, 24). Wershler's comment suggests that the digital context nullifies the possibility of dirty concrete poetry since even the "nastiest" poem, once scanned into the computer, becomes enveloped within the computer system and displayed upon the clean, transparent surface of the monitor. Computers,

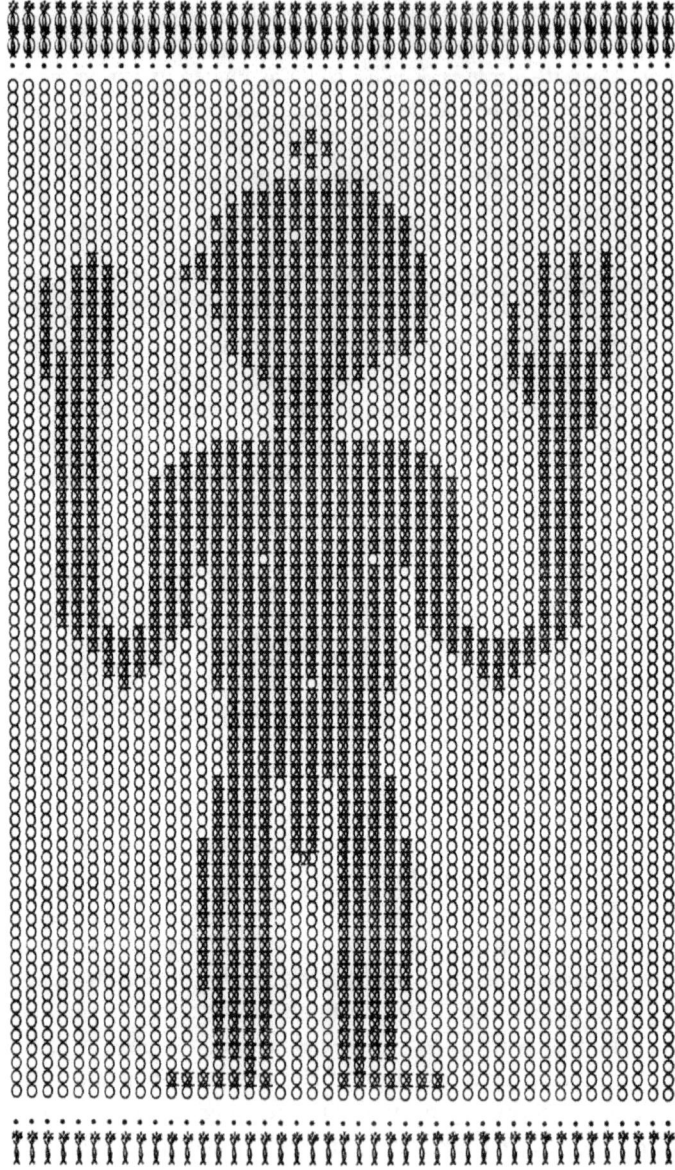

Figure 11.1. Typewritten portrait by bill bissett.

Source: bissett 1982.

too, pose some difficulties for poets when it comes to misusing or hacking the physical medium as poets like bissett and Nichol have done in the past. As C. T. Funkhouser has similarly suggested, "even if the poet-programmer wishes to instill disorder, the process calls for prescribed stylistic elements" (Funkhouser 2007, 20). Combined, these comments seem to indicate that technologically mediated "dirty" concrete poetry may no longer be possible because of the conditions imposed by those technologies; however, their comments could also be interpreted as a challenge to new-media poets to find ways of reconfiguring the possibilities of a dirty aesthetic while using digital media.

WITHIN THE TURN TOWARD CANADIAN DIGITAL POETRY

In "Toward a Theory of Canadian Digital Poetics," literary critic Dani Spinosa outlines the characteristics of three kinds of digital poetries: (1) "print books that use digital technology as integral to their productions (i.e., rather than simply as a word processor)," (2) "print books that have a supplemental born-digital element," and (3) "language-based performance that relies heavily on digital elements" (Spinosa 2017, 241), as demonstrated by the examples below. These categories delineate the types of digital poetry that make up the corpus of Canadian literature that, as Spinosa points out, "has *already turned digital*" (240). In other words, Spinosa suggests that Canadian poetry has been digital for a lot longer than previous scholarly discussions have admitted. With that being said, the Canadian poets in the 1960s through to the 1980s, living in the early moments of North America's rapid digitization, are more accurately located *within* the turn toward the digital, which is effectively described by literary and media critic Kate Eichorn, who documents the increasing prevalence of digital technologies in literary creation and describes how digital Canadian literatures pose generic definitional problems that complicate notions of literary nationalism (Eichorn 2015, 513). Following both Spinosa's and Eichorn's depictions of the turn and post-turn in Canadian digital literatures, I will now focus on specific poetries to understand the implications digital technologies had for poets as they emerged, especially for poets of the Canadian literary avant-garde.

Lionel Kearns is credited for having composed a foundational computer-adjacent poem with his "Birth of God/uniVerse," created in 1965 and published in 1969 in *By the Light of the Silvery McLune: Media Parables, Poems, Signs, Gestures, and Other Assaults on the Interface.* The poem might most closely approximate Spinosa's definition of a print work that uses digital technology, at least in the sense that the poem uses the foundational digits of binary computer code—ones and zeros—to create "a simple yet striking image that suggests the relationship that exists between text, image, and code in the new forms of contemporary expression" (Funkhouser 2007, 258). Kearns himself describes the work as a

> mathematical mandala embodying the perfect creative/destructive principle of the mutual interpenetration and balanced interdependence of opposites: one and zero, something and nothing, substance and void, being and oblivion, positive and negative, good and bad, spirit and flesh, black and white, yin and yang, male and female, thesis and antithesis, this and that—and all the possible dynamic relationships of these polarities, the simultaneous representations, of which are immediately obvious in the icons of sex, childbirth, and death. (Kearns 1969)

Kearns's ability to see a "mutual interdependence" of opposites is striking when considering remarks like Brand's and Baudrillard's above, both of whom saw digital media as oppositional to corporeality. Instead, Kearns's poem suggests that, if we were to extend his string of interrelated binaries, a new kind of interdependence between body and machine arrives with the computer.

Eichorn describes Kearns's *By the Light of the Silvery McLune* as reflecting "a simultaneous fascination with and skepticism about the new media arts" (Eichorn 2015, 517), as demonstrated by his "Kinetic Poem." Kearns writes:

> Now, I admit that this prototype model that you see on display
>   is something of a compromise as it has a live poet
>   concealed inside
> But I assure you this crudity will eventually be eliminated
> Because each machine is to be fully computerized
> And so be able to stand on its own two feet. (Kearns 1969, 19–24)

Kearns's tone is somewhat sneering as he parodies the possibility of advanced, automated computing. Yet, Kearns's comment is perhaps more prescient and insightful than a parodic reading might suggest. With its skepticism of computer advancements, "Kinetic Poem" captures the entanglement of humans and computers within the throes of poesis. Even when compared to our more advanced digital literatures of the twenty-first century, there always remains to be "a live poet / concealed inside" the poem (even though this poet is now better hidden under layers of applications). Despite the number of text generators, algorithms, procedures, and layers of code used in the composition of a digital poem, the poet always implements some degree of agency within the construction of the work as a programmer, editor, catalyst, etc. As a poem that may denigrate the possibility of a fully automated computer poetic, "Kinetic Poem" has thus far proven to be a prescient criticism of computational poetics.

Earle Birney's 1969 "Space Conquest : Computer Poem," in *Pnomes, Jukollages, and Other Stunzas* (from Nichol's Ganglia Press), was published in the same year as *By the Light of the Silvery McLune*. *Pnomes* foregrounds the materiality of the work with its expansive selection of paper, printing methods, and modes of interactivity. "Space Conquest," in particular, consists of twelve lines printed on chart-like, fan-fold computer paper. An aesthetic decision like this reminds the reader that the poem is created by a computer, thereby gesturing toward an alternative conception of authorship in the work of the poem. As DH scholar Dean Irvine points out, Birney—in the preamble to his reading of the poem at the Sir George Williams University Poetry Reading Series—"disclaimed responsibility for selecting the source texts" (2015) to partially sidestep his involvement with creating computer-based poetry. Instead, Birney is the editor of the computer-generated text. Conversely, Birney's comments on the creation of his computer poetry reveals the extensive human labour in creating the text:

> Well those ten lines, each of five syllables, came out of a computer at the University of Waterloo last week, into which we had programmed a hundred and eleven, the one hundred and eleven words of George Meredith's "Lucifer in Starlight," and the last thirty-three words of Archibald MacLeish's "The End of the World." Don't ask me why we picked those two poems, I had nothing to do with the picking of the poems. But some of us, two linguists, two linguist-isists, a mathematician, and myself, and masses of computers are producing this sort of poetry. It took point eight-three seconds, not even one second, to produce the hundred-some-odd lines, out of which I chose those ten. So you can see it doesn't take very long, once you've programmed the machine, to find the, you know, the entire text of *Hamlet*, but this is what we have done so far, we haven't put too much time on it yet. Some things that I haven't had computers write for me, although perhaps I might have, or should have. (Birney, 1968)

It is important to note that Birney worked with three persons to create his computer poems, a mathematician and two linguists, plus he drew the vocabulary of this poetry from two other poets (Meredith and MacLeish). The labour required for a single computer-based poem in this case appears to be much greater than the human involvement of a single analogue poem at the time. While Kearns's "Kinetic Poem" may playfully suggest the limits of computer-made poetry, Birney's poem, published in the same year, qualifies this joke to demonstrate that there is not one "live poet / concealed inside" but there is, in fact, a multitude.

Some poets and artists at the time held a more skeptical view of the role of the computer in writing and poetics. In 1967, poet and artist Roy Kiyooka writes a verse-letter to Max and Charlotte Bates (included in *Transcanada Letters*, edited by Glen Lowry) wherein, as Douglas Barbour notes, Kiyooka writes of the "art revolution in New York in a desire to battle for his own cultural community against all forces of what he saw as cultural decline" (Barbour 2001, 152). As part of this defence, Kiyooka exclaims: "Libidinous Dreams shall destroy / All yr IBM D-A-TA" (17). In this succinct and playful statement, Kiyooka expresses a seeming lack of faith in the possibilities presented by computers. The computer is another symptom of cultural decline. Instead, he claims that humankind's libidinous powers will eclipse the power of computers: "Long Live / Surrealism's Brute / Fantasias" (16). Notable here, too, is that one of Kiyooka's most celebrated works, *Stoned Gloves* (1971), is a book of poems and photographs that meditate on gloves fallen from workers' hands at Expo 1970 in Osaka, Japan. While the poem has no direct connection to digital media, *Stoned Gloves* is yet another text from this period concerned with the absence and presence of the body in poetry and art. Each pair of gloves, fallen to the ground, reminds readers of absent hands and labouring bodies. More pertinently, however, Kiyooka's comment in this verse letter to the Bateses puts more faith in psychosocial mechanisms than Baudrillard would in 1987. For Baudrillard, by 1987, the subject had already withdrawn from the world. Kiyooka, on the other hand, at least at this moment in 1967, believed that corporeality would triumph over the digital.

THE DIGITS AND THE DIGITAL: BISSETT'S IBM AND NICHOL'S APPLE IIE

Beyond these fleeting, early engagements with computers in poetry, bissett and Nichol composed books that divergently engage the rise of personal computers in the mid- to late twentieth century. Like Kiyooka, bissett, at least in the early 1970s, privileged the presence of the body in the poem over the computation of language. *IBM (saga uv th relees uv huuman spirit from compuewterr funckshuns)* is an oversized, staple-bound book that consists of handwritten and hand-drawn poems. The book loosely follows the progression of the alphabet from A to Z, and each page (sometimes pages) roughly corresponds to a different letter.

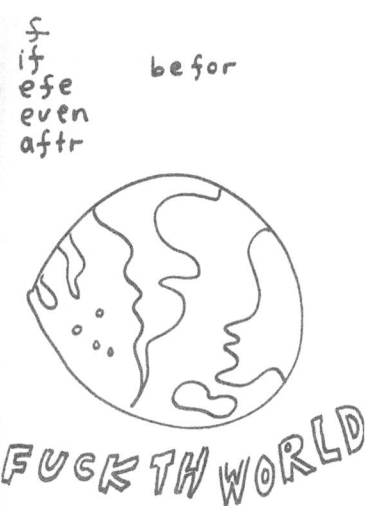

Figure 11.2. Hand-drawn poem with image by bill bissett.

Source: bissett 1971.

bissett explores a series of seemingly symbolic associations around each letter that correlates to their shape and sound (see figure 11.2 for an example). The book is a space wherein bissett, as freely as possible, spills onto the page in a way that is reminiscent of automatic modes of writing and drawing. In this way, IBM is a momentary rejection of digital communication machines and the promise they hold for the future of thought and expression.

The IBM in the title of bissett's book refers to the American multinational technology company whose un-abbreviated name is the International Business Machine Corporation, whose machines are implicated in the acceleration of capitalism in North America and were instrumental in planning and strategy for the military (see Betts's chapter in this collection for comments on militarism). bissett protested American foreign influence, the exploitation of life under capitalism, and positions himself as staunchly anti-war (see Daems 2010). IBM is a company associated with all these issues. bissett's title for the collection identifies the company as his target of critique; however, the title is also a scatological pun on a specific type of bodily "relees"—the BM also known as the bowel movement. In so doing, bissett's title privileges the body rather than the computer, which bissett sees as a means of divorcing the body from the mind, and that funnels human expression into the rigid grid of computational logic. The title, with its invocation of BM, is also indicative of an appeal to dirtiness, connecting back to the radical possibilities afforded by dirty concrete poetics.

Employing a dirty concrete aesthetic, much of bissett's poetry up to the early 1970s (and beyond) aggressively resists and rejects the standardization that overwhelmed the twentieth century as developments in information technologies accelerated. bissett tested the limits of the typewriter, finding ways to express his complex affects by relentlessly misusing the machine. bissett's IBM, however, is not a misuse of computer technology; rather it is a total rejection of the computer's systematization that is on its rise to prominence. Instead, bissett relies on the movements of his hand—leaving more direct traces of his body on the page. As also seen in the work of fellow poet Judith Copithorne, drawn poems can more effectively capture the pressure, movement, and speed of the hand on the page whereas the typewriter cannot do that with the same effect. Each page of IBM adheres to a free-flowing logic. bissett's IBM finds the computer and its interactions with the body to be a step too far toward displacing and concealing the body of the poet. The computer relies on specific patterns of code to function (much like conventional language); in response, bissett's IBM releases thought and feeling from those restrictions. For example, bissett writes a sequence for the letter C: "see / sea / c / si / eeee" (bissett 1971) and further elaborates his expression around the letter with a hand-drawn image of pine trees across a plain with the word "rescue" in a speech bubble. Or, in a sequence for the letter F bissett writes: "f / if / efe / even / aftr / befor / FUCK TH WORLD" (bissett 1971). It is this total rejection of "the world" as the incubation of codified logic that energizes the spirit of bissett's work. It is a rejection of the world as it was coming to be, and the search for a means of opening perception as computational logic takes hold, discarding the human body for the exigencies of the machine. The final pages of the book imagine the destruction of the computer:

th computer
bites th dust
    th dust makes
    th compuutor
    sick
    th compuutor
    dies
    zend (bissett 1971)

For bissett, what he perceives to be the erasure of the human form will eventually be destroyed. Wershler argues that bissett did eventually succumb to the allure of the digital machine and that this shift in the material context of his work indicates a political shift. He writes:

> Since the closing of blewointmentpress and bissett's subsequent move to publish with Talonbooks, the format and content of his texts have slowly stabilized. The drawings, paintings, and typewriter concrete poems still appear, but have a sanitized feel within the perfect-bound, desktop published, properly literary digest-size confines of Talon's editions. Although his writing has been comprehended by a computerized environment, bissett has not continued to push against the limits of that field in the same way that his earlier work pushed against the limits of earlier publishing technologies (the typewriter, letraset, mimeographs and small printing presses). (1997, 122)

However, bissett's work remains incredibly idiosyncratic and his transition to Talonbooks was necessary. In part, he sold blewointmentpress to recoup financial losses incurred by blewointment due to protests made by Canadian MPs in the late 1970s that inspired the Canada Council to reduce funding for the press. bissett's work with Talon marks a continuation and transformation of his poetics. To this day, his distinctive orthography is evolving and some of his recent works experiment with the affordances of cut, copy, and paste computer functions for producing meditative, repetitive visual poems. However, the radical push behind bissett's media-based poetics has diminished somewhat. bissett has not, to my knowledge, learned advanced coding or computer hacking techniques, nor are those skills reflected in his recent work. Compared to his early work, his more recent visual work does not foreground materials or processes in the same way that his typewritten and mimeographed poetry once did.

While bissett's IBM reveals a reactionary position against digitization and computing technology, Nichol saw the computer as a new opportunity for poesis and textual embodiment. In the paratextual matter that accompanies the 5.25-inch diskette for his computer poem suite *First Screening*, Nichol writes: "What most surprised me in this process [of writing these poems] was how concerns that had been present for me in the mid-60s', issues of composition and content i was confronting while working with my early concrete poems, suddenly found a new focus. In fact, i was finally in a position to create those filmic effects that i hadn't had the patience or skill to animate at that time" (1984). In Nichol's

Journeying & the Returns (sometimes also referred to as bp), Nichol tried to bring filmic and kinetic effects into his visual poems. For "Wild Thing," a three-inch by two-inch, single-stapled booklet, pages must be flipped in quick succession by the reader to activate the kinetic quality of the poem. In doing so, the letter shapes appear as though they are moving animatedly across the pages. The letters L-O-V-E morph into one another as they move across the pages. Similarly, "Journey to Cold Mountain," a title that implies movement through travel, is enacted by the reader as they flip pages that increase and decrease in size as though moving up and down a mountainside. These works anticipate Nichol's later, kinetic poetry and, with their requirement of specific movements of the hand, bridge Nichol's interest in poetry as a place always intended for digits.

First Screening, written on an Apple IIe in BASIC, opens with an alternative vision of the reading process. First Screening demands that the body move differently than a typical book may demand: there are no pages to turn to read the work, one must insert the floppy disk into the computer drive, open the file, and initiate the sequence on the screen. In this way, First Screening shares an affinity for unconventional textual interactivity just like Nichol's early work in Journeying & the Returns, which similarly ask the reader to change their reading habits—especially, the movement of their hands—to fully engage the text.

In First Screening, his "ANY OF YOUR LIP: a silent sound poem for sean o'huigin" highlights a complex body/computer relationship through visuality, sound, and movement (see figure 11.3). As Emerson also points out, the dedication is indicative of an absent presence of the body invoked by the idea of a sound poem that has no sound. Following the work of composer John Cage, however, we know that silence is not the absence of sound. So, while First Screening may not have sound programmed into it, sound is an integral part of the poem. In its programmed sequence, the poem consists of single words flashing individually in the centre of the screen. "MOUTH" is the central word, anchoring the rest of the poem. "MOUTH" becomes "mouth," "myth," "maze," "mate," and "amaze," all of which carry the same "m" sound. The poem ends with the repeated flash of "ing," a suffix that generally denotes an instance of a particular action. To be "mouthing" a word or phrase, for example, is to move your lips as if you are saying it. This poem might encourage interactivity wherein the viewer may say the words on the screen aloud, turning this "silent sound poem" into the score for a sound poem. The variation between lowercase and uppercase spellings may indicate fluctuations in volume and the tempo of the poem is indicated by the programmed transitions of the piece. However, I suggest that the poem, as a "silent" piece, is inviting the viewer to mouth the words (not saying them aloud) in synch with the poem sequence. This is a poem made of the sounds of the body. Instead of the sounds of the words, one will hear the sounds of the body during the act of silent enunciation or mouthing. What the viewer hears, then, is the sound of the mouth as an assemblage of muscle, tissue, and saliva in process—the subtle movements of salivary liquid, clicking and pops of the jaw, and the smack of the lips. In this way, Nichol's computer-based "ANY OF YOUR LIP" brings the body into the room in real time and in synchronicity with the machinations of the programmed poem.

Interactivity is pushed further in "Off-Screen Romance," a poem that must be initiated by the user to be viewed. Using the Remark function of BASIC (which allows the coder to leave an explanatory note) Nichol types: "For the curious viewer/reader, there's an 'off-screen romance' at 1748. You just have to tune into the programme" (1984). The user then types the command "RUN 1748" to initiate Nichol's most sophisticated poem sequence, a code-choreographed homage to the celebrities Fred Astaire and Ginger Rogers, known for their dancing. By typing the run sequence, the user/reader momentarily becomes part of the writing process by typing command lines, deepening their involvement with the computer. Further, "Off-Screen Romance" foregrounds the body and movement since the names of the dancers—Fred and Ginger—move across the page rather than in real time. In this way, the computer comes closer to registering the kinetic aspects of the body and its movements more effectively than any typewritten page-based poem. In "Off-Screen Romance" the names Fred and Ginger dance the screen, as indicated by the rapid blinking of the letters of their name. The user's eyes follow the movements of their names across the screen as they dance. These movements of the eye, of course, do not follow the regular linear, left-to-right movement of conventional poetry or prose. Only with its programmed, cinematic capabilities could a poem like "Off-Screen Romance" capture the complex and synchronized movements of dancers.

When examining Nichol's code for First Screening, it turns out that this poem suite requires even more direct involvement with the materials and processes of the computer-based poetry. The twelfth poem does not appear on-screen; rather it is embedded in the code itself, which can be viewed on the computer by, again, following the Remark function. This is, as Emerson points out, "one of the first works of codework, or literary writing that is code but not necessarily executable" (2014, 69). The poem is, like other on-screen pieces like "Letter," a permutational poem that again plays with the Remark function. Breaking the word "REMARK" into "REM" and "ARK," Nichol puns and pivots in various directions, including an allusion to the Biblical myth of the flood and Noah's ark:

```
3900   REM   ARK
3905   REM   BOAT
3910   REM   AIN
3915   REM   RAIN RAIN RAIN RAIN RAIN RAIN RAIN RAIN
       RAIN RAIN RAIN RAIN RAIN RAIN RAIN RAIN RAIN
       RAIN RAIN RAIN RAIN RAIN RAIN RAIN RAIN RAIN
       RAIN RAIN RAIN RAIN RAIN RAIN RAIN RAIN RAIN
       RAIN RAIN RAIN RAIN RAIN
3920   REM   BOAT
3925   REM   ARK
3930   REM   BOW
3935   REM   ARC
4000   END
```

Such a poem only approximates the possibility of a computer-based dirty concrete. It is not dirty in its aesthetic, but it does foreground the poem's materials and, by requiring the user to glimpse at the code, the processes of the poem's creation. The poem thus turns the computer from "transparent carriers of meaning to objects meaningful in themselves" (Emerson 2014, 126). This back-end poem also reveals Nichol's hands in the work of coding the poem. In this way, this hidden twelfth poem of the suite reveals the possibilities and limits of the writing device much in the same way that dirty concrete poetry did for the typewriter.

CONCLUSIONS: COMPUTER POETRY AND BELONGING

As indicated by a close reading of these poetries by Canadian poets working in the late twentieth century, it seems that Brand's comments on the body's relationship to the computer that opens this chapter require far more nuanced attention than his article offers. Both Nichol's and bissett's computer-oriented poems apprehend the promise of digital media with differing perceptions of how it engages and constructs the body. While IBM stages a rejection of computational logic for the way bissett perceives its denigration of the body, Nichol's *First Screening* embraces the potentialities of the Apple IIe and its new forms of user interaction and movement through language. In these two positions, I find not only significant conceptions of the body/computer relationship, I also find gestures toward their conceptions of digital media and its promises for social relations. For bissett in 1972—like Brand, Baudrillard, and Kiyooka—the computer results in the removal of the body, thus removing the poet from the social horizon of collective activity. Instead, bissett, with his distrust of multinational communication companies like IBM, privileges a kind of writing that on its surface seems more immediate and intimate than the writing that takes place in computational environments. Nichol, on the other hand, sees new possibilities for collective formation in computational poetics. At the start of his writing life, Nichol, in one of his most demonstrably material works posited the possibility of a "new humanism," in concordance with the new material affordances of new media—"there's a new humanism afoot that will one day touch the core of the world" (Nichol 1967)—and suggests that, in the 1960s, "we have come up against the problem, the actual fact, of diversification, of finding as many exits as possible from the self (language/communication exits) in order to form as many entrances as possible for the other.... The other is the loved one and the other is the key" (Nichol 1967). The language Nichol chooses here is significant. Exits and entrances are words that designate placeness—buildings and spaces for gathering. The poem is a site of potential for collective organization. So, when Nichol embraces the digital medium in the 1980s, he extends his purview from the 1960s in search of more entrances and exits. He offers more space and more openings and exits for persons to come into, out of, and linger in the poem. Nichol died too early to continue experiments with computers for poetry; however, as a foundational diskette of digital poems, *First Screening* continues to participate in ever-expanding dialogues regarding digital media and poetry as they exceed immediate social contexts. *First Screening*, then, participates in Nichol's lifelong project of continually expanding the poem as a site of potential collective belonging, using technology to expand the possibilities for others to enter and exit the poem.

NOTE

1.   OED Online, "digital, *n.* and *adj.*," https://www.oed.com/view/Entry/52611.

REFERENCES

Barbour, Douglas. 2001. "Reading Kiyooka's *transcanada letters.*" In *Lyric/Anti-lyric: Essays on Contemporary Poetry*, 145–56. Edmonton, Alb.: NeWest Press.

Baudrillard, Jean. 1987. *The Ecstasy of Communication*. Translated by Bernard Schutze and Caroline Schutze. New York: Semiotext(e).

Birney, Earle. 1969. *Pnomes, Jukollages, and Other Stunzas*. Toronto, Ont.: Ganglia Press.

———. n.d. "Earle Birney at SGWU, 1968." *SpokenWeb* (blog). Reading by the author, February 23, 1968. https://montreal.spokenweb.ca/sgw-poetry-readings/earle-birney-at-sgwu-1968.

bissett, bill. 1966. *We Sleep Inside Each Other All.* Toronto, Ont.: Ganglia Press.

———. 1971. *IBM (saga uv th relees uv huuman spirit from compuewterr funckshuns).* Vancouver, B.C.: blewointmentpress.

———. 1982. *Ready for Framing.* Vancouver, B.C.: blewointmentpress.

Brand, Stewart. 1972. "Spacewar: Fanatic Life and Symbolic Death." *Rolling Stone,* December 7, 1972.

Ceruzzi, Paul E. 2003. *A History of Modern Computing.* Cambridge, Mass.: The MIT Press.

Daems, Jim. 2010. "'i wish war wud fuck off': bill bissett's Critique of the Military-Cultural Complex." *TOPIA: Canadian Journal of Cultural Studies* 23–24: 368–80. https://doi.org/10.3138/topia.23-24.368.

Drucker, Johanna. 2009. "Graphesis and Code." In *Speclab: Digital Aesthetics and Projects InSpeculative Computing,* 133–143. Chicago Ill.: University of Chicago Press. https://doi.org/10.7208/chicago/9780226165097.001.0001.

Eichorn, Kate. 2015. "The Digital Turn in Canadian and Québécois Literature." In *The Oxford Handbook of Canadian Literature,* edited by Cynthia Sugars, 512–24. New York: Oxford University Press.

Emerson, Lori. 2014. *Reading Writing Interfaces: From the Digital to the Bookbound.* Minneapolis: University of Minnesota Press. https://doi.org/10.5749/minnesota/9780816691258.001.0001.

Funkhouser, C. T. 2007. *Prehistoric Digital Poetry: An Archaeology of Forms, 1959–1995.* Tuscaloosa: University of Alabama Press.

Hayles, N. Katherine. 1999. *How We Became Posthuman: Virtual Bodies in Cybernetics, Literature, and Informatics.* Chicago, Ill.: University of Chicago Press. https://doi.org/10.7208/chicago/9780226321394.001.0001.

———. 2002. *Writing Machine.* Cambridge, Mass.: The MIT Press.

Hilder, Jamie. 2016. *Designed Words in a Designed World: The International Concrete PoetryMovement, 1955–1971.* Montréal, Que., and Kingston, Ont.: McGill-Queen's University Press.

Irvine, Dean. 2015. "MISSION CONTROL: An Operator's Manual for Compulibratories." *Amodern* 4 (March). http://amodern.net/article/mission-control/.

Kearns, Lionel. 1969. *By the Light of the Silvery McLune: Media Parables, Poems, Signs, Gestures, and Other Assaults on the Interface.* Vancouver, B.C.: The Daylight Press.

Kelly, Kevin. 1994. *Out of Control: The New Biology of Machines, Social Systems, and the Economic World.* New York: Basic Books.

Kiyooka, Roy. 2005. *Transcanada Letters.* Edited by Glen Lowry. Edmonton, Alb.: NeWest Press.

———. 1971. *Stoned Gloves.* Toronto, Ont.: Coach House Books.

McLuhan, Marshall. 1964. *Understanding Media.* New York: Signet Books.

————. 1969. *The Gutenberg Galaxy: The Making of Typographic Man*. New York: Signet Books.

Munster, Anna. 2006. *Materializing New Media: Embodiment in Information Aesthetics*. Lebanon, N.H.: University of New England Press.

Nelson, Sharon. 1997. "A Just Measure: breath, line, body in the work of bill bissett." *The Capilano Review* 2, no. 23 (Fall): 35–50.

Nichol, bp. 1967. *Journeying & the Return*. Toronto, Ont.: Coach House Books.

————. 1984. *First Screening: Computer Poems*. Emulated version. Vispo.com. Edited by Jim Andrews. https://www.vispo.com/bp/emulatedversion.htm.

————. 2002. *Meanwhile: The Critical Writings of bpNichol*. Edited by Roy Miki. Vancouver, B.C.: Talonbooks.

Olson, Charles. 1950. "Projective Verse by Charles Olson." *Poetry Foundation*, October 13, 2009. https://www.poetryfoundation.org/articles/69406/projective-verse.

Spinosa, Dani. 2017. "Toward a Theory of Digital Canadian Poetics." *Studies in Canadian Literature* 42, no. 2: 237–55.

Stefans, Brian Kim, and Darren Wershler. 2003. "potentially suitable for running in a loop." In *Fashionable Noise: On Digital Poetics*, 15–38. Berkeley, Calif.: Atelos.

Wershler, Darren. 1997. "Vertical Excess: *What Fuckan Theory* and bill bissett's ConcretePoetics." *The Capilano Review* 2, no. 23 (Fall): 115–23.

CHAPTER 12

Nations of Touch: The Politics of Electronic Literature as Digital Humanities

Dani Spinosa

While they may appear at first to be likely bedfellows, there has always been some tension, in theory and in practice, between electronic literature and digital humanities (DH). Certainly, both share similar concerns regarding the digital as a medium, and there are clear intersections between the two fields in their goals and their methodologies. But, as of now, the relationship between digital humanities and electronic literature remains unclear or undefined. This chapter looks primarily to the ways that both electronic literature and digital humanities, particularly within Canadian critical frameworks, negotiate their skepticism of national borders (and, indeed, their clear interference with, or disregard of, these borders) to argue for the intrinsic relationship between these fields. Focusing on how these fields are approached by the academy writ large, I argue here that in Canada the open-access dissemination of electronic literature, the intrinsic global nature of the genre, and the interactivity of the born-digital text bring to the surface a digital humanities that could be, and should be, interested in opening, blurring, or breaking down national borders rather than making them distinct.

This chapter looks to the role of the production and study of electronic literature within the larger field of DH in the academy. In examining these related but distinct fields of study, I aim to reveal the ways in which digital media as a tool has both the potential for radical rethinking of the political conditions of authorship, textual dissemination and access, and readerly engagement while, at the same time, being mindful and critical of the ways that those same tools can be actively used for oppressive, elitist, and exclusionary goals. In other words, this chapter calls upon scholars of the digital humanities and electronic literature alike to think politically about access, commodification, and datafication, and to avoid a data fetishism that threatens to erase the relations of *people* in digital scholarly work. By looking briefly at three case studies—W. Mark Sutherland's poetry machine, *Code X* (published as a CD-ROM in 2002 by Coach House Books, then as an interactive website in 2009); Andrew Campana's generative work, "Automation"; and the Electronic Literature Organization's (ELO) production of the three existing volumes of the *Electronic Literature Collection* (ELC)—I hope to demonstrate ways that networked connectivity can act as a community-building space working against the exclusionary and limiting practices of genre, nation, and border, and to demonstrate that this community building is central to the parallel political potentials in digital humanities and electronic literature. To do so, I first work to situate electronic literary production within and as distinct from DH practices more generally.

I don't want to suggest that authors, critics, and academics have historically treated electronic literature and digital humanities separately. They have not.[1] The relationship between these two fields is hard to pin down, largely because neither field can be neatly delimited by historical, generic, or geographical markers. Instead, as Scott Rettberg observes in "Electronic Literature as Digital Humanities," "both 'electronic literature' and 'digital humanities' are loosely defined not by their attachment to a historic period or genre but by a general exploratory engagement with the contemporary technological apparatus" (2015, 127). If both fields of study are defined by their engagement with technology, broadly construed, then it would seem that they should be related fields.

DIGITAL HUMANITIES AND THE CORPORATE UNIVERSITY

Digital humanities has tended to define itself primarily as technology-based criticism of traditional objects of humanities inquiry, and that text is most often not born digital. Conversely, scholarship of electronic literary study has most often been print-based, or at least traditionally formatted, scholarship of born-digital or transmedial literary texts. Because of the technological nature of electronic literary study, these scholarly works have tended toward open access and networked production and dissemination, but they are not "digital humanities" insofar as the digital format is not integral to the scholarship itself. Some venues for electronic literary scholarship, most notably the Electronic Literature Organization's journal, *electronic book review*, are working now to expand the kind of scholarship they produce, with their newest publication, the *Digital Review*, calling for "multimodal, computational and/or interactive essay[s]" (*ebr* 2019) for publication. In addition to these space of overlap between electronic literature (also "e-lit") and the digital humanities, I want to argue that both are primarily and strategically interested in disrupting borders—of genre, of nation, of the line between literary text and scholarship—and that disruption is enhanced when DH scholarship and electronic literary production work in tandem.

To argue that part of what binds e-lit and digital humanities is both fields' proclivity for critiquing genre and nation requires, first, that I address the clear problem of this conversation: that the very tools that enable digital creative and scholarly projects to critique such closed structures as genre and nation are the *same* tools which allow the digital medium to be so readily co-opted by larger corporate or nationalist organizations. Indeed, the easy dissemination of open-access digital work, the facility of access and distribution, and the gestures toward interactivity and ergodic or haptic engagement make it possible for DH tools and e-literary works to bypass cultural gatekeepers. Electronic literature can be, and often *is*, self-published, open access, and circulated internationally rather than governed by publishers and editors who select when and how a work is disseminated. But those same tools can easily be used by larger corporate, nationalist, or academic-administrative interests to further the ever-increasing corporatization of humanities departments. And indeed, as more and more of our scholarly and pedagogical work moves online because of the current pandemic (at the time of writing this chapter), DH scholars and electronic literature scholars alike are increasingly suspicious of this corporatization and datafication of academic work generally.

Part of the problem, it would seem, is that administration and funding bodies tend to encourage the production of DH tools that are readily marketable and sellable. As Daniel Allington, Sarah Brouillette, and David Golumbia argue in their "Neoliberal Tools (and Archive): A Political History of the Digital Humanities," digital humanities' adoption of these methods helped the neo-liberal university to bypass the difficulties of commodifying humanities scholarship. That is, whereas traditional humanities scholarship tends to rely "on painstaking individual scholarship and produc[e] forms of knowledge with less immediate economic application," humanities computing tools often "provid[e] a model for humanities teaching and research that appears to overcome these perceived limitations" (Allington, Brouillette, and Golumbia 2016). In other words, many of the tools for humanities computing have resulted in a vein of DH scholarship that eschews the kinds of interpretive work and communities, as well as the politicization of texts, digital and otherwise, that have been hallmarks of critical humanities scholarship. This digital depoliticization is the cost of the digital humanities' desire to produce data sets, visualizations, and graphs that are easily marketable to funding bodies and tenure committees alike. As a result, Allington, Brouillette, and Golumbia argue that "Digital Humanities has played a leading role in the corporatist restructuring of the humanities." Now that we are apparently in the third wave of the digital humanities, the tools of humanities computing offer scholars the opportunity to return to some of the more useful elements of print-based criticism with new eyes and a plethora of resources presented in the early, quantitative days of the digital humanities without neglecting the importance of experiential and affective analysis.[2] Of course, Allington, Brouillette, and Golumbia's work, and their 2016 essay in particular, has come under harsh criticism for its tendency to generalize the digital humanities; they notably miss, for example, the often critical and sometimes quite radical political digital humanities that has been carried out across universities and research labs. While many DH projects continue to raise important political questions about how universities produce and distribute scholarly work, DH tools continue to be used to support corporate interests in the university. But it has also become increasingly clear that neo-liberalism and capital cannot and will not be defeated by humanities reading practices, interpretive strategies, and digital inquiry.

IS ELECTRONIC LITERATURE DIGITAL HUMANITIES?

So, where does electronic literature fit into this difficult negotiation of the role of the digital humanities in the academy? Here, I want to argue that electronic literature offers some ways of looking at the incorporation of the digital into literary studies that shows us how we may continue to harness the powers of the digital in the humanities to make us continually aware of how these tools can be co-opted. I do not want to present the born-digital or e-literary text as utopian or inherently radical; such a claim is clearly false. Instead, I want to point only to what electronic literature, as it has been practised and presented so far, offers us some glimpses about what networked computing *can do* for literary studies in the academy. One of the most important of these features, as highlighted by Alex Saum-Pascual's work on digital humanities and electronic literature pedagogy, is the way that electronic literature intervenes in traditional conceptions of the literary, challenging English departments to accept the digital as a viable literary medium. For Saum-Pascual, "the reading and writing of digital works, mostly distributed online—together with the work of a community of collaborative e-lit producers—has challenged established *literary* concepts such as 'author,' 'authorship,' 'work,' and even the act of reading itself" (2017). While Saum-Pascual's interest here is in the ways that electronic literature rethinks authorial and readerly practices, the same tools she points to—collaboration and open, digital dissemination—similarly work to challenge established *national* concepts as well, particularly as they pertain to the literary.

In other words, electronic literary production and study have a simultaneously uneasy relationship with the English department and with national literatures.

For Saum-Pascual, by virtue of its ease of dissemination, and by the communities required to produce it, electronic literature in general resists national literatures; "e-lit," she argues, "may take us back to a time before national languages and literary traditions" (2017). In other words, electronic literature lets us imagine a time before a need for unification and standardization of literary language(s) to reach a larger audience produced clearly definable literary languages and national literary traditions.

ELECTRONIC LITERATURE AS ANTI-/INTER-NATIONAL

In the introduction to their pivotal collection, *Electronic Literature Communities* (2015), Scott Rettberg and Patricia Tomaszek argue that electronic literature has long depended on the global network and wide-reaching networked communities. They argue that "[b]ecause the electronic literature community is intrinsically intertwined with the global network, the development of e-lit has been more international in nature than many literary movements that preceded it" (6). This is not to say that electronic literature ignores that national boundaries exist; indeed, despite the international nature of electronic literary production, Rettberg and Tomaszek point out that "it is still the case that many communities are emerging from and are responsive to national and language-based literary traditions" (6). Rather than imagining a utopian networked society, Rettberg and Tomaszek's collection demonstrates that electronic literature can and has worked to disrupt and blur national literary traditions through collaboration and exchange. In his article on the subject for *Digital Humanities Quarterly*, Rettberg points to the invaluable intersections and conversations happening between "the French, Spanish, German, Dutch, Brazilian, Scandinavian, English, American, and Canadian electronic literature communities" (2015), who interact through the international Electronic Literature Organization. The ELO, while based in the United States, has an international board and whose major outputs (the *Electronic Literature Collection*, the open-access and crowdsourced Electronic Literature Directory, and the annual conference) are decidedly international.

As demonstrated, for example, by the work presented at the ELO's 2018 Arabic Electronic Literature Conference in Dubai, the academic reception and study of electronic literature may have had its roots in anglocentric and Western-normative mindsets, recent research into the issues of access, bias, and normativity in the field have dramatically opened up our conceptions of an e-literary canon facilitated by networked computing.[3] We might argue, too, that we see a clear divide between the way that departmental administration views DH projects and the way those same organizations view electronic literature. For the most part, administrators see great interest in DH scholarly projects because they produce quantifiable tools for study and lend themselves to easy enumeration in tenure files, and, therefore, the common perception is that these works garner funding. However, these departments do not typically see similar use value in electronic literature; for the most part, literature and creative-writing departments tend to avoid or, worse, malign electronic literature for being too simple or without the intellectual rigour of print-based work.

ACTIVE READERSHIPS AND ERGODICS

This is obviously not the case. In fact, much of the complexity of born-digital literary works lies in the ergodic nature of many works of electronic literature, which encourages an engagement with the reader or user. According to Espen Aarseth's *Cybertext*:

> In ergodic literature, nontrivial effort is required to allow the reader to traverse the text. If ergodic literature is to make sense as a concept, there must also be nonergodic literature, where the effort to traverse the text is trivial, with no extranoematic responsibilities placed on the reader except (for example) eye movement and the periodic or arbitrary turning of pages. (1997, 1)

The line between ergodic and non-ergodic text is necessarily fraught. Demonstrating this, the Wikipedia entry on "ergodic literature" laments the distinction by observing: "Under [this] definition [...] *Finnegans Wake*, the *Critique of Pure Reason*, and *Being and Time* are considered nonergodic literature as they require only 'trivial [...] effort to traverse the text[s].' A stack of stained and mouldering newspapers, on the other hand, is ergodic literature" (2019). Unlike whoever wrote this Wikipedia entry, I do not want to lament this divide. Instead, I want to argue that the reader alienation that coincides with much print-based experimental work is a unilateral power structure in which the author retains a level of authority and power over their audience, an authority Joyce, Kant, and Heidegger would likely enjoy. I want to begin thinking about electronic literature, in its turn to the ergodic and to the reader, as a way of proposing community and active engagement at the limits of larger structures of power and control.

While the ergodic may initially have been interested in engagement and interactivity, its definition prioritizes the *difficulty* or *effort* involved in navigating a work of electronic literature that is distinctly not open, to say the least. Instead, this chapter argues that the digital humanities writ large, and electronic literature as one manifestation of the digital humanities, needs to start to consider the ease with which one can navigate and engage with an e-literary text as an issue of access, and one that works to dismantle the elitist—or, at the very least, the occasionally prohibitive—nature of exegetical reading processes. To read these electronic literary works as opening requires, too, that we as readers and as scholars recognize that those same structures can be elitist or inhibiting. Digital media are tools, which means that the digital is not a radical reformation of genre, nation, or textual practice. As a reminder of this, I often return to Loss Pequeño Glazier's indispensable *Digital Poetics: The Making of E-Poetries* (2002), which reminds us on the first page that

> we have not arrived at a place but at an awareness of the conditions of texts. Such an arrival includes recognizing that the conditions that have characterized the making of innovative poetry in the twentieth century have a powerful relevance to such works in twenty-first-century media. That is, poets are making poetry with the same focus on method, visual dynamics, and materiality; what has expanded are the materials with which one can work. (1)

Ultimately, electronic literature as a genre embraces a new awareness of the interconnectivity of nations and communities. Electronic literature as a genre invites readers into that community. While this is important in terms of access, and in terms of opening literary production and reception, it's also one of the same tools that can be co-opted for larger institutional structures.

Electronic literature can, and for some readers, I suppose, *does* encourage a kind of postcritical or apolitical reading process. However, as electronic literature moves to interactive, ergodic, and generative models that do not directly encourage hermeneutical activity, they continue to encourage an active, engaged readership, which is necessarily politicized. The links between digital humanities and electronic literature are made most apparent here, in the politics of digital practices like archivization and data production (despite the claims some make toward their apolitical nature) and the refusal of hermeneutics in the generative or "unreadable" e-lit text. I argue here, through the case studies of Sutherland's *Code X*, Campana's "Automation," and the production of the ELO's *Electronic Literature Collections*, that a refusal of authorial control and exegetical reading practices is indeed a politics that demonstrates a shared goal of digital humanities and e-lit can be—should be—a recognition of the workers, the humans, involved in this writing, and that political import of the relations between them.

CASE STUDY ONE: W. MARK SUTHERLAND'S *CODE X*

I would like to start with Sutherland's born-digital poem/poem generator, *Code X* (first exhibit 2002, current digital edition 2020), to discuss how some of the work by Canadian writers uses the affordances of digital technology to create freely accessible work that transcends national and genre boundaries and does so in a distinctly political fashion, arguing that the disruption of these borders is an inherently political act by way of its reaching out to its audience. Sutherland argues that in *Code X*, as in most ergodic texts, "the reader/viewer/listener/performer is actually a reader/writer/collaborator by participating directly in a process to create the poem(s). The writer/artist, on the other hand, shapes the conditions for this collaboration and is therefore equal parts poet and programmer" (Sutherland and Spinosa 2017). *Code X* is a born-digital application that allows users to create their own sound poetry. Each key places a typewriter-font collection of dispersed letters on the screen while at the same time queuing an audio track. The visual appearance of the work, a black screen with white and red Courier text, bears no small resemblance to early concrete and typewriter poetics.

In *Code X*, the "readers" become engagers, players, or "performers" who make some interesting agential choices in the text. In the information page that accompanies the web-based version of *Code X*, Sutherland specifically uses the term "performers" to describe the audience of his work, suggesting that the text is subsumed by, or at least less important than, the audience to which it is addressed. Sutherland points to the fact that *Code X* is a sandbox in which its audience can play and produce theoretically infinite permutations of the work's performance but is at the same time a fairly closed system. The performance of *Code X* will, of course, look and sound different depending on who is interacting with it, what letters they choose, with what speed or pattern they type, what hardware is used to engage with the piece, how long it is used, and whether the work is left to lapse into its "random" mode (see figure 12.1).

But, as long as the performer types each letter of the alphabet at some point during a session, the result is the same, the appearance of Sutherland's prewritten final paragraph. As the information page tells us, "Code X is housed within a self-referential paragraph containing every letter of the alphabet" (Sutherland, n.d.). Moreover, while the order, overlap, and frequency of the sounds may vary, each letter typed will play the same "10 second phonetic improvisation" that

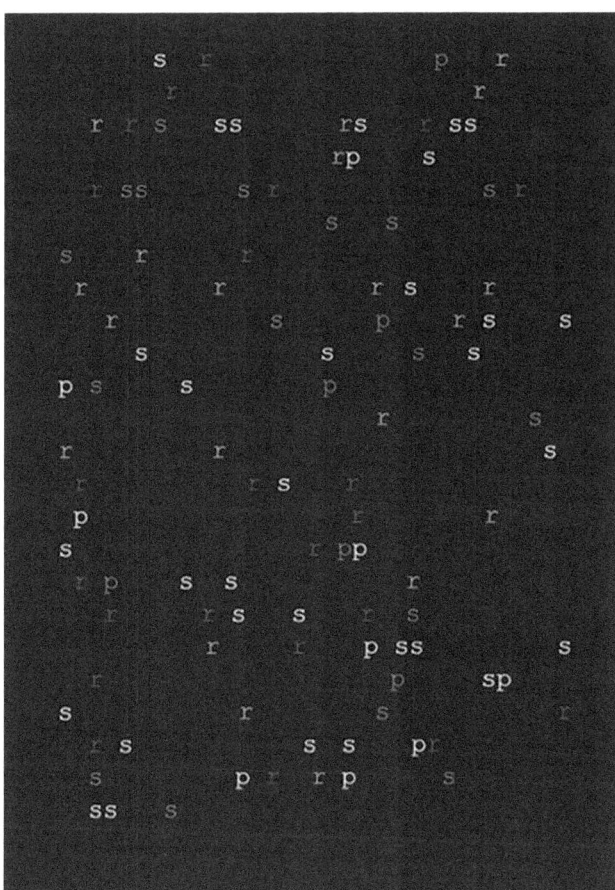

Figure 12.1. *Code X* run in random mode.

Source: Sutherland, n.d.

Sutherland recorded for each separate letter. Regardless, the two fundamental elements of the work—the paragraph and the recorded sounds—never change. Thus, while the role of the audience is agential, interactive, and integrative in many ways, only the process differs; the ultimate outcome remains fundamentally constant. In engaging with—or performing—*Code X*, we slightly alter the text. The voices and visuals produced by our interaction are predetermined, and though they look random when only a few letters are activated, they ultimately form a pre-written textual "whole." In fact, the "self-referential" paragraph that is ostensibly the work's conclusion speaks of the reading process of the digital text as leading to the end goal of making adequate and substantial meaning from the text at hand (or cursor). It reads in part, "reading was a road a car a mnemonic mechanism driving towards form and meaning" (Sutherland, n.d.), positioning the text as a kind of teleological red herring, where no real interpretive meaning is ever produced. The fact that the poem reverts to an automated mode suggests, on one level, that perhaps the individual performer is not even required for the text to run. If the activity of the performer can be and is performed by an automated algorithmic function that continues indefinitely and does not repeat itself, then our engagement with the work as performers is still integrative, but it is not necessary. And yet the random mode can never "replicate interactivity" with complete success: the automated function reveals letters too slowly and thus letters fade into the black background before the full paragraph can be revealed.

Unlike the individual performer, the random, automated function of *Code X* will never reveal that paragraph in full; this paragraph can only be achieved by an individual knowingly typing each letter of the alphabet at least once in a short-enough time span. The terminus of the full paragraph and the automated function might be taken to suggest a "truth" or a "fact" of the text that is initially hidden from the viewer, an endpoint that results from an alphabetically exhaustive use of the work that the reader can reveal but cannot intervene in. However, the case is not that simple, due in large part to the many red words revealed within other words. The word "ode" is revealed as red text in "code" with the letter C remaining in white. Almost every word of the final paragraph uses the differentiation of red and white to reveal words within words, which gesture toward the multiple and individual-specific readings that are contained within (and resist the limits of) the arbitrary whole of this final paragraph. Furthermore, the performer is not only featured in the final paragraph but is even, through these words within words, addressed—although referred to in the third person. By way of the "hi" salutation that is hidden within the first word "while," the performer is invited into the reading practice. Sutherland's opening, "while staring at the computer the abecedarian catalogued every key," is an obviously self-referential statement that describes his composition of the paragraph itself, which contains every letter of the alphabet and is a kind of catalogue of all letters (see figure 12.2). The final paragraph, when complete, makes cohesive semantic sense, despite the occasional awkward syntax; by contrast, the red words contained within other words do not unite to form logical sense. Despite the promise of arriving finally at "form and meaning" that the paragraph offers, these words within words

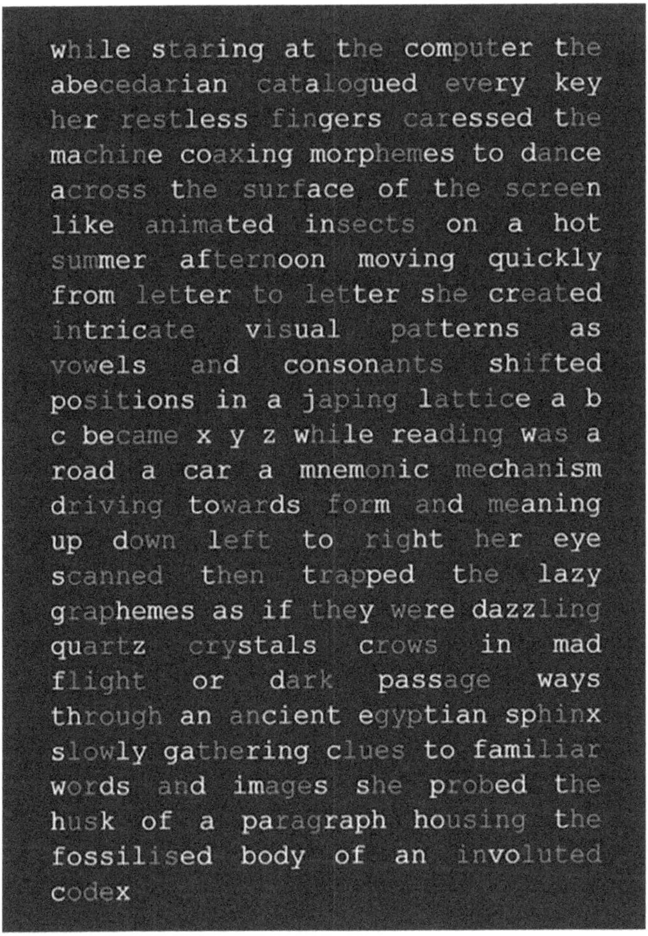

while staring at the computer the
abecedarian catalogued every key
her restless fingers caressed the
machine coaxing morphemes to dance
across the surface of the screen
like animated insects on a hot
summer afternoon moving quickly
from letter to letter she created
intricate visual patterns as
vowels and consonants shifted
positions in a japing lattice a b
c became x y z while reading was a
road a car a mnemonic mechanism
driving towards form and meaning
up down left to right her eye
scanned then trapped the lazy
graphemes as if they were dazzling
quartz crystals crows in mad
flight or dark passage ways
through an ancient egyptian sphinx
slowly gathering clues to familiar
words and images she probed the
husk of a paragraph housing the
fossilised body of an involuted
codex

Figure 12.2. The full final paragraph of *Code X*.

Source: Sutherland, n.d.

suggest instead the arbitrariness and inadequacy of the sense at which we might arrive. The final paragraph positions the creative performer at the crossroads of sense and nonsense, where any permutation of these letters can and should be used, but any permutation would result in the same outcome: the dissolution of those larger structures that typically govern meaning making and reading processes.

CASE STUDY TWO: ANDREW CAMPANA, "AUTOMATION"

While Sutherland's work is both a text and a tool for further textual production, I'd like to argue that a similar reaching out to an active audience can happen even when the electronic text in question is non-interactive. To do so, I'd like to look for a moment at Campana's "Automation," a generative electronic poem that flaunts its international nature and its anti-exclusionary sensibilities. As Campana's introductory note to the work indicates, the work itself is situated in the communal space of public transportation and draws parallels between various international experiences of public transit by drawing on one universal feature: the repetitive nature of announcements throughout transit systems. Campana tells us that "Automation" was "[i]nspired by the endlessly repeated automated announcements in Tokyo train and subway stations" and is explicitly international, existing in two versions: "'自動化' in its Japanese version, and 'Automation' in English" (Campana 2014, para. 1). While they are quite different, Sutherland's and Campana's texts do share one important feature: both texts employ some level of generation in the reading process. But, unlike Sutherland's, which requires reader engagement on some level to ultimately produce the final, teleological paragraph, Campana's "Automation," as the name suggests, will continue running indefinitely, producing new lines using Campana's predetermined (but rather large) lexicon. Campana explains the poem's process on the poem's website:

> It uses the syntax of the familiar [...]. "The doors on platform 1 are closing. Please be careful." Every 8 seconds, a script generates a new line by randomly selecting the platform number, subject, verb, and exhortation from a preset list. It displays the result on the screen and then generates a new line. Browsers capable of speech synthesis will also read the text aloud in either English or Japanese. (2014, para. 2)

Thus, each iteration of "Automation" will be different, but will follow the same basic syntactic structure. For example, my most recent run-through of the poem, as of my writing this, reads as follows:

> The last train is visible on Platform 33. Please cooperate.
> The express train is gathering on Platform 8. Please confirm.
> The rapid train is arriving on Platform 29. Please instruct us.
> The doors are laughing on Platform 7. Please confirm.
> The tunnel is arriving on Platform 23. Please be calm.

Clearly, some of these lines appear just as banal as the typical subway announcements one might hear in any major city centre. "The last train is visible on Platform 33. Please cooperate," could very well be a subway announcement here in Toronto where I write this chapter. But, because of the random nature of the poem's generation, and owing to the choices Campana made when selecting the lexicon from which the generating code can select, some of the lines verge into the surreal ("The doors are laughing on Platform 7. Please confirm.") or the potentially frightening ("The tunnel is arriving on Platform 23. Please be calm."). "Automation" works to defamiliarize the banality of these transit announcements, juxtaposing this quotidian repetition with moments of absurdity, humour, and, in some cases, an impossibility of logical meaning. What I find most interesting about this project is the sheer internationality of it; these might be subway announcements in London or Paris or New York or Toronto.

And yet, there is something uniquely Japanese about Campana's initial concept for "Automation," which is, after all, inspired specifically by the Tokyo subway and by Japanese poetic structures. Moreover, Campana has his doctorate in Japanese literature from Harvard and currently teaches in the Department of Asian Studies at Cornell. In their entry on the work for the Electronic Literature Directory, Marta Deyrup and Mouannes Hojairi argue that "Campana has created a multimedia work that reflects upon the rigidity and punctuality of the Japanese transportation system (and perhaps Japanese culture)" (2018). But "Automation"'s internationality is precisely in the concomitant specificity and universality of the work. As Deyrup and Hojairi also point out, while "the text of this work is generated in both the Japanese and the English languages simultaneously...each of the two texts follows the semantic structure of that language creating in the viewer a different experience." This issue of translation, altering the semantic structures of the announcements, was highlighted when the work was included and translated as a part of the Renderings project, an initiative headed by Nick Montfort that worked on translation computational literature from other languages into English and beyond in the Trope Tank at MIT in 2014.

CASE STUDY THREE: THE ELECTRONIC LITERATURE COLLECTIONS

Finally, I want to consider the production of the first three volumes of the ELO's electronic literature anthologies as examples of scholarly work that bridges the gap between DH practices of archiving, curating, and producing datasets and the work of producing and reading electronic literature. Published in October 2006, the first volume of the ELO's *Electronic Literature Collection* anthologizes 60 representative and various works of electronic literature and digital poetry. This first curated volume was selected and edited by four editors, N. Katherine Hayles, Nick Montfort, Scott Rettberg, and Stephanie Strickland, all of whom are major scholars and writers in the fields of digital humanities and electronic literature. As the first major anthology with the designation of "electronic literature," the first ELC contains many vital and pioneering works by e-lit mainstays such as M. D. Coverley, Maria Mencia, J. R. Carpenter, Deena Larsen, Robert Kendall, John Cayley, Michael Joyce, and Shelley Jackson.

A CD-ROM with identical contents was published along with the online collection so that this first volume might "reach the broadest audience possible and to provide for reading, classroom use, sharing, and reference on and off the network" (Hayles et al. 2006). Many of the works collected in this first iteration of the ELC are now considered to be foundational works in their genre, or of electronic literature in general, which was itself a political action, asserting that these various generative works, animated graphics, and text-based games were, in fact, literary products worthy of reading and study like any print-based work. The first volume of the ELC is available to view and navigate in four different organizational formats. The main index presents the texts by way of hyperlinked thumbnail graphics, but the works are also sorted variously by author and title alphabetically, and then categorized by keywords, which are also defined on the keyword index page. This keyword index page, while not exhaustive, is one of the first glossary-style pages looking to define the major terms of e-literary study—for example, kinetic, codework, installation, hypertext—for ease of scholarship.

Even though this first ELC was produced nearly fifteen years ago, the keyword page continues to be a useful resource for defining and clarifying e-literary terms. My work with the Electronic Literature Directory is to continue this kind of glossary-style work to encourage and facilitate more scholarship into the field, and to render more accessible these technological terms which can prove alienating to non-specialists. Additionally, the keyword feature demonstrates how varied the first volume of the ELC proved to be in its representation e-literary formats and styles. The second volume maintained this keyword feature alongside the title and author sorting. Interestingly, the third added country and language to its indexing pages, to encourage the study and reading of e-lit works produced in languages other than English, and in countries outside of the Western tradition.

These three volumes of the ELC demonstrate that the work of curation and archiving—the ELC also provides stable hosting for digital works that require it—need not be an elitist or exclusionary practice. The open call for editors for the fourth volume of the ELC demonstrates an interest in further opening this discussion; potential editors were asked to self-nominate, with the required qualifications stating only that "[n]ominees must have a PhD or terminal degree (such as an MFA) or a substantive body of elit work to be considered" (Marino 2019), with no clarification of what constitutes a "substantive body of elit work." Thus, we might argue that the DH work of archiving, curating, editing, and producing metadata for these e-literary works is starting to open. ELC 4 will continue to be open access and will continue to prioritize a variety of works in different languages, genres, and styles by international authors at various stages in their careers. The work that still needs to be done, however, is to help funding bodies, departments, and tenure and hiring committees to recognize this work as valuable knowledge production.

CONCLUSIONS

I want to close by considering the ergodic in digital poetics as an extension of Donna Haraway's theory of the touch, the point at which Haraway's theories of the cyborg and her work in critical animal studies meet. Haraway starts *When Species Meet* with a question: "Whom and what do I touch when I touch my dog?" (2007, 3). She considers touch as the primary site of the encounter that invites us into ethical connection with the Other; otherness, the comments section seems to assure us, can be reinforced by the digital and the separation suggested by hardware. Nonetheless, Haraway insists that touch, even with the digital intervening, "ramifies and shapes accountability. Accountability, caring for, being affected, and entering into responsibility [...]. Touch does not make one small; it peppers its partners with attachment sites for world making. Touch, regard, looking back, becoming with—all these make us responsible in unpredictable ways for which worlds take shape" (35).

It is fitting, then, that one other way of talking about the ergodic is to consider the haptic tendencies of digital technology. Haraway's touching is the meeting point between Aarseth's "ergodic" (etymologically the "pathway" of the "work") and the haptic (etymologically, "to touch," from the same root as tactile). So, what and whom do I touch when I touch this text? I touch out into this network, and by ease of entry and use, new numbers of readers, writers, users, and producers can touch back. The new ergodics is an ergodics of ease, of touch; the effort involved is not trying to make sense out of Joyce or Kant but of the difficulties, the strains of reaching out. The digital does not revolutionize this process, it merely makes those conditions of touch more apparent, brings us to—to return to Glazier—"an awareness of the conditions of texts," of their production and relation. And, as the ELO works on the production of ELC 4, I am excited about the direction, the ways that electronic literature and the use of the digital in humanities departments can expand what we do. But we need to remain acutely aware, too, of how quickly those tools become weapons if held that way.

NOTES

1. There has regularly been inclusion of e-literary study in prominent DH venues and publications. The Digital Humanities Summer Institute (DHSI), held annually at the University of Victoria, always offers at least one course in the production and study of electronic literature, and in 2016 the DHSI hosted the ELO's annual conference. The *Digital Humanities Quarterly* occasionally publishes essays on electronic literature, and most edited collections or journal special issues on digital humanities pay some attention (as is the case with this collection) to the intersection between electronic literature and the digital humanities.

2. The "Digital Humanities Manifesto 2.0" by Todd Presner, Jeffrey Schnapp, Peter Lunenfeld, et al., argues that the earliest DH projects, its "first wave," were "quantitative, mobilizing the search and retrieval powers of the database, automating corpus linguistics, stacking hypercards into critical arrays." The second wave of the digital humanities emerged as "qualitative, interpretive, experiential, emotive, [and] generative in character" (Presner et al. 2009).

3. Because electronic literature is a field in its infancy, it is even more important to address the fissures and gaps in whom we read or study. Rettberg points to this when he writes for *Digital Humanities Quarterly* that though the international community of electronic literature writers, artists, and scholars "don't necessarily speak the same languages, we are all becoming increasingly aware of each other's work. The field of electronic literature is a network of networks, and we are only beginning to learn how to work together" (Rettberg 2015). In a similar vein, then, while I was unable to attend the Dubai conference because of finances and my adjunct position's lack of institutional support, I was able, through DH practices, to review and compile some of the papers presented at this conference in a gathering for the *electronic book review*; in the introduction to this gathering, I argue that the papers presented, and the various formats for open-access presentation and dissemination of this work, "prioritizes the communicative aspects of digital scholarship, reminding readers that this scholarship is meant as discursive and dynamic rather than static" (Spinosa 2018). The *ebr* gathering on the Arabic Electronic Literature Conference demonstrates that electronic literary production and study can work with larger DH practices to open canons, nations, and genres of literary to encourage discursive scholarly practices, to disturb Western-normative reading and writing practices, and to work against paywalls and other economic or location-based barriers to access.

REFERENCES

Aarseth, Espen J. 1997. *Cybertext: Perspectives on Ergodic Literature*. Baltimore, Md.: Johns Hopkins University Press.

Allington, Daniel, Sarah Brouillette, and David Golumbia. 2016. "Neoliberal Tools (and Archives): A Political History of Digital Humanities." *LA Review of Books*, May 1, 2016. https://lareviewofbooks.org/article/neoliberal-tools-archives-political-history-digital-humanities/.

Campana, Andrew. 2014. "Automation." Printer's Devil Review. https ://pdrjournal. org/arts/Andrew_Campana/automation.html.

Deyrup, Marta, and Mouannes Hojairi. 2018. "Automation." Electronic Literature Directory, June 6, 2018. http://directory.eliterature.org/individual-work/4978.

Ebr (*electronic book review*). 2019. "Call for Submissions." https://electronicbookreview. com/call-for-submissions-the-digital-review/.

Glazier, Loss Pequeño. 2002. *Digital Poetics: The Making of E-Poetries*. Tuscaloosa: University of Alabama Press.

Haraway, Donna. 2007. *When Species Meet*. Minneapolis: University of Minnesota Press.

Hayles, N. Katherine, Nick Montfort, Scott Rettberg, and Stephanie Strickland, eds. 2006. *Electronic Literature Collection: Volume One*. Electronic Literature Organization. http://www.collection.eliterature.org/1/.

Marino, Mark. 2019. "Call for Nominations for ELC 4 Editors." Electronic Literature Organization. July 15, 2019. https://eliterature.org/2019/07/call-for-nominations-for-elc-4-editors/.

Presner, Todd, Jeffrey Schnapp, Peter Lunenfeld, et. al. 2009. "The Digital Humanities Manifesto 2.0." *UCLA Humanities*. https://www.humanitiesblast.com/manifesto/Manifesto_V2.pdf.

Rettberg, Scott. 2009. "Communitizing Electronic Literature." *Digital Humanities Quarterly* 3, no. 2. http://www.digitalhumanities.org/dhq/vol/3/2/000046/000046.html.

———. 2015. "Electronic Literature as Digital Humanities." *A New Companion to Digital Humanities*. https://doi.org/10.1002/9781118680605.

Rettberg, Scott, and Patricia Tomaszek. 2015. "Networks of Creativity: Electronic Literature Communities." In *Electronic Literature Communities*, edited by Scott Rettberg, Patricia Tomaszek, and Sandy Baldwin. Morgantown: West Virginia University Press.

Saum-Pascual, Alex. 2017. "Teaching Electronic Literature as Digital Humanities: A Proposal." *Digital Humanities Quarterly* 11, no. 3. http://digitalhumanities. org:8081/dhq/vol/11/3/000314/000314.html.

Spinosa, Dani. 2018. "Essays from the Arabic E-lit Conference." *Electronic book review*. https://electronicbookreview.com/gathering/essays-from-the-arabic-e-lit-conference/.

Sutherland, W. Mark. n.d., *Code X*. Accessed June 16, 2022. https://www.wmarksutherland.com/code-x.

Sutherland, W. Mark, and Dani Spinosa. 2017. "In Digital Ether: W. Mark Sutherland in Correspondence with Dani Spinosa." *Jacket* 2. September 21, 2017. https://jacket2.org/interviews/digital-ether.

Wikipedia. 2019. "Ergodic Literature." Wikimedia Foundation. Accessed June 8, 2019. https://en.wikipedia.org/wiki/Ergodic_literature.

CHAPTER 13

Stop Words

Klara du Plessis

"Stop Words" is a set of five short poems based on five sets of commonly used stop words gleaned from topic modelling and computational literary studies: articles, pronouns, prepositions, conjugations of to be, and negation. Each poem is transformed using sound-wave-visualization software, resulting in five visual poems or images. Combining the precision of research technology with the spontaneity of creative practice, these images function as interpretative graphical material—visualizing poetry as simultaneously scientific quantitative and affective qualitative diagrams. "Stop Words" is also an intervention in the humanities/DH schismatic debate, digitally supplementing the traditional verbal dimension of poetry while celebrating an interpretative, subjective, digitally generated product.

STOP WORDS

Technologically speaking, stop words are a set of words strategically compiled to be consistently removed from a particular text or grouping of texts for computational analysis. The intention is to remove high-density, high-probability terms that can slow down or skew the retrieval of data or research outcomes; the assumption is that these words are negligible in relation to the overall semantic meaning of the text or text grouping. For "Stop Words," I found a commonly used list of stop words from the Natural Language Toolkit website (NLTK Project 2019), a platform for building programs to work with language data. This list was roughly divided into articles, pronouns, question words, conjugations of to be, conjugations of to have, conjunctions, adverbs, negation, prepositions, and verb forms signalling tense.

As a poet and trained close reader of literature, the notion that any word can be devoid of relevance rings false. Particularly laden, existential words like forms of to be are fundamental to the ontology of a text. Whether action happens in the house or on the roof of the house are distinctions that could completely change a narrative trajectory. Whether a knife or that knife is used in the murder mystery has fetishistic significance. For me, it seems clear that every word in a text contributes innately to its overall meaning and that the act of consistently removing certain words, no matter how small, might equally skew search results as not removing them at all. The difference is, of course, that the excision of words during computational analysis is quantitative in scope, while the silencing of words that I am mourning is more closely aligned with a qualitative interpretation of that text.

A study by Alexandra Schofield, Mans Magnusson, and David Mimno, "Pulling Out the Stops: Rethinking Stop Word Removal for Topic Models," queries the utility of stop words altogether. After running the same texts through computational software with and without lists of stop words, these scholars noticed little difference in the search results. They contend that the amount of labour required to compile a relevant list of stop words is not commensurate with the degree of nuance prompted in the results; they argue that "although stop word removal clearly affects which word types appear as most probable terms in topics [...] this improvement is superficial, and that topic inference benefits little from the practice of removing stop words beyond very frequent terms" (2017, 1). That is, while I argue for the exclusion of stop words on the grounds of the inherent significance of all words, stop words could also, on the contrary, be functionally meaningless in terms of modifying computational search results.

As a corrective to the dismissal of stop words, "Stop Words" celebrates their relevance by employing them as inspiration for the following set of five poems. "Stop Words" further transforms the term's technological relevance into an invocation of a cessation of language. The poems' words on the page are transformed into audio, which is, in turn, translated into visual representation. That is, the words are the raw material mobilizing the project as a whole; at the same time, the words are stunted, mutated to the point of unrecognizability. Significantly, Susan Brown, in her essay "Evolving Digital Modes of Scholarly Production," underscores the fact that "digital technologies [...] are producing a shift from textuality to visuality. Voice and sound technologies [...] are shifting dissemination toward orality" (2011, 8). "Stop Words" literalizes this progression, emphasizing the interrelationship between the verbal, sonic, and visual, especially with the flexibility and facility of a digital context.

POEM 1

**Tiny determination
of terms:
squat plasticity of articles, resting squarely on the ledge of better meaning**

a
an
the

POEM 2

Pronoun sound clap

I
me
my
myself

we
our
ours
ourselves

you
your
yours
yourself
yourselves

he
him
his
himself

she
her
hers
herself

it
its
itself

they
their
theirs
themselves

POEM 3

**Hypermasculinity in a structure, kink
in the hyphen,
in the sense of positionality**
            **governing a word
a word governing**

as
until
while
after
into
of
at
by
for
with
about
against
between
through
during
before
above
below
down
from

to
up
in
out
on
off
over
under

POEM 4

**In the soft wardrobe of being,
suture an ego timidly /
boldly**

am is are was were be been
being

POEM 5

**Refusal, self-motivated resolution, trajectory ends,
whip smart knot tied back into knowing**

no
nor
not

PRAXIS

After reading and recording the five short poems line by line on GarageBand, I input each line/audio file (see. figure 13.1) into sound-wave-visualization software. Experimenting, I used three different softwares—Drift, Praat, and a free online app called convert.ing-now.com. Once each audio file had been visualized (see figure 13.2 for an example), I played—played with each software's features, manipulated and maximized the visual potential of the graphic. The intention was never to warp the actual sound wave, but to maintain its integrity in relation to the original poem and to its recording. At the same time, this representation of the sound wave shifted the focus to its aesthetic effect as an image or artwork in its own right. Once satisfied with the graphic, I screenshot it. Since some of the poems included three to five lines, I further played with Microsoft Word, combining the graphics of the lines into a new whole.

I created Image I (see figure 13.3.) using Drift. Drift is a downloadable software that works in conjunction with Gentle. I first uploaded a sound file along with its transcript and Gentle generated a phonetic breakdown of the words heard. I then uploaded the same sound file and transcript onto Drift. Drift visualized the file's sound wave while also inserting the words into the graphic to display the relation between the text and the visualization. I decided to use Drift for the first poem's image exactly because of the residual inclusion of text. It seemed

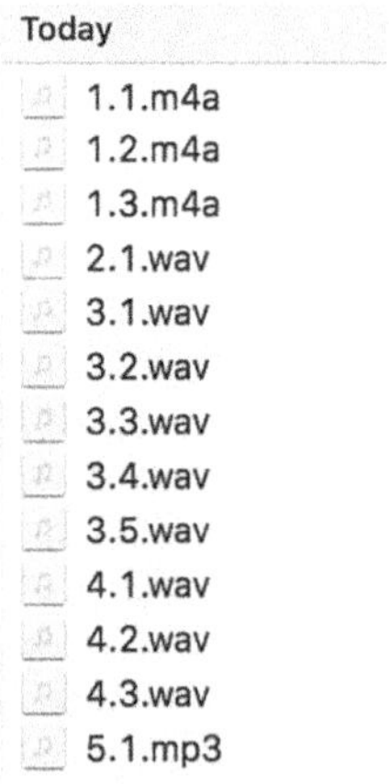

Today

1.1.m4a
1.2.m4a
1.3.m4a
2.1.wav
3.1.wav
3.2.wav
3.3.wav
3.4.wav
3.5.wav
4.1.wav
4.2.wav
4.3.wav
5.1.mp3

Figure 13.1. "Stop Words" audio files.

Source: Klara du Plessis.

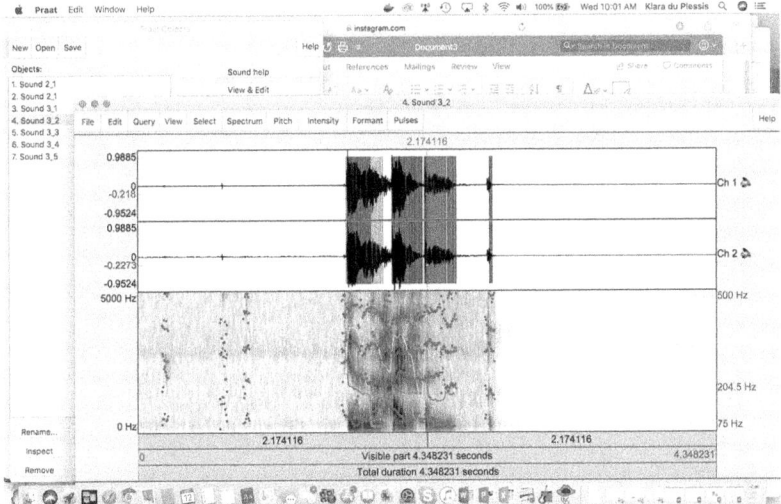

Figure 13.2. Play rendered in Praat.

Source: Klara du Plessis.

appropriate for "Stop Words"—as a project that, among other goals, is concerned with *stopping words* in the sense of visually recontextualizing language—to begin at a point where the actual words of the written poem are still recognizably present. The final line of the poem was too long to screenshot as a single image and was divided into two image files—this was fortuitous, especially when I visibly flattened the visual of the phrase "squat plasticity of articles."

Images II and III (figures 13.4 and 13.5) were created using Praat. Praat is also a downloadable software specifically intended for literary study; *praat* means "to speak" in Dutch and Afrikaans. Praat transforms an audio file into a sound-wave form and a spectrogram. Both can be further manipulated to display pitch, vocal intensity, formant, and sound pulses in different colours. Image II showcases formant in red—vocal frequency that determines the phonetic quality of vowel enunciation.[1] Image III uses only the spectrograms of each poem line. I like the brutalist aesthetic of the spectrogram, its monochromatic but highly affective visual rendering. The spectrogram seems architectural in a way that aligns well with the phrase "Hypermasculinity in a structure" and the institutional logic of the repeated word "governing." By angling the spectrograms vertically, they resemble a concrete construction or enclosure even more starkly.

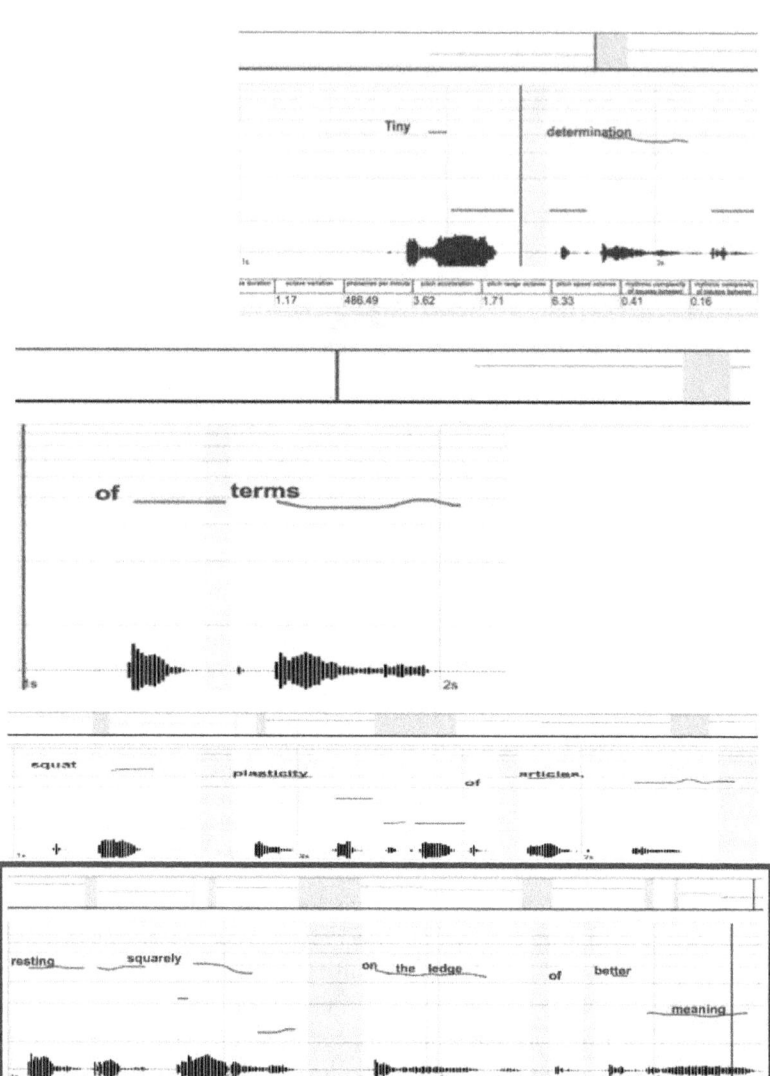

Figure 13.3. Image I: Audio file visualization made using Drift.

Source: Klara du Plessis.

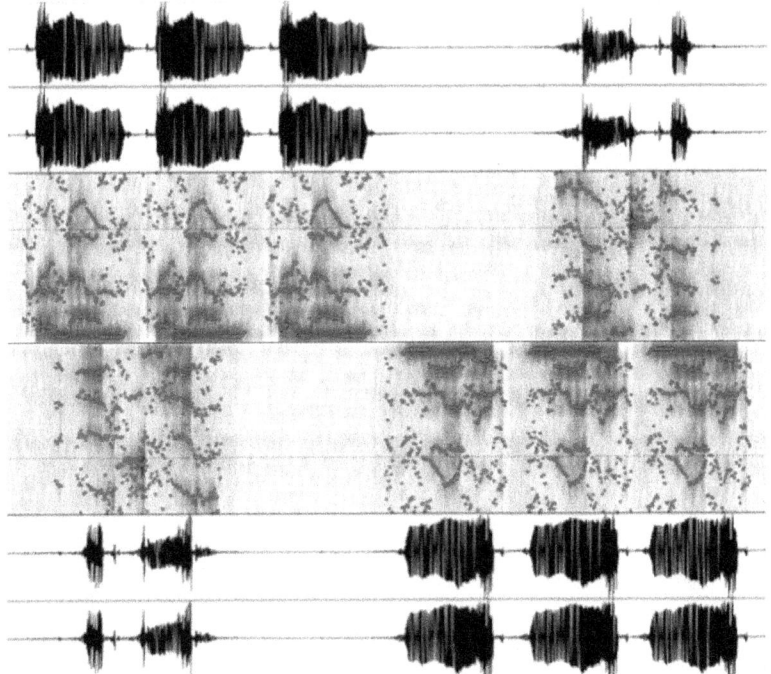

Figure 13.4. Image II: Audio file visualization made using Praat.

Source: Klara du Plessis.

Images IV and V (figures 13.6 and 13.7) were created using convert.ing.now. com. This is a basic app, which transforms audio into either a sound-wave form or a visually stunning, colourful spectrogram. Apart from being able to select the sound-wave colour palette, this app includes no further tools to highlight features of the sound wave. I have used this app before for images in my poetry collection *Hell Light Flesh* (du Plessis 2020) and appreciate the simplicity of its design; the sound-wave form has a succinct minimalism to it, which lends itself to being further crafted into a larger, often geometric, composition, as can be seen in Image IV. The rectangular border of Image IV formally mirrors "the soft wardrobe of being" of Poem 4, suggestive of lyrical interiority and a sense of self. Image V features both the spectrogram and sound-wave form of Poem 5. By far the boldest image, its bright-red and cutting spectral peaks signal negation,

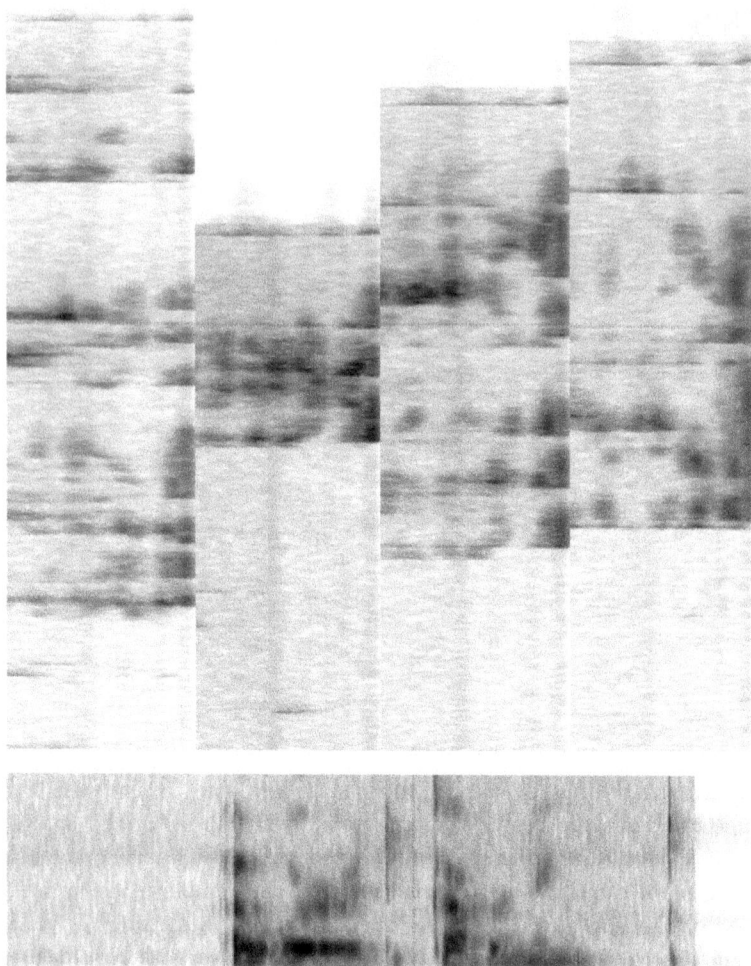

Figure 13.5. Image III: Audio file visualization made using Praat.

Source: Klara du Plessis.

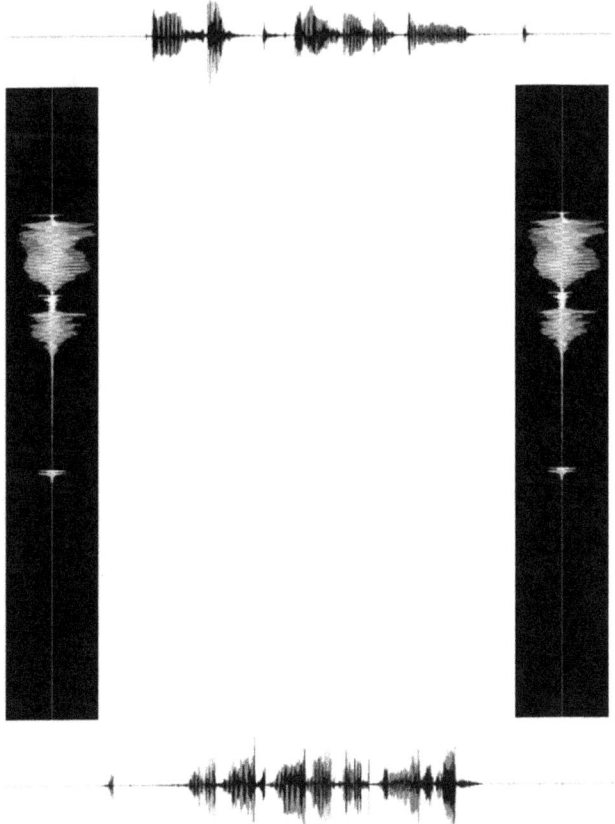

Figure 13.6.   Image IV: Audio file visualization made using convert.ing-now.com.

Source: Klara du Plessis.

danger, cessation. This final image visualizes an austere but audacious world devoid of words, yet highly evocative on a non-verbal, experiential plane.

The five images progress from the first readable graphic, which still includes words, through a transformation of horizontal graphics to vertical and construct-ive, generative constellations, to the final bold, unreadable but highly affective representation. While playfully showcasing different possibilities of sound-wave visualization and pushing these beyond their purely functional roles into the terrain of creative, interpretative media, the set of images also collectively

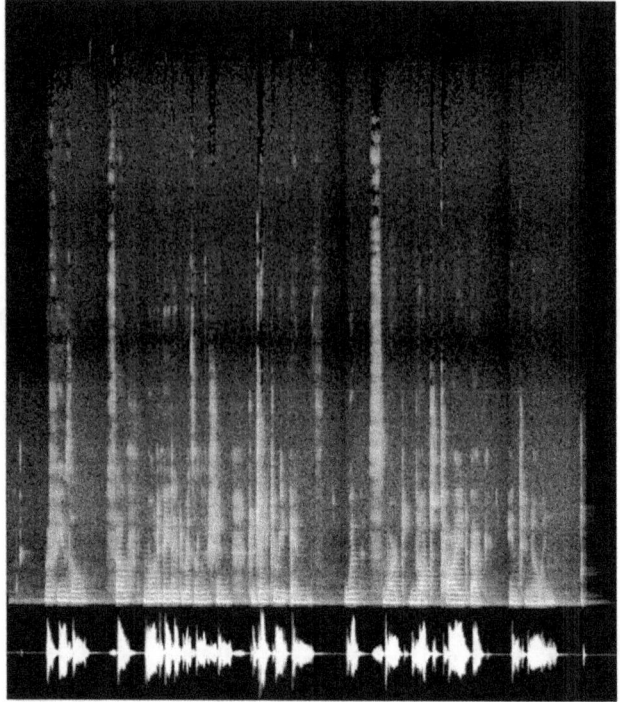

Figure 13.7. Image V: Audio file visualization made using convert.ing-now.com.
Source: Klara du Plessis.

maintains a consistently abstract, minimalist, and print-like aesthetic with a general greyscale and red colour scheme.

ART DATA

In her essay "Humanities Approaches to Graphical Display," Johanna Drucker (who, apart from her considerable scholarly bibliography on interpretative visualization, is also a visual artist) calls for an exploration of new, qualitative methods for visualizing research results. Her essay experiments with graphical information that would be more closely aligned with the exegetic nature of humanistic inquiry itself. She suggests, "humanistic methods are counter to the idea of reliably repeatable experiments or standard metrics that assume observer independent phenomena. By definition, a humanistic approach is centred in the experiential, subjective conditions of interpretation" (2011, 5–6). Although Drucker focuses on graphs, "Stop Words" expands her argument to include digital visualizations of poetic material, emphasizing the instability of a quantitative/qualitative divide—visualizations of both textual and audio materials are simultaneously a priori renditions and emotive, interpretative displays.

In other words, the set of five images that constitute "Stop Words" are simultaneously accurate graphical representations of the original set of five poems and evocative, creative readings of those texts. While recontextualized, reoriented, and transformed into independent artworks, the sound-wave visualizations themselves have not been warped, adjusted, or rendered quantitatively different from the sounding of the matching poems; if removed from the artworks, the sound-wave visualizations could, hypothetically, be reverse engineered back to the poems' words. In the faithfulness of their depiction of the poems, the visualizations *are* literature; each sound-wave form is a poetic line; each image is the stanza of a poem. At the same time, each image is a faithful, even scientific, research representation. As such, the visualizations encompass "the critical distance between the phenomenal world and its interpretation" (Drucker 2011, 1), straddling the extremes of realist representation and subjective rendering. Take the difference between figures 13.8 and 13.9, for example. While both sound-wave forms are precise and unaltered visualizations of poetic lines, the functional black sound wave on a white background signals a textbook graph,

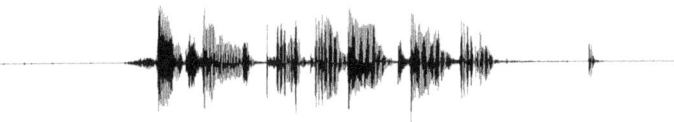

Figure 13.8. Sound-wave form of the second line from "Stop Words" Poem 4.
Source: Klara du Plessis.

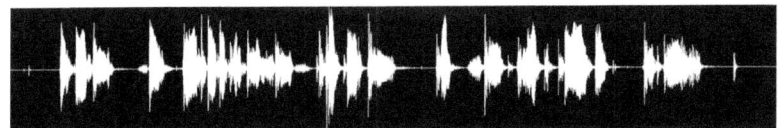

Figure 13.9. Sound wave form of "Stop Words" Poem 5.
Source: Klara du Plessis.

information that has been rationally represented and can be cognitively understood. In contrast, the bold white on sombre black sound wave opens to the world of indeterminate interpretation, expanding allusively to frigidity, snow, otherness, apocalypse....

In thinking through the schism between quantitative and qualitative graphics, Drucker generatively contrasts what she calls "data" and "capta." She explains: "Differences in the etymological roots of the terms data and capta make the distinction between constructivist and realists approaches clear. *Capta* is 'taken' actively while *data* is assumed to be a 'given' able to be recorded and observed" (2011, 2). Drucker underscores the kinetic nature of information over the passivity implied by its static and supposedly quantifiable rendition: "Humanistic inquiry acknowledges the situated, partial, and constitutive character of knowledge production, the recognition that knowledge is constructed, *taken*, not simply given as a natural representation of pre-existing fact" (2). There is thus an iterative quality to interpretation, which includes progression, flux, and an acceptance of change. If the assumption is that data can be reproduced endlessly to create the same result each time of research mapped out on a pair of axes, then the visualization of interpretative knowledge would be a shapeshifting mirage that modulates its appearance with the ebb and flow of thought. Capta represents "relational information" (6), taking and transforming increments of interpretation, grasping at the edge of an articulation, which constantly pushes and expands its own receding horizon line of creative mental activity.

"Stop Words" merges data and capta, a kind of creative art data, which offers a constant fluctuation of give and take. While "Stop Words" literally illustrates a poem according to measurable attributes like pitch, vocal intensity, and more, those same illustrations can also—in a counterintuitive gesture—take on divergent forms. Contrast figures 13.9 and 13.10, for example. Although both are accurate visualizations (sound-wave form and spectrogram) of Poem 5, of the same information, they are clearly radically different in shape, colouration, and formal construction, and equally distinct in terms of emotive intensity. Arguably, these different visualizations of textual material can be representative of the subjective nature of humanistic interpretation itself. Despite the reiterated fact of their positivist accuracy, the range of possibility of the sound-wave visualizations (even before their creative integration into the set of five images) models as a convex radiation of literary interpretations. Figure 13.9 mirrors the extra-long protracted line of Poem 5, emphasizing refusal and the strength inherent to saying *no*. Figure 13.10, in contrast, has a bruise-like, variegated quality focusing more keenly on the "whip smart" hurt of negation.

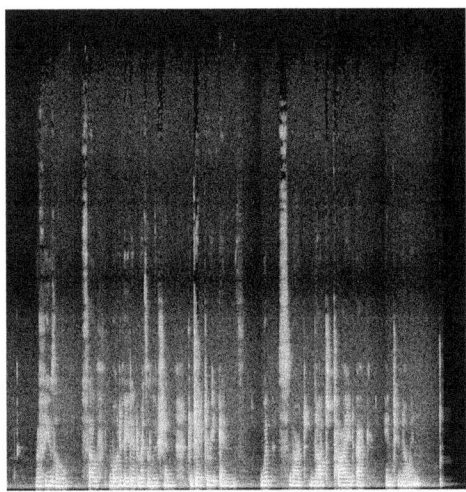

Figure 13.10. Bold spectrogram of "Stop Words" Poem 5.

Source: Klara du Plessis.

"Stop Words" gleans a technological toolkit from the digital humanities and orients it toward a creative, literary and fine-arts product. On the one hand, one might contend that the sets of software used in this process are just a re-navigation of other artistic media harnessed to experimental image creation; the apparent novelty of DH methodologies simply leads to the visual deforming of a text—been there, done that. On the other hand, however, "Stop Words" goes beyond a DH means of production: it stages a theoretical intervention, shifting the expectations from the statistical outcomes of, for example, distant reading and other clinical, computational approaches to an acceptance of indeterminate, visceral results. This is a merging of traditional humanities and DH research, a recalibration of scientific fixity to a variegated art data of new media. The novelty of the digital humanities in relation to past research methods is less relevant than the combination of digital humanities and past research methods. Together they can be applied to strengthen research methodologies and render them contemporary.

Interpretative readings of art data slip and slide along the subjective axes of more traditional poetic analysis. The sound-wave visualizations become formal iterations of the poems, equally scientifically determinate and vulnerable to differing exegetic stances as the form of a strophe typed out on a page. As digital translations of poems into the visual domain, these sound-wave forms and spectrograms conspire not to stop words but to endlessly protract the potentiality of those words' meaning.

NOTE

1. Full-colour versions of the images in this chapter are available in the open-access edition of the volume on the University of Ottawa Press website: https://press. uottawa.ca.

REFERENCES

Brown, Susan. 2011. "Don't Mind the Gap: Evolving Digital Modes of Scholarly Production across the Digital Humanities Divide." In *Retooling the Humanities: The Culture of Research in Canadian Universities*, edited by Daniel Coleman and Smaro Kamboureli, 203–31. Edmonton: University of Alberta Press.

Drucker, Johanna. 2011. "Humanities Approaches to Graphical Display." *Digital Humanities Quarterly* 5, no. 1. http://www.digitalhumanities.org/dhq/vol/5/1 /000091/000091.html.

du Plessis, Klara. 2020. *Hell Light Flesh*. Windsor, Ont.: Palimpsest Press.

NLTK Project. 2019. "Natural Language Toolkit." Documentation. https://www.nltk. org.

Schofield, Alexandra, Mans Magnusson, and David Mimno. 2017. "Pulling Out the Stops: Rethinking Stop Word Removal for Topic Models." *Association of Computational Linguistics Anthology* 2 (April): 432–36. https://doi.org/10.18653/ v1/E17-2069.

Part 3

Digital Canadian Archives

CHAPTER 14

Wages Due Both Then and Now

Pascale Dangoisse, Constance Crompton, and
Michelle Schwartz

C aring is work; consciousness-raising is work; fucking is work. Individually, people who do this work know, well, just how much work it is. It took feminist analysis to reveal the broad social repercussions of under-valuing that work. This chapter takes as its jumping-off point, as the opening line suggests, the Canadian Wages Due Lesbians song "Fucking is Work" (Agger et al. 1975). Wages Due Lesbians' (WDL) provocations, like the slogan "fucking is work," have a long and specific history; indeed gay liberationist movements were not developed *sui generis* but, instead, have a distinct lineage: in the case of WDL, through Italian Marxist feminism and the Wages for Housework cam-paign, which shaped the issues that Canadian lesbians could and did take up, shaping in turn the way that Canadian lesbian activism grew and the actions it espoused. This chapter will not, however, be devoted to detailed Marxist-feminist analysis of women's unpaid work under capitalism but rather to what Canadian digital humanities can learn from the concrete political action of WDL activ-ists. Wages Due Lesbians navigated the heteronormative underpinnings of the Marxist-feminist call for an equivalent to universal basic income for Canadian women, and harnessed a wide variety of media and outreach activities to share their analysis. We write from the perspective as researchers and workers on the Lesbian and Gay Liberation in Canada (LGLC) project: it is the perspectives of the activists that we study that have been transformative for us, and that, we expect, have much to offer in this volume's reflection on the way we work in the digital humanities in Canada.

The LGLC project started as a traditional humanities project. It is built on the aggregation and synthesis of archival material, augmented with further research, offered to readers online with digital affordances beyond those pos-sible in print. In some ways the project's challenges are purely technical: the LGLC extends traditional humanities chronology development into the realm of the digital humanities through the TEI-XML (Text Encoding Initiative-Extensible Markup Language) encoding that underpins it, and through the humanities-based investigation of graph databases and lightweight JavaScript development to turn private TEI-XML–based analysis into a public-history website, lglc.ca. But it has also been the testing ground for experiments in how to digitally represent personhood, social change, and identity shifts. It is these types of experiments that move the digital humanities away from purely technical pursuits of online publishing, database design, or corpus analysis. As is often the case in humanities scholarship, the research team has been inspired by the historical self-positioning of the activists we study, to consider and iteratively reconsider our own history-making practices.

In this chapter, we outline how the LGLC project came into being and how the political organizing tactics of the feminist organization WDL have shaped our own digital research and publication strategies. The work of the digital humanities is, in the final analysis, humanities work.

The digital humanities in Canada need to further develop tools and avenues to make our research empowering to others. While the LGLC project has been fairly traditional in its outreach, through the public history lglc.ca website and academic articles and conferences, we have also been researching the outreach methods of 1970s activists, as we start to experiment with film nights, Wikipedia edit-a-thons, volunteering, and other activities that do not count in the neo-liberal university, but that might extend the reach of the project in useful ways. This chapter offers the insights we have gathered from WDL, whose fonds are part of the University of Ottawa's Canadian Women's Movement Archives, and reflects on the way their activist principles can shape the outreach the LGLC team undertakes. Wages Due Lesbians' consciousness-raising work both within and outside of the mainstream feminist movement, and the battles that the group fought on behalf of lesbian mothers in Canada, has made us reflect on our labour practices within the LGLC project as well as within academia. Like WDL, the LGLC project is explicitly non-capitalist, while working inside a capitalist system. Like many DH projects, it relies on the labour conditions and practices built into the modern academy, many of which have a long tradition in the humanities, others of which are the product of more recent exigencies. In order to best draw out the provocations that the WDL publications encourage readers to take up, it is worth considering their genesis before turning to labour in Canadian DH projects in general and in the LGLC project in particular.

WAGES DUE

In 1967, the Canadian federal government established the Royal Commission on the Status of Women. Following the completion of the commission's report in 1970, the National Action Committee on the Status of Women was formed as an umbrella organization pushing for "the implementation of the recommendations of the Royal Commission on the Status of Women" (Scala, Montpetit, and Fortier 2005, 587). The decade that followed was a time of great momentum for the women's movements in Canada. By 1977, the Canadian Human Rights Act declared that no discrimination could be made on the basis of sex, race, or religion. Things were starting to move: activists and their supporters saw the time was ripe to push forward with demands that would ensure women's social equity. One of these groups was the Wages for Housework Committee, Canadian chapter, or WfH.

The International Wages for Housework Campaign (IWFHC) grew out of the International Feminist Collective, a Marxist-feminist women's organization based in Padua, Italy, in 1972. The IWFHC's political stance and motivations were a response to Mariarosa Dalla Costa and Selma James's pamphlet *The Power of Women and the Subversion of the Community* (1972), which identified a relationship between women's oppression and capitalism. As the capitalist approach to participation in society is based on a highly stratified method of measuring value and productivity through wages, the authors argued that unwaged women were therefore excluded from full participation in society (Dalla Costa and James 1972, 79–86; Rousseau 2015, 366). In the capitalist system, women who are relegated to housework cannot be considered "productive" as they do not "produce" something that has market value. Women's unpaid domestic labour is, however, critical to the working of capitalism since it supports and makes possible men's labour and prepares children to be future labourers. Further, Wages for Housework activists grasped, in real and practical terms, how capitalism creates power imbalances and struggles that reach into the confines of the bedroom and shape workers' sexual lives (Bezanson 2006). The activities of working-class women in the home (cleaning, nursing, educating, cooking, emotional support, sexual labour, etc.) are functional for capitalism: these activities are all performed to ensure reproduction and well-being in the current class structure, making it easier for working-class men to engage in working-class jobs. The IWFHC argued, therefore, that all women's labour should be valued as equal to men's labour.

The Toronto committee of Wages for Housework became an umbrella organization, supporting various subgroups that emerged organically to cater to more specific challenges faced by different groups of women. One of these subgroups was WDL. Formed in 1973 to "give women a choice outside of poverty or relationship to men" (Rousseau 2015, 366; WDL 1973), the IWFHC's Marxist-feminist approach served as the guiding light for WDL's organizational perspectives and core activism. WDL focused almost exclusively on the question of income, taking on projects such as organizing opposition to government cuts to Canadian family-allowance payments. Wages Due Lesbians was also instrumental in creating and developing the Lesbian Mothers Defense Fund, which ran from 1978 to 1987, and supported lesbian mothers financially, legally, and emotionally during child-custody battles.

Wages Due Lesbians was fairly inclusive for a group organized in the 1970s, when matters of class and race were largely neglected or ignored in many feminist organizations. Wages for Housework and WDL put social reproduction theory into action: in their view, that women were prohibited from obtaining an equitable wage because of the gendered construction of their role and labour was the link between all women, regardless of their initial wealth or lack thereof, their sexual orientation, ability, or ethnicity: "if women do not have money, they are not able to exist as good, neoliberal subjects. The focus on wages/wagelessness means that Wages Due was successful in finding common links between the struggles of diverse groups of women" (Rousseau 2015). Furthermore, Wages for Housework activists made bids to connect their movement to others; for example,

> we wanted to make these few comments on the attitude of revolt that is steadily spreading among children and youth, especially from the working class and particularly the Black people, because we believe this to be intimately connected with the explosion of the women's movement and something which the women's movement itself must take into account. (Dalla Costa and James 1972, 10)

Wages Due Lesbians members protested at the Supreme Court of Canada on July 13, 1977, and framed their motivations as those of solidarity across historically marginalized groups, albeit using terms that are objectionable today:

> One of the most violent punishment lesbian women face for stepping out of line is the loss of the custody of our children. Like prostitutes, welfare women, immigrants, disabled women, prisoners and mental patients— we have our children taken away every day. Fifty people, who knew that our fight is also theirs, joined Wages Due's picket. [...] They came from the Women's Counselling Referral and Education Centre, the Law Union of Ontario, the Community Homophile Association of Toronto, Prisoner's Rights and many other groups [...]. Among the speakers were: Florence

Sims of Black Women for Wages for Housework, Anne Walker of Wages Due Lesbians in London, England, and Judy Ramirez for the Immigrant Women's centre. (WDL 1977)

Wages Due Lesbians' focus on wage issues also affected their relationship with other organizations, straight women, other lesbians, and men. By directing their attention to the capitalist system, WDL was able to move away from making the movement about men and to move beyond the "simple" division of gender and labour; a task not easily achieved within the mainstream liberal feminist movement. Working in solidarity with other groups helped the organization gain respect outside of activist circles, as well as much-needed visibility and media coverage (including the *Toronto Sun*, the *Los Angeles Times*, *Éditions du Remue-Ménage*, and *Chatelaine*), at conferences, and next to larger or mainstream organizations (including Gay Alliance Toward Equality, Gays of Ottawa, and the National Gay Rights Coalition). Wages Due Lesbians members argued that their specific experience would help strengthen the feminist movement, as lesbians are the ultimate manifestation of dependence to men in a capitalist society:

Right now a lot of lesbians and other single women find themselves being forced to look for a man. Women who want to come out as lesbians can't afford to abandon what little security marriage offers. Why should we have to depend on a man? None of us, lesbians or straight, want to be pushed into a relationship because we can't afford to be on our own. (Raymond 1976)

Lesbianism thus became a means to challenge not only the way in which society has woven together womanhood and housework but also the heteronormative family structure as the only viable financial option for women under capitalism. Wages Due Lesbians' argument reframes lesbianism: no longer restricted to women loving women in an intimate or sexual sense, lesbianism becomes a political stance as well. Wages Due Lesbians members were also thinking about lesbians' place and space in the socio-political context of the time. Through open discussions at conferences or through writing to each other about their views, they attempted to bestow new power on lesbianism.[1] By using a socio-political perspective to analyze their relationship to men and labour, WDL was able to speak openly about previously taboo subjects such as sexuality, intimacy, and care.

Correspondence between members of WDL and other organizations documents the debate within the group as well as the personal identity struggles of group members. Lesbian women were trying to understand not just their "place" in society but also within the larger and "straight" Wages for Housework movement, as women in relationship to and with other women, and as lesbian women in relation to gay men. The WDL correspondence is intimate, showing signs of nuance, questioning, of making one's way. Women discussed coming out or staying closeted because they feared for their personal safety and feared workplace harassment and dismissal. Members spoke of women supporting other women, and women caring for other women. They discuss the inability to be free or caring in relationships predicated on financial dependence. Finally, women talked about coming out in a very broad sense. Coming out as a lesbian was not only about sexual orientation but encapsulated a sense of women taking care of other women, as only women could understand the experiences and needs of women. This type of coming out was a political, albeit occasionally transphobic, way of taking a stance against heteronormative relationships in which "a power [imbalance] precludes any possibility of affection and intimacy" (Dalla Costa and James 1972).

Wages Due Lesbians was not immune to divisions in the lesbian feminist movement in the face of expanding political approaches to oppression (WDL 1975a). Because WDL's struggle was about the power imbalances created through wages, their struggle was not against men per se, but rather against political and economic social stratification. In their view, the separatist actions taken by more radical lesbian groups meant increased isolation from society, and that is what, according to WDL, was functional for the state:

> We, by our previous discussions and our paper, brought an understanding to the rest of the women in the network. And a power because we could point out our powerlessness as lesbian women and how the state has used our differences as lesbians to divide women from each other and to keep us and other women powerless. (WDL 1976–1977)

According to WDL, the state benefits from the power imbalances created by capitalism and supports the segregation and isolation of women who fail to conform to the heteronormative division of labour or relationships that buttress the system. The members of WDL were not going to promote further isolation; they wanted to be part of society, accepted for who they were, and paid for their labour. In contrast to their separatist contemporaries, WDL continued on with their overarching perspective on collaboration and solidarity across marginalized groups.

Wages Due Lesbian's activities encompassed traditional events such as rallies and protests, pub and café meetings, and discussions and workshops, but its members also used more innovative ways to disseminate their message. As some of Canada's first out lesbians, WDL organized door-to-door activities, dances, and conferences. Moreover, they took advantage of every media channel to push their mandate forward: petitions, publications, newsletters, books, pamphlets, posters, mainstream newspaper and magazine interviews, and, most notably, music. Indeed, the WDL group perhaps achieved the most recognition for their song "Fucking is Work," which complemented the paper of the same title. The paper articulates two ideas that defined WDL: "Wages for Housework recognizes that doing cleaning, raising children, taking care of men is not women's biological destiny. Lesbianism recognizes that heterosexual love and marriage is [sic] not women's biological destiny. Both are definitions of women's roles by the state and for the advantage of the state" (WDL 1975b).[2]

From a twenty-first-century vantage point, what is most striking about this opening passage is its refusal of biological arguments. This refusal not only signals a central difference between radical and socialist feminist thought (where radical feminism advocated separatism in response to their perception of men's innately oppressive and violent behaviour), it also reveals the difference between WDL's approach and the "born that way" approach to contemporary gay-rights claims. For WDL, lesbianism can be a choice, and one that is no less worthy of social and economic standing for being a choice. To promote this point of view, "Fucking is Work" was unapologetic about using "vulgar" language to help gain visibility. This subversive work on sexual labour was central to demonstrating how the capitalist system reaches into the most private sphere of the home— the bedroom.

Though based out of Toronto and London, Ontario, WDL carried out activities elsewhere in Canada. One far-reaching initiative was CORA, the Women's Liberation Bookmobile (1975). CORA, a school bus converted into a travelling library, was named for suffragist E. Cora Hind. Judith Quinlan and Ellen Woodworth, both members of Wages Due Lesbians, quit their jobs in order to drive the bus, filled with feminist books about the history of women in Canada, birth control, daycare, and food pricing, through small towns in southern and central Ontario, where such books were hard to access (Hurst 1974, E6). WDL also published and distributed newsletters and pamphlets that had a larger reach than their home base of Toronto and London; for example, in Ontario the WDL material was distributed in Bolton, Trenton, and Kingston. In this way, they were able to connect with closeted or isolated women that were scared or alone. Many women reached out in letters to WDL, often with gratitude, and sometimes, sadly, to be removed from a mailing list for fear of being outed.

Before it disbanded, in 1984, WDL members worked to dismantle many divisive barriers through their continued efforts: geographical (urban/rural), political (Marxism in the capitalist system), economic (class), gendered (straight/ lesbian), ethnic (racialization). Before Kimberlé Crenshaw (1989) coined the term "intersectionality," WDL understood and addressed the complexities of the many intersecting experiences of women. Wages Due Lesbians, through their astonishing number of publications and events, made it possible for women of most walks of life to imagine a life without men, without the consistent oppression of the heteronormative capitalist system.

It is through lived experiences that WDL members understood the world and how to change it. The deep thinking about their unique situation was done behind the scenes, among themselves, in letters and at conferences, while the public-facing WDL was affirmative, strong, dynamic, and innovative. They found new channels of communication to make their voices heard (through comedic acts, songs, and bookmobiles). They skillfully gained support from other organizations. They took part in the mainstream activities of social movements but they also worked to represent people with multiple intersecting marginalized identities to give voice not only to themselves but to other underrepresented people.

LESBIAN AND GAY LIBERATION IN CANADA UNDER THE HOOD

Wages Due Lesbians were innovative, using traditional and non-traditional methods to promote their socio-political analysis and to encourage consciousness-raising. The LGLC project has reached an important milestone, and needs to innovate too. The team has completed the remediation of two print chronologies and is now faced by the need to plan next steps, including outreach and change through traditional and non-traditional means. Even if the team will not immediately turn to provocative song writing to help bolster the message of the LGLC— that gay liberation was not a solely urban movement and that gay liberation has a specific politic that is in danger of being forgotten in the current right-of-centre political and cultural moment—the team needs to look to its own subject matter to both find gaps and silences in the historical record and to find innovative ways to ensure it is not forgotten.

At the centre of the project is *Lesbian and Gay Liberation in Canada: A Selected Annotated Chronology* by our collaborator Donald McLeod (1996, 2016; see figure 14.1), the head of book and serials acquisitions at the University of Toronto

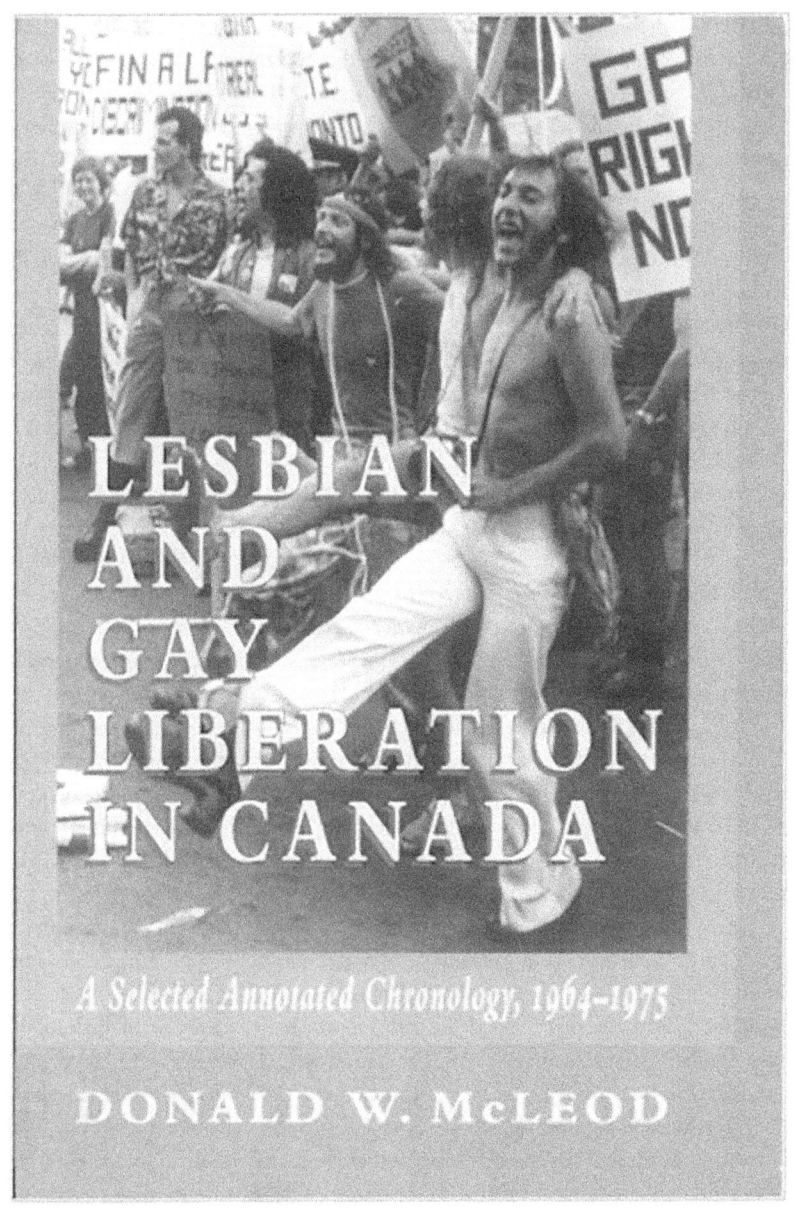

Figure 14.1. *Lesbian and Gay Liberation in Canada.*
Source: Donald McLeod.

Libraries. The book's two volumes span from 1964 to 1981 and consist of 3,100 events that cover everything from book launches to bar closings, including rallies, raids, letter-writing campaigns, and more. McLeod created these chronological entries by methodically reading through periodicals and other archival material, noting events captured in articles, advertisements, and letters to the editor. Many of the entries include more than one citation—in addition to reading through the gay-liberation press, McLeod turned to mainstream newspapers and periodicals to record how events were covered, and indeed occasionally incited by, the mainstream press.

The DH contribution to the project does not just hinge on the digitization of those 3,100 events (though that particular facet of the scholarship is covered in this section) but in creating ancillary material that enriches the text. We have collected supplementary information on the periodicals, people, places, and organizations in the text to better show the interconnectedness of the gay-liberation movement in Canada. As a result, the database that underpins the project houses an additional 31,000 records, and is now accompanied by a prosopography, or collective biography, which is online in beta form at prosopography.lglc.ca, which houses further biographical information for over 3,000 people involved in the Canadian gay-liberation movement as text and RDF (Resource Description Framework), the format of the Semantic Web (a machine-readable, interoperable version of the Web).

The LGLC is a DH project built using humanities methods and workflows, undertaken by the project principal investigators and, at the time of writing, 16 research assistants.[2] At the heart of the LGLC's digital work is plain text encoded in TEI-XML, the language of the Text Encoding Initiative. This format is easier to archive than a database or complex front-end, making TEI-XML an ideal format for long-term preservation. We have conducted a number of analyses on the resulting TEI-encoded corpus, looking for patterns and trends in the movement, working to understand the dynamics of gender, geography, mobility, and more in the development of gay liberation in Canada.

The original version of the lglc.ca web app was built very much on the humanities model: a number of research assistants working closely with the principal investigators, whose role it was to teach the research assistants the skills they would need for the project. We worked through lessons on XSLT, Cypher, Bootstrap.js, node.js, virtual machine architecture, and others to create a searchable graph database out the TEI-XML; to put a node.js front-end on the database; and to serve it to the Web.

Now that we have encoded the original chronologies and published them online as a searchable database, we need to move beyond creating ancillary material to adding new events to the chronology by seeking out events activists not covered in the original text. While not exactly uneven, the LGLC material reflects its sources: the events have been drawn principally from English-language periodicals and ephemera collected by the ArQuives (formerly Canadian Lesbian and Gay Archives), the Archives gaies du Québec, the BC Gay and Lesbian Archives, the Canadian Women's Movement Archives, the Glenbow Museum's archives, the Toronto Reference Library, Robarts Library at the University of Toronto, and beyond. The original LGLC events skewed toward both anglophone activities and toward those recorded in readily archivable material, which tend to bear witness to events featuring urban white men: the opening of men's bars, bathhouse raids, action by elected officials.

To attempt to rectify this first shortcoming, the bias toward anglophone content, the team has engaged with more francophone primary source material than was possible in *Lesbian and Gay Liberation in Canada*. The second shortcoming, the project's focus on men, has been a bit harder to address, and is what has garnered our interest in WDL.

It takes a tremendous amount of labour to work through the periodicals and other archived material (to say nothing of the acquisition, deaccession, description, and maintenance work of the archivists themselves) that has informed the project. And yet, we have discovered, archival research is also not alone sufficient for this recovery work. The men involved in the gay-liberation movement tended to be wealthier than women, with more print media, services, and spaces available to them. As a result, they have left more of a trace in the archival record. We are thus left with the challenge of recovering information about events beyond the protests, bar openings, raids, elections, and other events that end up in newspapers and periodicals, to record information about the consciousness-raising groups, dry dances, bookmobile tours, hotlines, and other lesbian-organized interventions and events.

Our recovery work is twofold. A part of the LGLC team is dedicated to reviewing oral testimony and ephemera by Canadian lesbians to capture information about the consciousness-raising meetings, potlucks, conferences, and women's dances that were not publicized in the gay or straight press.[3] While the *Toronto Sun* and *Chatelaine* may have led us to WDL, it is our attempts to use organizational records held in the archives to get at events that are more ineffable than those that turn up in our usual periodical sources that have given the LGLC team a more complete understanding of the history of WDL, their political tactics, and their effect on gay organizing in Canada.

THINKING THROUGH WAGES DUE

Research into WDL's innovative and generous approach to activism has changed the way LGLC team members work, write, and teach. Wages Due Lesbians was explicitly non-separatist in its approach, working to change Canadian society and legal structures through alliances and strategic collaboration in cities and less urban environments, rather than encouraging women to find separate space on womyn's lands in western Canada and the United States as many radical groups did. While its roots were Marxist, WDL recognized the intersectional nature of women's oppression, better perhaps even than its parent organization, Wages for Housework. Like the WDL's pamphlets, bookmobile, and local events, our work in the digital humanities needs to implement outreach strategies beyond traditional academic publishing. We need to develop practical ways of using humanities research to empower people, whether that happens on the Internet, at local events, or in classrooms. In the spirit of "fucking is work," one team member dedicates multiple hours of the class she teaches to discussing the link between capitalism, neo-liberalism, sex, labour, and the private sphere. She also volunteers at her local elementary school to teach children about representation. The encoding process itself also becomes a space for subversive feminist and political practices (Schilperoort 2015; Taylor 2013). Great effort was placed on naming authors of documents to add the silenced voices of the marginalized women who were either active in WDL or connected to it. If the name of the authors could not be found, or if the author of the document wished to be anonymous, the document was still integrated into the project by using a specific attribute. Sometimes, organization name was used as author of the document. The naming practices are an important political tool for the project as it can show how numerous lives were touched by WDL activism. A particular difficulty arose when trying to capture precise dates of documents or events. As "feminist markup can be understood as a spectrum instead of in terms of a concrete definition" (Schilperoort 2015), rather than dismissing the events that did not conform to the encoding standards the LGLC project allows for uncertainty in date markup. Further, as the lesbian battle for equality somewhat extended beyond the original LGLC project timeline, events from after 1981 related to activism that started before 1981 are now part of the database, and the team is working on the expansion of the project's range to 1985. Using feminist markup practices as a political tool, and understanding how encoding can be understood as a spectrum, helped fuse the subversive and political stance of WDL with DH practices. As Taylor (2013) points out, "by building digital collections, humanities scholars develop techniques for making explicit the structure and semantics of texts; make information available to be used for research, education, and personal enrichment; and enable users to interact with information in dynamic ways" (181).

The LGLC project is just that: it has had many outputs, including journal articles, book chapters, conference papers, and Wikipedia edit-a-thons, but the furthest reaching—and from a DH perspective, the most exciting—has been the development of lglc.ca, the public-history website that allows readers to navigate the events recorded in lesbian and gay liberation in Canada. We are currently enriching it with contextual essays and information about people, places, and organizations to allow readers to understand the movement and its contexts without relying solely on chronological organization. We plan to add lesson plans mapped to provincial curriculum. We will also continue to convert the project's underlying TEI-XML into linked data for better integration of events our team has uncovered in other history-based digital projects.

While DH has been criticized as a tool of the neo-liberalization of the university, there is compelling evidence that these critiques rely on cherry-picking from outside the field (Grusin 2014). DH scholarship, particularly in Canada, is undertaken primarily by humanities scholars, and, occasionally in collaboration with computer-science and social-science scholars. While the digital skills that students in DH courses may develop are often marketed to students, parents, and others in a neo-liberal way, actual DH teaching and research rely on humanities approaches to technology (Sayers 2011, 279–300). As Lefebvre (1971) points out, capitalism only continues to function by exploiting imbalances in power and colonizing new non-capitalist spaces, not just geographically but also in terms of time and demographics too (consider, for example, how much productive labour we engage in on social media or how many Canadian cities have 24-hour grocery stores. Our leisure time and the night are ripe for colonization by capital, as is the university). The university is one such space. The way we construct and share our research findings and creation can, in a small measure, be a way to push back against the capitalist colonization of the university.

In the final analysis, the group's work exacted a heavy toll from the organization and its core members. As the organizers struggled to make ends meet, they still invested their time and money in the cause, leaving little to sustain themselves. And as the organizers worked tirelessly to make their case heard, the organizations' processes, logistics, and planning suffered a great deal. Wages Due Lesbians did not have a succession plan—its members worked unceasingly on women's immediate problems. After a dozen years of work that exhausted members emotionally and financially, WDL disbanded. Unless their theory, history, and practice are shared, their work will be left to be redone in every generation. Digital humanities can be at the forefront of sharing this historical knowledge beyond traditional academic venues to keep each generation from having to fight the same fight and do the same consciousness-raising over again.

The online work of LGLC project is in part designed to help prevent the erosion of historical memory that undermines ongoing cultural critique. This work takes place in a university context. Universities have, like unions, religious groups, and other organizations, been a counterweight to capitalist organizations: the goal of the university is to create new knowledge for the public good, rather than to maximize profits. While parts of this mission may be eroded by lack of funds and other supports, digital humanities can and should be part of that counterweight.

NOTES

1. The material traces of these organizing efforts may be viewed online at the University of Ottawa's Canadian Women's Movement Archives https://biblio.uottawa.ca/en/archives-and-special-collections/womens-archives and the Rise Up: Digital Archive of Feminist Activism https://riseupfeministarchive.ca/.

2. The LGLC project has had the good fortune to be funded by the Social Science and Humanities Council of Canada. This has not only allowed us to train students in humanities and DH research methods and practices, but also to develop ways to share the knowledge we've developed with the Canadian public. These students—Caitlin Voth, Jessica Bonney, Sarah Lane, Raymon Sandhu, Anderson Tuguinay, Travis White, Stefanie Martin, Cole Mash, Seamus Riordan-Short, Rebecca Desjarlais, Nadine Boulay, Pascale Dangoisse, Alice Defours, Candice Lipski, Oxana Pilenko, and Ewan Matthews—have been invaluable to the project's success. The LGLC's public-history websites, lglc.ca and prosopography.lglc.ca, are the fruit of the team's effort to make the project publicly accessible. The work was done by the project team, with advice from John Simpson at Compute Canada and from the University of British Columbia's research computing expert, Wade Kavier. While this was in keeping with humanities ways of working and training, it posed security risks to the website, as none of the team members had the systems administration and app-security knowledge to keep the web app secure and uncorrupted. We invested in a professional redesign of the app in 2018, and the site is now hosted by Toronto Metropolitan University Library, with support from the Centre for Digital Humanities at Toronto Metropolitan University and the Humanities Data Lab at the University of Ottawa.

3. We will not address our research in this area in depth here, but we recommend that readers interested in women's oral testimony turn to the Archives of Lesbian Oral Testimony, created by our collaborator on the LGLC project, Elise Chenier: https://alotarchives.org/.

REFERENCES

Agger, Ellen, Lorna Boschman, Judith Quinlan, Boo Watson, and Ellen Woodsworth. 1975. "Fucking is Work." Wages Due Lesbians Fonds. 10-027-S1-F4. University of Ottawa Archives and Special Collections.

Bezanson, Kate. 2006. "Gender and the Limits of Social Capital." *Canadian Review of Sociology* 43, no. 4: 427–43. https://doi.org/10.1111/j.1755-618X.2006.tb01142.x.

Crenshaw, Kimberlé. 1989. "Demarginalizing the Intersection of Race and Sex: A Black Feminist Critique of Antidiscrimination Doctrine, Feminist Theory and Antiracist Politics." *University of Chicago Legal Forum* 1: 139–67.

Dalla Costa, Mariarosa, and Selma James. 2017. "The Power of Women and the Subversion of the Community." 1972. In *Class: The Anthology*, edited by Stanley Aronowitz and Michael J. Roberts, 79–86. Hoboken, N.J.: Wiley-Blackwell. https://doi.org/10.1002/9781119395485.ch7.

Grusin, Richard. 2014. "The Dark Side of Digital Humanities: Dispatches from Two Recent MLA Conventions." *differences: A Journal of Feminist Cultural Studies* 25, no. 1: 79–92. https://doi.org/10.1215/10407391-2420009.

Hurst, Lynda. 1974. "Bookmobile Takes Word to Small-Town Feminists." *Toronto Star*, May 30, 1974, E6.

Lefebvre, Henri. 1971. *Everyday Life in the Modern World*. London, U.K.: Penguin.

McLeod, Donald W. 1996. *Lesbian and Gay Liberation in Canada: A Selected Annotated Chronology, 1964–1975*. Toronto, Ont.: ECW Press/Homewood Books.

———. 2016. *Lesbian and Gay Liberation in Canada: A Selected Annotated Chronology, 1976–1981*. Toronto, Ont.: ECW Press/Homewood Books.

Raymond, V. 1976. "Money Needed: Local Lesbians Told 'Coming Out' Risky." *Ottawa Citizen*, June 30, 1976.

Rousseau, Christina. 2015. "Wages Due Lesbians: Visibility and Feminist Organizing in 1970s Canada." *Gender, Work & Organization* 22, no. 4: 364–74. https://doi.org/10.1111/gwao.12092.

Sayers, Jentery. 2011. "Tinker-Centric Pedagogy in Literature and Language Classrooms." In *Collaborative Approaches to the Digital in English Studies*, edited by Laura McGrath, 279–300. Logan: Computers and Composition Digital Press/Utah State University Press.

Scala, Francesca, Éric Montpetit, and Isabelle Fortier. 2005. "The NAC's Organizational Practices and the Politics of Assisted Reproductive Technologies in Canada."

In *Canadian Journal of Political Science* 38, no. 3: 581–604. https://doi.org/10.1017/S0008423905003574.

Schilperoort, Hannah M. 2015. "Feminist Markup and Meaningful Text Analysis in Digital Literary Archives." *Library Philosophy and Practice*, 1228. https://digitalcommons.unl.edu/libphilprac/1228/.

Taylor, Susan Gail 2013. "Collaborative Approaches to the Digital in English Studies." *Computers and Composition* 30, no. 3: 180–82. https://doi.org/10.1016/j.compcom.2013.06.003.

WDL (Wages Due Lesbians). 1973. "Wages Due Lesbians—Circular Letter." Wages Due Lesbians Fonds. 10-027-S1-F2. University of Ottawa Archives and Special Collections.

———.1975a. "For Betty." Wages Due Lesbians Fonds. 10-027-S1-F3. University of Ottawa Archives and Special Collections.

———. 1975b. "Fucking is Work." *The Activist* 15, no. 2: n.p.

———. 1976–1977. "Debate on Wages Due Being a Lesbian Group." Wages Due Lesbians Fonds. 10-027-S1-F3. University of Ottawa Archives and Special Collections.

———. 1977. "Lesbians on the Move." *Wages for Housework: Campaign Bulletin* 2, no. 1 (Fall): 2. Wages Due Lesbians Fonds. 10-027-S1-F9. University of Ottawa Archives and Special Collections.

Women's Liberation Bookmobile. 1975. "The Adventures of CORA the Bookmobile." The Women's Liberation Bookmobile Fonds. Fonds 10-015. University of Ottawa Archives and Special Collections.

CHAPTER 15

Analog Thrills, Digital Spills:
On the Fred Wah Digital Archive Version 2.0

Deanna Fong and Ryan Fitzpatrick

Since the 1960s, poet Fred Wah has been central to a series of important moments and conversations in North American poetry and Canadian literature. There is a wide arc from Fred Wah, University of British Columbia student, to Fred Wah, Parliamentary Poet Laureate. As a younger poet, Wah was involved with the seminal poetry newsletter TISH (and other little magazines) and in cross-border dialogues with key American poets (Charles Olson, Robert Creeley, Allen Ginsberg, Denise Levertov, etc.). Over the course of his career, living in Vancouver, Buffalo, Nelson, and Calgary, Wah's ongoing work has intersected with and shaped conversations around racialization and hybridity, ecology and ecopoetics, and the relationship between politics and poetic form. In the introduction to Wah's collected works, *Scree* (2015), editor Jeff Derksen gestures to the political potential of these community affiliations, suggesting that "these coalitions—which I see as a joining of poetic communities with the urgency of a social question of a political moment—are assertions of a social and aesthetic grouping" (Derksen 2015, 11). Derksen makes it clear that Wah's work is relational and coalitional—that it works with and responds to social contexts across geographical scales.

With our work on the Fred Wah Digital Archive (fredwah.ca), we have tried to offer a glimpse into the sliding social contexts of Wah's work in some way. The archive collects print work by and about Wah: his books, chapbooks, journal publications, as well as his editorial work, and secondary sources about his writing. It anticipates ongoing plans to include audio, video, and unpublished archival material. It aims for something comprehensive, even as it acknowledges the impossibility of any total view of Wah's work. Wah's poetic oeuvre is marked by its attention to public discourse, its interest in the improvisatory potentials of poetic language, and its devotion to the analog thrills of small-press publishing. The work of the Wah archive, then, is invested in a combination of public discourse and circulation: the ways his work has been historically performed and received, and the networks and communities Wah works within and responds to. In this chapter, we discuss the strategies we've employed to illuminate the relational and coalitional valences of Wah's work within the material substrate of the archive.

Like Wah's work, the archive itself has a similarly shifting context, passing between hands and between institutional homes. While we think through our relationally driven redesign of the archive, in this chapter we also ask how our reflections on the sociality of Wah's poetics have shaped our simultaneous reflections on our own shared labour on his archive—a shared relationality that is continuous with the coalitional nature Derksen reads in Wah's oeuvre. In particular, we want to grapple with the messy historical and institutional pressures our archival work contends with, asking how affect and sociality inflect issues of sustainability, credit, and labour that are key in the digital humanities. In our work on Wah, we've had to contend with the hot and cold realities of care in the neo-liberal university: can we get enough people to *care* about the work of one poet, however canonical, to sustain a project even as we find ourselves only temporarily attached to it, no matter how much *we* care, because of our own precarious positions within the resource-granting university?

1.0 TO 2.0, OR IN ADVANCE OF THE NEXT ITERATION

Our version of the Wah archive is entangled with another attempt to catalogue and map Wah's work, building on previous work by Susan Rudy—both her bibliography (in manuscript form) and an earlier version of the website. The Fred Wah Digital Archive emerges from a print manuscript compiled by Susan Rudy in 2009 that she translated into a digital archive. Styled after Roy Miki's *A Record of Writing: An Annotated and Illustrated Bibliography of George Bowering* (1990), Rudy's unpublished manuscript "'Loose Change': Fred Wah, A Life in Writing" compiles Wah's work up to 2009.

Her descriptive bibliography is print-centred, though it includes entries for a number of audio- and videotapes. She collects and categorizes Wah's work by publication type—a section for books, another for work in anthologies, another for work in periodicals—with a short section that collects his editorial work (including the numerous small magazines he worked on) and another for critical and academic work about Wah and his work. Even in its analog form, the bibliography shows the limitations of slotting the messy social activities of writing and publishing into discrete categories. For example, how to describe a handwritten Christmas card, addressed to a friend, which contains an early draft of a poem that appears in a subsequent book of poems? In which section does it belong—correspondence, manuscripts, or occasional works? In addition to the vagaries of bibliographical classification, such artifacts are subject to much looser rules of archival description than books or journals, meaning that more training in archival protocols and rigorous process documentation are necessary. Rudy's bibliography is invested in collecting not only Wah's book-length works but also the social substrate that precedes published versions—that is, the material history of a text as it moves from artifact to artifact, evolving as it comes into contact with other actors and texts. It is interested in tracing variance, revision, and dialogue with other writers. In a similar vein, an appendix to the bibliography (compiled by Calgary poet-academic Derek Beaulieu) lists the titles of Wah's poems, indexing the various publications in which they appear across time. As one might imagine, this list quickly overflows its container, as bibliography entries pile up and as title variants proliferate.

Seeing the potential in DH, in 2009 Rudy turned the ship of her paper bibliography into the first iteration of the Fred Wah Digital Archive with the help of Darren Wershler, Bill Kennedy, and a small group of graduate students at both the University of Calgary and York University under the banner of the research project Artmob. The site was originally developed on Drupal 5 (an open-source content-management system); federal funding allowed the team to create a custom interface for the archive's multimedia content. Shifting the bibliographic information to a digital working environment where users could access, organize, and interact with full-text scans of Wah's work, the interface largely transferred the bibliography's concerns, absorbing both its thoughtful categories and hard limitations. Searching through it, you could access the bibliographic information for each entry, including a scan that reproduced Wah's work. We can catch a glimpse of the archive's structure via the Internet Archive's Wayback Machine (see figure 15.1). Rudy's introduction places the site between the history of the bibliography and the present tense of the notice board, both aiming to be a complete repository of Wah's work while also providing news about "what Fred Wah is doing right now" (Artmob 2012).

Clicking through, the site traces the contours of Rudy's bibliography with its interest in sociality. For example, an entry for an untitled poem describes it as "one of seven poems in *Open Letter* 12.2 (2004) selected by Louis Cabri for a reading at the Alley Alley Home Free conference." This poem is one entry out of seven that comprises item D178 in the bibliography, which collects the "Alley Alley Home Free" special issue of *Open Letter*.[1] We're reminded that the poem is from Wah's *Pictograms from the Interior of BC* (1975) and was subsequently collected in his selected poems *Loki Is Buried at Smoky Creek* (1980). We're given dates and a Modern Language Association (MLA) citation (see figure 15.2).

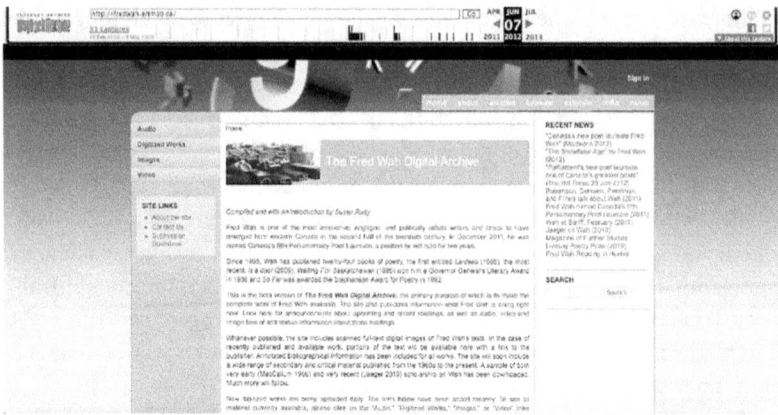

Figure 15.1.   Homepage of the Fred Wah Digital Archive, version 1.0.

Source: Fred Wah Digital Archive 1.0, The Wayback Machine.

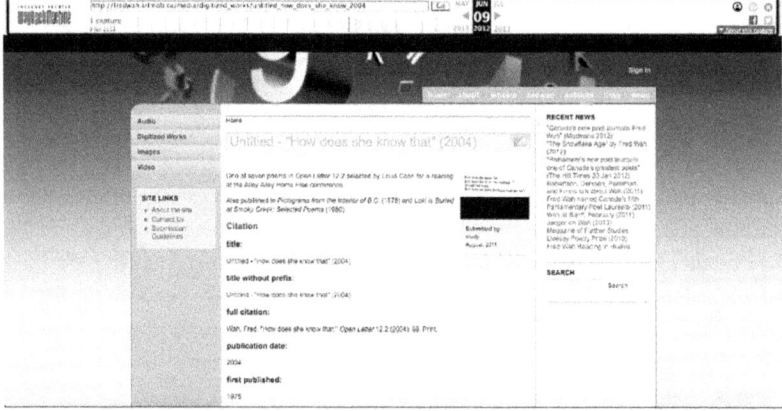

Figure 15.2.   Bibliographical entry on the Fred Wah Digital Archive, version 1.0.

Source: Fred Wah Digital Archive 1.0, The Wayback Machine.

There are a lot of questions that a scholar could leap from or into occasioned by this entry, pushed into readings of older books or the comparatively recent relations of "Alley Alley Home Free"'s Festschrift, both on the page in the form of the special issue (if you can track down a copy in the stacks) or off (if you can chat down the folks involved or watch the filmed conference panels). Indeed, our archival work, described in detail below, intends to point user-scholars toward such information by making the connections between artifacts, events, and people visible.

The early site held massive potential, we think, to open up critical discussions of Wah's work and poetics more generally. But when the grant ended and investigators moved on to other projects, the site wasn't migrated to successive versions of Drupal and the content went dark. In the summer of 2014, we began working on the second iteration of the Fred Wah Digital Archive. The new version was led by graduate students: the two of us, at different stages of our doctoral degrees in English at Simon Fraser University, and our colleague Janey Dodd, who was transitioning from her master's at SFU to her PhD at the University of British Columbia (UBC). A network began to assemble around the project as Fong approached both the SFU library, through contemporary literature collections librarian Tony Power, and Derksen, the graduate chair in English (and a former student of Wah's) for technical and financial support. The network expanded into the library, the English Department, and the wider professional networks around the digital humanities in Canada. In the library, the project was supported by labour and input from copyright officer Don Taylor, digital projects librarian Rebecca Dowson, archivists Nailisa Tanner and Melanie Hardbattle, systems librarians Mark Jordan and Janice Banser, digital initiatives librarian Ian Song, and the acting dean of library services, Brian Owen; in the English Department, Professors Clint Burnham and Christine Kim; and in the larger DH community, Professor Susan Brown of the Canadian Writing Research Collaboratory, poet-critic Erín Moure, Professor Karis Shearer at UBC Okanagan, geographers Nick Hedley and Andre Iwanchuk, and Bill Kennedy (who worked alongside Wershler on the design for the earlier version of the site).

The materials from the original site came to us as both a ZIP file containing PDFs and images sent by Kennedy and two large boxes from Rudy containing her archival records of both the bibliography and the site. Faced with the messy contents of the original iteration of the site and archive alongside, it became clear that neither source was a mirror of the other. Each contained an incomplete set of files when compared with the paper bibliography, which was also incomplete when compared with Wah's complete oeuvre. We grappled with unwieldy, overlapping, incomplete, redundant, and/or sometimes opaque information, not only in the metadata that accompanied the files but also within the artifacts themselves, drawn from Rudy's archive and SFU Special Collections and Rare Books. We found many items that weren't accounted for and had trouble finding items that were. We found versions of artifacts that matched the content described in the bibliography, but not the physical description (different sizes, covers, page counts—all a product of hands-on, small-run print production). We puzzled over how to insert new archival records into a fixed alphanumerical index (A1, A2, A3, etc.) organized chronologically—should we preserve the order of the original bibliography by adding addenda (A1a, A1b, etc.) or should we start anew? All these questions prompted a deep investigation into our investments in, and desires for, the archive, asking us to consider what information we wanted to make salient, and where we would have to compromise for the sake of clarity and useability.

<h2 style="text-align:center">TRACKING PUBLICNESS</h2>

While we were salvaging the files from the original site, we had the opportunity to reconsider the concept and design of the archive while also trying to remain faithful to the labour of Rudy, Wershler, Kennedy, and their teams. Key to our reconsiderations was an interest in the social contexts and conditions of Wah's poetry, especially as our archival work began to entangle us within that context— this work another entry in the bibliography. Negotiating with Rudy's bibliography, focused on individual publications as points of engagement, we began to wonder about the social connections only visible when you turned the archival substrate just right: could we read Wah's network of publishers, collaborators, and co-conspirators as a text in its own right? How would the legibility of this network inflect the reading and reception of Wah's creative work? Given that sociality and place feature so prominently in his writing, how could the archive, as a digital tool, do the editorial heavy lifting of an annotated edition while opening opportunities for serendipitous discovery and casual browsing?

Because Wah's work famously sprouts out of the coterie contexts of 1960s Vancouver, specifically around Tish and UBC, we saw a real value for accounting for the relational soup his work continually bubbled in, both for academics looking to situate Wah and his work historically and for non-academic readers looking to follow the threads of Wah's career. We also saw an opportunity to counter the fetishization that starves our understandings of individual books by stripping them of both the labour that goes into making them and the wider social contexts they circulate in. We began to ask a whole host of questions about Wah's relationships to others that led us to look for strategies to include information about the community formations that sprung up around his work. We found answers in the textual and paratextual information that accreted around books and journals—in tables of contents, on acknowledgements pages, in blurbs and reviews, in fields of citation.

Our questions in building the 2.0 archive echo those of the social turn in bibliography in the 1990s (championed by textual scholars such as Jerome McGann, John Bryant, and D. F. McKenzie) that calls for a holistic integration of material form, social and historical transmission, and textual criticism when considering the codex. McKenzie's "The Book as Expressive Form" (1999) redefines bibliography as the "sociology of texts," opposing the notion of a "pure" analytical bibliography that produces empirical evidence about a book's authorship, place of provenance, or date of production, while remaining scientifically detached from the interpretation of its content. Instead, bibliography should open itself to history, "account[ing] for non-book texts, their physical forms, textual versions, technical transmission, institutional control, their perceived meanings, and social effects" (13). In other words, it should consider the book as, and within, a network. While McKenzie's work provides a crucial turn within the discipline, it is nonetheless still subject to a major limitation: the acknowledgement of a book's socio-historical context services the goal of analyzing it as a "record of cultural change" (13), whose meaning exists primarily in the past.

In translating a chronological bibliography to a modular database, we realized the ways that Wah's poetic practice emerged from, was embedded in, and was productive of a range of historical, geographical, and social contexts. Wah's work troubles the primacy of the individual codex as the unit for analyzing the production, consumption, and circulation of texts, which became immediately apparent when we were faced with the unwieldy contents of the 1.0 site and archive. In between our conversations with Wah and our time spent holding the print material in our hands, we wanted to find room in the metadata for the information that exceeded Rudy's bibliography, especially as we ran into moments like this passage from Wah's 1972 book, *Tree*:

> Biography: It started between Mike and I just after I moved out to South Slocan from Buffalo. Then Stan started listening to them when he came up from Vancouver. Last fall Derryll said he and Michael would like to print them on their new press up in Argenta. So a week ago Gladys and Lars arrived at South Slocan and so did Derryll and Shirley. Derryll said he was ready to print, so. Brian arrived Monday night and he and I came up to Argenta Tuesday. We ordered the paper that afternoon. Wednesday morning I talked with Bird about doing some drawings for the book and so that started then. Gladys M. also arrived on Wednesday to work on her magazine, *Hamill's Last Stand*. Brian gave a reading that night and I started working on the typewriter Shirley had found for us over in Meadow Creek.

Yesterday I talked about Love in the World Problems class in the Argenta Friends School and started typing plates. Today, Pauline, Jenefer, Erika, Gladys, and Lars arrived. Michael and Derryll are up the hill printing. Bird is working on the title page drawing. The house is full and the sun's coming out over the head of Kootenay Lake. It's 4:30 and that's about it. April 28, 1972.

In this passage, the production of the book object is rendered visible. Not only does Wah provide a narrative of the book's material production by listing names and roles, he also places the sociality of that production at the heart of the very geography that he writes about in the rest of *Tree*. The pieces (the people, the printing materials and equipment, the lake and the sun) assemble into a milieu larger than but inclusive of the book. No longer can we simply question how the book will act upon, or transmit to, the reader, we must also consider the ways it is, even at the moment of its inception, already enmeshed in a social and spatial production.

We're also in a unique situation because of the heavy involvement of the archive's subject in its production. When the original site went offline, Wah approached Fong to reboot the archive on Wershler's suggestion.[2] Wah and Fong met at Beans, a small café on Cambie Street, where they discussed his desires for and concerns about the project, including issues around access, design, and sustainability. Leaping from café to café, she met with Rudy at Peyton and Byrne (a café in the British Library) to discuss what she wanted to see happen with her project as we rebooted it. Our archive dealt with history, this was clear, but it was also part of an ongoing assemblage in the present—the archive as more than a record of writing, but an intervention into how that writing is received and circulated. In this sense, the Wah archive had to negotiate a doubled publicness: both the publicness of Wah's writing and the archive itself as part of Wah's public face in the present. This doubleness asks us to be ethically responsible to the various stakeholders whose lives are still unfolding in the present. These stakeholders include Wah as the author of the work that is the site's content, Rudy as the original author of the bibliography, and Wershler and Kennedy as the authors of the infrastructure of the 1.0 site. We wanted to make sure that everyone's contributions were recognized and not risk effacing the intellectual and technical labour that formed the basis of our inherited project.

So, in the summer of 2015, as we sat in SFU Special Collections to sort through the archive, supplementing it with material from SFU's collection, we concluded that Rudy's bibliography needed to be opened up, moved from a fixed text to a living one. We added new entries and edited old ones. We changed the bibliographic numbers that organized things (and thought about abandoning that system altogether).

We found ourselves drawn in different ways to this naming of the very public networks of literary production. We wanted to credit the community labour embedded in both Wah's work and our own. We began recognizing this labour in the banal everyday practices of the poetry world: in acknowledgements pages, in biographical notes, in the lists of whom Wah was published by or alongside. We became invested in the ways the public networks of literary production could be credited and made visible through the work of the digital archive.

The question for us as designers of the archive, then, is how to represent or materialize that larger network in a way that is useful for readers, researchers, and scholars of Wah's work. How to diagram the machine, especially when its parts are constantly on the move? Immediately, we were struck by the question of what to reproduce. With a single-author codex this was an easy decision, but what about an anthology or a journal? How much of that object was important in recording Wah's place in a literary milieu? We decided to include as much contextualizing information as possible, rescanning texts to include features like covers, tables of contents, lists of contributors, and biographical notes (see figures 15.3 and 15.4).[3]

Most importantly, we began to consider the ways our digital frame might help us catch information the paper bibliography didn't, information that was easier seen when wading through the stacks of books and journals armed with our understandings of the ways literary communities often work. "Publicness" is at the heart of Wah's work and we wanted to extend the work of Rudy's bibliography to provide a better view to social relations and circulations that happened off the page. We wanted to enable readers and researchers to follow the threads that connected different entries by tracking a poem's publication history or who Wah was being published alongside or by.

To construct the architecture of this context, we added to the metadata set in a way that acknowledges the larger productive and receptive networks around Wah's oeuvre (see figures 15.5 to 15.8):

(1) An "associations" taxonomy, which maps a network of relationships involved in the production of the text (editors, artists, co-contributors, etc.), the materiality of the artifact (printers, technicians, etc.), and Wah's personal and literary network (friends, collaborators, influences, references, teachers, students, etc.). While these associations are listed as they

appear in codices, they also become indexical terms that offer alternative pathways to navigate the archive. For example, Roy Miki, a long-time friend and co-conspirator of Wah's, has a number of relationships to many of his texts: they are published by, edited by, and dedicated to him; they reference and respond to his work, creating an ongoing dialogue between texts, across time. Clicking on or searching for Miki's name brings up a complete set of the works he's attached to. Additionally, a bibliographical entry can have any number of associations such that relations of editing, design, affiliation, dedication, and co-publication are all represented.

Fred Wah Digital Archive

A digital repository and bibliography

≡ Menu

Home - West Coast Line 31.3

West Coast Line 31.3

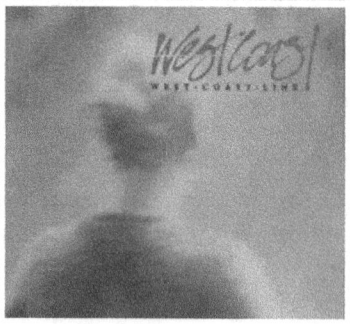

Section D
Bibliographic ID: D164
Year: 1997-1998
Genre:
Essay
Format:
Periodical
Periodical title:
West Coast Line
Publication number: 31.3
Page number(s): 72-84
Titles:
Speak my Language: Racing the Lyric Poetic
MLA Citation:
Wah, Fred. "Speak my Language: Racing the Lyric Poetic." *West Coast Line* 30.3 (1997-1998): 72-84. Print.

Published or edited by:
Roy Miki
Artist for:
Karen Moe
Published alongside:
Marie Annharte Baker
Jodey Castricano
Susan Clark
Heather Fitzgerald
Peter Hudson
Karlyn Koh
Ashok Mathur
Erin Moure
Mark Nakada
Sonia Smee

Figure 15.3   Figure 15.4

Figure 15.3 and Figure 15.4. Bibliographical entry on the Fred Wah Digital Archive, version 2.0.

Source: Fred Wah Digital Archive 2.0.

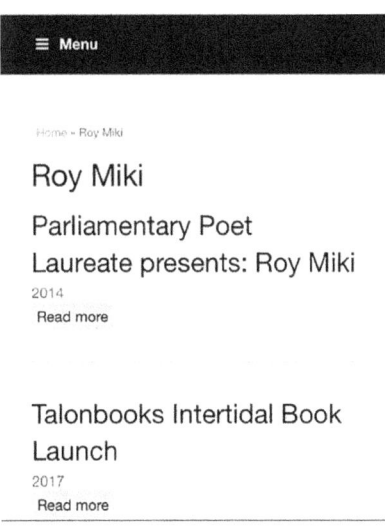

Figure 15.5   Figure 15.6

Figure 15.5 and Figure 15.6. Associations entry on the Fred Wah Digital Archive, version 2.0.

Source: Fred Wah Digital Archive 2.0.

(2) A "related works" node reference that interlinks instantiations of a work (whether partial, complete) across formats (critical articles both by and about Wah, anthologies, journals, ephemera, books, etc.). On a more granular level, an index of titles allows the user to cross-reference different versions of a given piece, tracking variations across multiple edits, as well as circulation in different venues. In this manner, we expand the work of the appendix of titles given in Rudy's original bibliography, making it into a searchable, indexed record of writing.

(3) Two "geographical location" vectors that allow us to trace (a) the place where the item was published, and (b) where Wah lived at the time of publication.[4] Plotting this geographical metadata allows us to trace

currents of textual circulation, adding another layer of relational informa-tion that illuminates literary sociality in a spatial sense. In our future work on the archive, we are interested in the different ways we might visualize this dynamic geographical information, such as timeline-integrated mapping.

Through context-rich metadata sets like these, the digital archive makes visible networks of affiliation, camaraderie, and influence, both between contemporaries

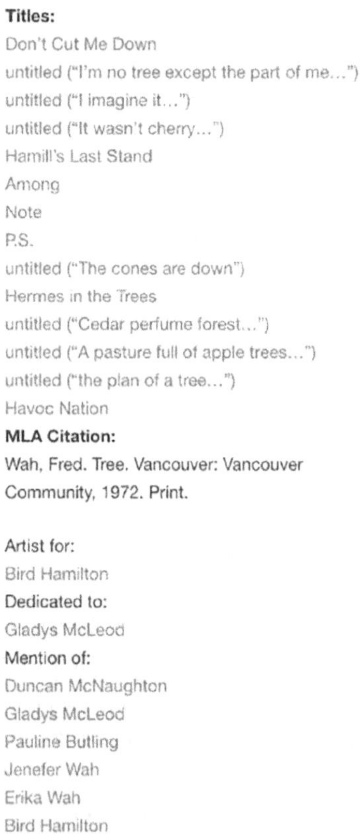

Titles:
Don't Cut Me Down
untitled ("I'm no tree except the part of me…")
untitled ("I imagine it…")
untitled ("It wasn't cherry…")
Hamill's Last Stand
Among
Note
P.S.
untitled ("The cones are down")
Hermes in the Trees
untitled ("Cedar perfume forest…")
untitled ("A pasture full of apple trees…")
untitled ("the plan of a tree…")
Havoc Nation
MLA Citation:
Wah, Fred. Tree. Vancouver: Vancouver
Community, 1972. Print.

Artist for:
Bird Hamilton
Dedicated to:
Gladys McLeod
Mention of:
Duncan McNaughton
Gladys McLeod
Pauline Butling
Jenefer Wah
Erika Wah
Bird Hamilton

Home » Don't Cut Me Down

Don't Cut Me Down

Fred Wah launches Scree at the Western Front in Vancouver, BC (full-length)

2015
Read more

Poetry Reading - March 8, 1979

1979
Read more

Boundary 2 3.1

1974
Read more

Tree

1972
Read more

Figure 15.7 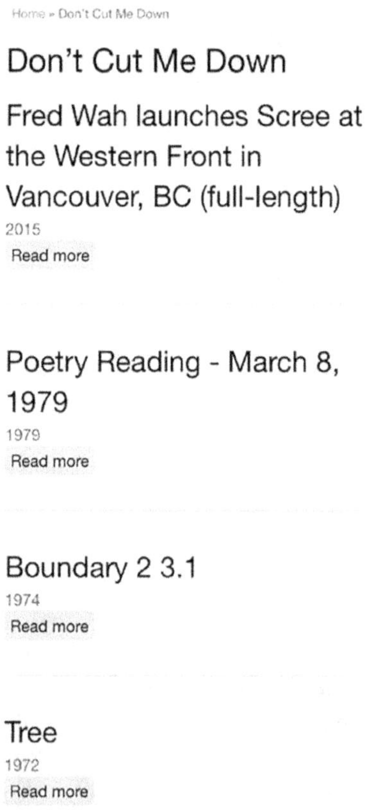   Figure 15.8

Figure 15.7 and Figure 15.8. List of poems in *Tree* and the connected page of related works for "Don't Cut Me Down" on the Fred Wah Digital Archive, version 2.0.

Source: Fred Wah Digital Archive 2.0.

and generations. The exciting part of this project lies in making more public the kinds of knowledge that would otherwise exist through our relationships with one another, in the memories we hold and the stories we share. It goes a long way to make social dynamics an object of study in their own right. Freed from the codex as the primary unit of study, the composition of the literary field shifts to include different kinds of contributions and labour, like inspiration, design, editorial work, reception, and other forms of support.

TIGHTENING CHANGE, LOOSE SUSTAINABILITY

In one sense, the archive allows us to "scale up" to a formal plane above the individual codex where we can make connections between different parts of the field legible. In another, the archive itself is subject to the disciplining mechanisms of other structures, particularly, in our case, the university, with its fraught relationship to authorship, labour, and funding. We want to conclude our contribution by turning back to the processes still at work—the fragile, tenuously held networks at the heart of the archive's current iteration. If a primary concern of this project is *sociality itself*, how do we sustain that project when the agents driving it change over time as they convocate, retire, die, etc.? In particular, how might our student-initiated project organize around already-existing institutional hierarchies without getting caught in a utopian fantasy that through the project those hierarchies disappear, or that the agents involved don't need compensation for their labour?

As we write this, we've already drifted off the project's playing field into postdocs and precarious labour pools, though we're still connected to wider literary and academic communities animated at the heart of the project. If the standard roles for students are as paid research assistants, who, in many cases, relinquish their claim to intellectual property once their contribution to the project is done, or as a principal investigator who receives limited-duration funding, what space is available outside these two positions? All this to ask: how do you ensure a project's sustainability outside the usual funding models that maintains an ethical relationship to labour and intellectual property for all parties involved? And what role does sociality play in the development and sustenance of those ethical relationships?

This question of sustainability is the topic of discussion in a 2011 special issue of *Profession*, where contributors view sustainability as one of the limitations to evaluating digital scholarship. In his paper "On Creating a Useable Future," McGann proposes shifting the focus of sustainability in DH initiatives from project funding to the actors involved.[5] He calls DH scholars to arms, challenging them to fulfill their duty as cultural custodians by being active agents in digital design and circulation. He insists that "the scholar's interests ought to be determining ones—perhaps, if there is such a thing, *the* determining ones" (2011, 24), suggesting that the other actors in the digital humanities align their interests and labour to fit the scholar's. McGann laudably intends to protect scholarly exchange from the encroachment of for-profit publishing; however, in the process, he imagines a system where he and other scholars are displaced from their positions at the top of the academic pyramid by an influx of "out-of-department" professionals. But his stance fails to meaningfully acknowledge the rest of his network, leaving a list of individuals who are never named.

In the same issue of *Profession*, Bethany Nowviskie's paper "Where Credit Is Due" resists the single-scholar model McGann leans into, arguing against lone-wolf scholarship where the other professionals around a digital project would act as publishers might for a print volume. Instead, she argues in favour of a greater acknowledgement of a project's production network. Nowviskie poses this question of authorial network in affective terms when she asks: "Can we imagine collaborations in which not only faculty members but also named librarians, administrators, non-tenure-track researchers, and technologists begin to *feel* a private as well as professional stake?" (2011, 170; emphasis added).

What does it mean for the participants in a project to *feel*—to both carry and share an affective stake in a project? If we turn to the model proposed by Wah and his contemporaries who bolstered the Canadian DIY publishing scene in the 1960s and 1970s, the answer is that collaborators invest in the co-creation of a product not for the end game of financial or cultural capital but, rather, for the reciprocity that comes from being part of a community. Such promise is apparent in Wah's "biography" in *Tree*: the book is nothing if not a record of comings and goings, gestures of care, and time freely given to others as a powerful anticommercial gesture. To a certain extent, we believe that some of this ethos can be (and currently is) fostered in DH initiatives. This is especially true of projects in which all participants feel a sense of ownership over their own labour, and feel the impact of that labour as contributing to a larger whole.[6]

At the same time, we must not commit the mistake of using affect to mask unethical labour practices, assuming that participation in a community is reward enough in itself when the community in question may be subject to differences of experience and power. Though communities might be maintained through care for each other and for the work, they are also subject to institutional pressures. This is especially true in DH projects that emerge from the structures of the academy, where funding models often engender hierarchies between project leads and co-applicants (usually tenured faculty) and other stakeholders such as community collaborators and student researchers. In this case, it is essential that these other stakeholders be compensated for their work with more than just good feelings; however, this is not to suggest that the powerful effects of community participation somehow be antithetical to financial compensation. The opposite, in fact, is true: if participants are compensated for their labour *and* encouraged to become affective stakeholders in the project, then the likelier they are to carry that work forward into their own intellectual and professional lives. This has certainly been the case for us.

We need to keep sociality in sight when working on projects like this. Although the project has been supported institutionally at SFU through the English Department, the library, the Digital Humanities Innovation Lab, and the Scholarly Digitization Fund, what ultimately has held the project together is that we care about one another, about Wah, and about his work. As we trudge forward in our careers, we continue to find renewed opportunities and points of interest to return to the archive. Some of these are opportunities we've sought out ourselves—for example, Fong's postdoctoral involvement with the SpokenWeb project, which promises new methods to support the audio and video components of the archive. And other opportunities have cropped up circumstantially as a result of our embeddedness in different institutional and writerly networks—like the occasion to write this chapter after spending time with Paul Barrett at various conferences and events. As the networks around the archive slide and shift, new possibilities can open up to engage with the work, but at the same time can make it difficult to carry forward. There's a certain comfort, however, to the idea that the archive may not always exist in its current form or may one day merely exist as a record of our labour. The comfort comes from leaving a thoughtful record of sociality in which names are named, labour is accounted for, and relations acknowledged.

NOTES

1. The "Alley Alley Home Free" issue of *Open Letter* was edited by Frank Davey, Nicole Markotić, and Susan Rudy (2004) to collect contributions from the 2003 conference of the same name held at the University of Calgary that acted as a Festschrift for Pauline Butling and Fred Wah on the occasion of Fred's retirement. The poems in the issue were performed by Fred at a public event during the conference, but curated by Fred's student and friend Louis Cabri, who went on to edit an edition of Fred's selected poems in 2009 for Wilfrid Laurier University Press. Louis and Nicole stay at Fred and Pauline's house in Vancouver every summer.

2. In editing this text, we made the shift from referring to Wah, Wershler, Rudy, and ourselves by our first names to our surnames. It felt strange to speak about this project so formally since we also saw ourselves embedded in the sort of productive/professional friend map that Wah writes about in *Tree*. To us, this seems to be the conundrum at the heart of the archive: the competing demands for, on the one hand, historical objectivity, clarity, and distance from the archival subject; on the other, a fuzzy scholarly *process* that involves conversation, verbal agreement, affective management, and other interactions that play out on a first-name basis. The point is not to land firmly on one side or the other, but to talk about and name the tensions specific to this kind of digital archival work that involves living subjects, friends, colleagues, allies, etc.

3. Our approach to copyright was informed by conversations with SFU's copyright officer, Don Taylor. We reproduced the front matter of books (tables of content, publication information, etc.) as these are not considered protectable works. Where copyright was held by journals, we sent letters of permission to reproduce Wah's work. We also sent letters of permission to the copyright holders of any artwork that was necessary to reproduce, such as cover artwork. Finally, we added a disclaimer to the site that the material that appears on the site is used for the purposes of academic research and critical study, and appears with the permission of the author(s).

4. When we were conceptualizing what information would be useful to readers and researchers, we began to think about Wah's movements across the North American continent and how that shifted the communities he participated in. Wah in 1960s Vancouver wrote in a different milieu to that in Calgary in the 1990s. However, we chose to include where he lived at the time of publication because the data on this are clear-cut, whereas the timelines of a book's composition can be protracted and messy, happening over multiple geographies. In due consideration, further study would be rewarding in collaboration with Wah, to figure out the compositional timelines and geographies for all his books.

5. McGann's article develops from his introductory keynote to the Online Humanities Scholarship: The Shape of Things to Come conference, held at Rice University in 2010. The piece was originally titled "Sustainability: The Elephant in the Room" and appeared in the published proceedings from the conference.

6. For a discussion of affect and student labour in collaborative DH environments, see Anderson et al. (2016).

REFERENCES

Anderson, Katrina, Lindsay Bannister, Janey Dodd, Deanna Fong, Michelle Levy, and Lindsay Seatter. 2016. "Student Labour and Training in the Digital Humanities." *Digital Humanities Quarterly* 10, no. 1. http://www.digitalhumanities.org/dhq/vol/10/1/000233/000233.html.

Artmob. 2012. Fred Wah Digital Archive, version 1.0. https://web.archive.org/web/20120607231612/http://fredwah.artmob.ca/.

Davey, Frank, Nicole Markotić, and Susan Rudy. 2004. "Alley Alley Home Free." Special issue, *Open Letter* 12, no. 2.

Derksen, Jeff. 2015. "Reader's Manual: An Introduction to the Poetics and Contexts of Fred Wah's Early Poetry." In *Scree: The Collected Earlier Poems, 1962–1991*, edited by Jeff Derksen, 1–16. Vancouver, B.C.: Talonbooks.

McGann, Jerome. 2011. "On Creating a Useable Future." *Profession* 14: 182–95. https://www.mlajournals.org/doi/pdf/10.1632/prof.2011.2011.1.182.

McKenzie, D. F. 1999. "The Book as Expressive Form." In *Bibliography and the Sociology of Texts*, 9–30. Cambridge, U.K.: Cambridge University Press. https://doi.org/10.1017/CBO9780511483226.

Miki, Roy. 1990. *A Record of Writing: An Annotated and Illustrated Bibliography of George Bowering*. Vancouver, B.C.: Talonbooks.

Nowviskie, Bethany. 2011. "Where Credit Is Due: Preconditions for the Evaluation of Collaborative Digital Scholarship." *Profession* 14: 169–81. https://www.mlajournals.org/doi/pdf/10.1632/prof.2011.2011.1.169.

Rudy, Susan. 2009. "'Loose Change': Fred Wah, A Life in Writing." Unpublished bibliography. Wah, Fred. 1972. *Tree*. Vancouver, B.C.: Vancouver Community Press.

———. 1975. *Pictograms from the Interior of BC*. Vancouver, B.C.: Talonbooks.

———. 1980. *Loki Is Buried at Smoky Creek: Selected Poems*. Edited by George Bowering. Vancouver, B.C.: Talonbooks.

———. 2015. *Scree: Collected Earlier Poems, 1962–1991*. Edited by Jeff Derksen. Vancouver, B.C.: Talonbooks.

CHAPTER 16

Humanizing the Archive: The Potential of Hip-hop Archives in the Digital Humanities

Mark V. Campbell

Since the late 2000s, hip-hop culture has become a subject of archival expansion at American universities, yet not of DH research. This is particularly surprising given the digital dimensions of much of hip-hop music and culture, not to mention the digital humanities' stated interest in "diversifying" and even "decolonizing" digital humanities. With both Harvard and Cornell establishing collections of historical hip-hop artifacts early on—and schools such as Tulane and the University of Houston newly following suit—the academy clearly finds something interesting in hip-hop's past.

The challenges of reading hip-hop in relation to the digital humanities are brought into focus via the work of Sylvia Wynter, particularly as Wynter's work examines the notion of the human that shadows the digital humanities. Her work attends to the Black lives that are often conspicuously absent or significantly underrepresented in the field yet captured in the emergence of hip-hop archives at Ivy League schools. Human-geography scholar Katherine McKittrick reminds us that disciplinary thinking and its classificatory systems are "produced in the shadows of biological determinism and colonialism" (2021, 38). Thus, when Wynter identifies Black life as an excessive disruption to the theorization of Man, I argue that it is equally disruptive to the digital humanities as well as to traditional archiving practices.

Wynter's immense intellectual project, one focused on finding a new language for human "forms of life," has dramatically shifted the intellectual terrain across several fields and disciplines (2021, 242). In proposing that we understand the human as both a biological entity and a cultural invention—one that relies on both a mythos and a bios, rather than simply an overdetermining biocentric notion—Wynter's work also challenges the foundations of the humanities. Wynter elaborates how the narrative, discursive, and cultural aspects of European Man are equally significant as the biological, scientific claims of Man. As such, Wynter's work does not merely offer a genealogical account of the narrative overrepresentation of Man but also challenges the relationship between this narrative of the human and its concomitant representation in visions of the humanities. Wynter's interrogations of Europe's Man, its provincialism, and its

rigid hierarchical orderings have made clear how considering Black cultural production does not merely open the boundaries of the humanities but also requires
a new definition of both the humanities and the idea of who counts as human.

Black diasporic populations disruptively live in excess of the white racial
imagination, as speaking (former) commodities who test the limits of the language of Western humanity. From the speculative imaginings of Afrofuturism to
the interdisciplinary innovations of hip-hop culture, Black cultural expressions
of what it means to be human is, as Wynter suggests, a praxis (McKittrick 2015).
This has implications for the digital humanities, particularly as hip-hop employs
digital technologies to challenge narrow visions of the human and the humanities
via a critique of notions of authorship, community, and globality. These practices
evoke what Wynter terms the "social imperative of black culture," which is to
invent a social order in which Black life accrues human value—beyond the replicating powers of the labour market's hegemonic value creation schemas (n.d.,
734). The social value created by hip-hop's creative art forms, from beatboxing
to turntablism, stand in stark contrast to the market notions of the human as
Homo oeconomicus. (Wynter and McKittrick 2015). This disruption of hierarchical
human ordering via social value created by hip-hop culture outside of the labour
market prompts us to look at the humanities and the limits of its analytical possibilities when Black life disappears from the conceptual frame. The limited framing of the human in the humanities presents an opportunity for us to examine
what hip-hop culture—particularly before its commodification following its first
hit record, by Kurtis Blow—might have to offer to the digital humanities now
that an archival turn is shedding light on its earliest years.

In what follows, I interrogate how hip-hop culture and its archivization might
disturb the digital humanities' poor track record around Blackness and reorient
trajectories in the field. The archiving of hip-hop can make plain the archive's
problematic formation to amplify and make audible Black life beyond ocular-
centric ways that assist the dominant culture in its hierarchical arrangements
of human life. Hip-hop disrupts the archive's historical linearity with multi-
temporal sonic innovations that challenge the coloniality of traditional archival
methods and activities. Hip-hop culture—including b-boying, DJing, and graffiti
art—interweaves and interconnects racialization, justice, geography, and power
in uniquely creative ways. Personal style and improvisatory skills contribute to a
sense of value that refuses to follow the hierarchy of racial schemas. The interconnection of these concerns is one way to enter an intersectional approach to
Black life that is sorely missing within the digital humanities. This intersectional approach reads the emergence of hip-hop archives in dialogue with the

institutionalization of the digital humanities in order to understand how the digital humanities have failed to contend with Black people and Black cultural practices. A critical, intersectional approach to digital humanities and hip-hop archives necessarily pulls librarians and archivists into a world in which social justice is a constant demand, and where anti-Black trauma and death haunt collections and archival practices.

As a "living archive," hip-hop's cultural dynamics play out on multiple frequencies; paying close attention to the archivization and archivability of hip-hop's sonic innovations and hyper-local practices allows me to excavate innovations and Black life beyond racial schemas. For example, reflecting on the practices of mixtape DJs in the 1990s—made possible through collections such as the Mixtape Museum archive project or the Houston Hip Hop Research Collection at the University of Houston—moves us toward an analysis of innovative distribution methods and DJ mixing strategies and away from a focus on race, Blackness, and the white racial imagination. This resonates with Stuart Hall's observation that the constitution of an archive coincides with an artistic movement's "new stage of self-consciousness" (2001, 89). Archiving thus enables a deep reflection on the seemingly fleeting moments of this youth-driven culture, returning to seminal moments, repressed histories, and innovative forms of expression and selfhood often rejected by the music industry's emphasis on newness and commodification.

In the Canadian context, the erasure and invisibility of Blackness identified by scholars across many fields—such as Charmaine Nelson in art history, George Elliot Clarke in English, McKittrick in geography, and Rinaldo Walcott in cultural studies—stands in stark contrast to our hyper-visibility in areas productive to the white racial imagination. So as the work of Philip S. S. Howard (2018) on blackface in Canada demonstrates, Blackness is consistently amplified by the white racial imagination, while spaces like Hogan's Alley, Nigger Rock, and Africville become sites of the erasure of Black life (McKittrick 2002). Under these circumstances of simultaneous erasure and fetishization, hip-hop in Canada wrestles with the nation's inequality through critiquing official multiculturalism, refusing state renditions of static "immigrant culture," and presenting other versions of Canada, both lyrically and visually.

My own work on digital archiving of Canadian hip-hop reveals the import of local cultural practices and largely unknown actors. In the archive of Canadian hip-hop, there are numerous Black Canadians of critical importance, like Master T, a video jockey who developed talent and provided a national stage for artists. Similarly, there are seminal venues and festivals, photographers, graphic designers, visual artists, and DJs on community radio. Demuth Flake is an exemplary figure who photographed much of Toronto's hip-hop culture in the 1990s; Odario Williams was responsible for the massively successful Peg City Holla held in Winnipeg from 1997 to 2006; Ebonnie Rowe who founded and continues to run, since 1995, a female-only performance event, the Honey Jam—an annual event that has put notable artists such as Nelly Furtado, Jully Black and Haviagh Mighty on stages early in their careers. Therefore, constituting an archive is also an act of amplifying and illuminating the cultural workers, artists, and audiences that made hip-hop in Canada possible. Centring these Black Canadians in exploring hip-hop's archival potentials is essential as it interrupts Canada's track record of disappearing sites of Black life. This form of archival work refuses preoccupation with the known historical actors, too often correlated with a market definition of success, or a preoccupation with "first" Black actors. For instance, the artistic and archival practices associated with mixtapes challenge notions of authorship, originality, and commodification. Mixtapes present the difficult taxonomical questions to the archival sciences and digital humanities, as they disrupt copyright, deprioritize the recorded album, centre Blackness in digital practices, and praise the DJ for technical innovations and abilities. The presence in the archive of many marginalized or ignored contributors to the culture, often obscured because these artists refused to reproduce familiar tropes legible to the white racial imagination, both decommodify Black life and help reveal how the limits of cultural infrastructure silence Black artists.

In what follows, I focus only on the sonic innovations found in hip-hop music, purposely sidestepping an ocular and biocentric focus of how hip-hop and, more specifically, rap music often gets taken up. Rather than disregarding race, my focus on remixing, sampling, and mixtapes intentionally disturbs the comfortable ways in which mainstream Western society accepts and fetishizes Black death, marginalization, and epidermically overdetermines Black life. I begin with a reading of hip-hop culture that takes Wynter's notion of a deciphering practice as a way to move beyond the aesthetic fetishization of hip-hop culture. I move to an exploration of the digital humanities and its inability to reckon with Blackness and its own anti-Black legacies. With specific attention to Marisa Parham's innovative connections of Blackness to digitality and Safiya Noble's work on racism and algorithms, I think through the potentialities the archiving of hip-hop presents to the digital humanities. In the final section, I focus on the specificities of hip-hop's artistic practices and how they propose we think of archiving and digital archives beyond its Eurocentric and biocentric underpinnings.

A DECIPHERING TURN

Parham reminds us that "Black studies-inflected frameworks are important to digital humanities because at the technical core of so much contemporary technological innovation we find literal and figurative resonance with histories, materialities, and other structuring realities of Black diasporic experiences— digitality" (2019). Centring Black life within the digital attempts to dislodge the ways in which digital life replicates social binaries and hierarchies; when Blackness enters the frame, as Simone Browne (2015) signals for us, the whiteness underpinning technologies reveals itself. In a related manner, Wynter's 1992 essay, "Rethinking Aesthetics: Notes Towards a Deciphering Practice," explores the role of aesthetics in demystifying power structures and presenting a new language from which to develop another social reality from the vantage point of the minoritized "others." Wynter's argument provides a language to speak of the way aesthetics can alter the current arrangements of power and knowledge so deeply embedded in middle-class, Western, bourgeoise notions of taste. For Wynter, Black popular music and culture "should induce counter-writing and a counter-politics of feeling" (270)—feeling that does not fall into an existing dialectic of good/bad, powerful/disempowered. Within the multidisciplinary practices in hip-hop culture, this means recognizing how hip-hop's innovations in dance, graffiti, rhyming, and DJing subvert middle-class notion of taste (as the reproductive function of securing middle-class life and values).

Attention to aesthetics, particularly in a moment of hip-hop culture's hypercommodification and popularity, offers an opportunity to explore the culture using a practice of deciphering. Investigating sonic innovations within hip-hop culture, sampling, mixtapes, or remixing, in terms of its aesthetic and politics, centres the culture's methods and creative practices over its consumptive and racial collusion with the status quo. Using a deciphering practice with hip-hop culture means interconnecting social institutions, policies, and Black geographies with a deep-listening practice and engaging hip-hop's signifying practices (such as ciphers) as a multifaceted behaviour-inducing text.

Wynter's deciphering practice reveals some of the ways in which hip-hop aesthetics influence individual behaviours, social realities, and terms of value that hip-hop artists and audiences develop and promote. Sonically, the aesthetic innovations found in hip-hop cultures—from sampling to mixtapes to remixing—induce and evidence a human modality that diverges from governing codes of taste. In these moments, made possible by sonic innovations in hip-hop, the racialized hierarchical schema of Man and Man's Other is reordered so that time, space, and sonic innovation intervene in the production of value or its negation as related to Black lives. Attention to these innovations within the digital humanities and archiving project presents an opportunity to question and diverge from the limited ways in which the human exists within the digital humanities.

Deciphering hip-hop practices reveal how sampling is more than merely using portions of existing recorded music to make new music. Specifically, sampling is a reflection of a former commodity creatively exceeding the social limitations imposed on objects—a reflective artistic practice that reverberates throughout post-middle-passage Black life. The act of sampling provides us with access to imagine how Black populations continually innovate new relations to constitutional, social, and extrajudicial restrictions. On multiple occasions, attempts were made in Caribbean colonial societies to ban the use of drums; today, copyright laws make the clearance of music samples prohibitively expensive. Read within a deciphering practice, sampling in hip-hop becomes more than just a violation of intellectual-property rights, it unravels a code of legal workings that continue to circumscribe and deny Afrosonic expressions of subjectivities, the legal personhood of African-descended peoples, and circumscribe property rights. Sampling also exposes the limits of the Western cultural enterprise as unprepared to expand definitions of music and music making invented by Afrodiasporic artists.

Hip-hop cultural aesthetics and methods pose nuanced questions of form, style, and politics, and we risk overlooking these methods and aesthetics when hip-hop is imagined to fit within existing and traditional archival practices and institutional archives. Without being immersed in hip-hop culture, archivists and librarians risk developing collections that fail to capture the innovative methods by which the dominance of racial schemas and discourses of marginalization are circumvented, reworked, or simply destroyed. To effectively understand hip-hop culture, archives must thoroughly investigate the methods and artistic practices that distinguish hip-hop as a cultural form.

A DH framework that takes hip-hop archives seriously needs more than simply an additive approach. Hip-hop archives cannot be reduced to a marginalized field that can simply be added on. Interpreting hip-hop archives through a deciphering practice reveals how they challenge existing codes, practices, and ideologies that make colonial archives possible. Roopika Risam suggests a foregrounding of difference using an intersectional approach is necessary in the digital humanities to resist the easy binaries that structure life of the Other (2015). As such, a hip-hop archive can be more than the accumulation of items preserved for scholars and collectors, they can become testaments to a set of differences that reorder the white racial imagination and the reifying of Western power and knowledge.

DH 3.0?

Unsurprisingly, there have been very few DH projects that have taken an interest in hip-hop culture. The debates in the digital humanities have been a sustained critique of who is doing the work versus who is talking about the work, the hack versus yack discussion. This false binary exposes the limits of the digital humanities, where orientation has not been toward social justice but, rather, a more insular experience, in which difference in its many forms threatened the formation of the field. Simply put, the field privileged its own formation to the neglect of the lived realities of equity-seeking groups. This trend is starting to change with the work of scholars such as Risam, Parham, Noble, Catherine Steele, and Kim Gallon. The nature of the human within the humanities remains, for much DH scholarship, unquestionably rooted in a Western liberal framework, one in which Others not fully deemed human remain outside of the frame. As Gallon makes clear when envisioning the work that needs to be done in the digital humanities, significant attention needs to be devoted to "how computational processes might reinforce the notion of a humanity developed out of racializing systems" (2016). The work of Browne, Noble, and others dismisses the neutrality of technology—a neutrality that resonates with the tenets of Western, European enlightenment thinking.

Black studies can bring to the digital humanities an orientation that prioritizes justice and the myriad ways in which Black life is consistently rubbing up against the limits of a European-conceived human. Black studies refuses to disentangle digital technologies from state violence and planetary environmental crises that shape our contemporary moment. Noble is clear when she insists, "we can no longer pretend that digital infrastructures are not linked to crises like global warming and impending ecological disasters" (2019). The deafening silence of the digital humanities, as body cameras and social media continue to capture state-sanctioned murders and the ensuing virality, is a detriment to the field and clearly anti-Black. It is only in the last few years—evidenced in Noble's work and elsewhere, such as Nehal El-Hadi's essay, "Death Undone," and Bethany Nowviskie's June 17, 2020, blog post—that Black life and Black studies are beginning to gain traction in the digital humanities. In fact, the failure of the digital humanities to engage with the sites where blackness and technology intersect, via Afrofuturist thought, for example, is a failure to move to an intersectional lens. As Risam reminds us, "Afrofuturism is an African American literary and artistic movement that foregrounds speculative approaches to displacement, belonging, and home for the African diaspora, that deeply structures the imaginative possibilities of black artists and scholars invested in a future of black life" (2015).

Furthermore, the overlaps between digital humanities, archival practices, and Black studies remain largely unexplored. Existing hip-hop collections and projects at several universities in the United States, such as collections at Harvard, Tulane's NOLA Hiphop and Bounce Archive, and the William & Mary Hip Hop Collection, have begun collecting documents and artifacts without a Black studies frame from which to foreground justice, and without an intersectional and interdisciplinary examination of Black life. Items and records are currently treated in a fashion that aligns with contemporary uses of archives—siloed and for the benefit of the academy, not those whose lives populate the archive's indexes. Relatedly, the works of scholars in the Black digital humanities have yet to have an impact on the existing hip-hop collections across various academic institutions in the United States. Put differently, Black studies encourage us to ask of hip-hop archives, how might Black life and liberation be enhanced by these entities?

ARCHIVING HIP-HOP

Hip-hop archives, both in their digital and brick-and-mortar incarnations, are increasing in number across American universities. In the South, New Orleans and Houston are home to three archives; in the West, Seattle is now home to a hip-hop archive; on the Eastern seaboard, Boston is home to two archives, while collections have also emerged in New York and Ithaca; in the Midwest, Virginia and Indiana are also home to hip-hop collections. In Canada, my own efforts focus on the digital-only Northside Hip Hop Archive, which includes content from Ontario, Saskatchewan, Quebec, Nova Scotia, and Manitoba. Hip-hop archives in the United States feature different scales of digital sophistication, and they also work in siloes, with varying relations to power, academia, and the hip-hop artists in their local community.

Within existing hip-hop archives, a tension exists between the archival presence of local hip-hop artists and commercially successful artists. The majority of hip-hop archives, including the Massachusetts Hip-Hop Archive, in partnership with the Boston Public Library, focus on local communities. Recognition of the most commercially successful hip-hop artists tends to align with Ivy-League aspirations, yet a hip-hop archive that does not archive its local artists remains unresponsive to its local community, refuting the same responsiveness that has made hip-hop culture concerned with its local environment. For some archives, like the University of Washington's Seattle Hip Hop Archive, the relatively small size of the local hip-hop community (compared to New York's or Houston's) is not a deterrent to ensuring it reflects its locality. The Seattle archive boasts more than 1,800 local hip-hop tracks. In a city like Houston, with a wealth of success-ful hip-hop artists on the national stage (beginning with the Geto Boys in 1991), two university archives hold a significant quantity of materials in impactful col-lections. Meanwhile, for other archives, such as the Cornell Hip Hop Collection, the relatively small size of the local hip-hop community means working closely with the collections bestowed by New York City and New Jersey residents.

The exclusion of local artists from hip-hop archives, primarily in the Hiphop Archive & Research Institute, at Harvard University, and to a lesser degree in Cornell's, is not simply a matter of what is missing or a gap in local connections. The representation of localness is an important facet of hip-hop culture—one that is disrupted and complicated by the shift to the digital and by traditional archiving practices (Forman 2002). Without local context and content, hip-hop archives risk producing an alienating narrative of hip-hop culture that stands in contrast to the allure of the locally produced. Part and parcel with representa-tion of local artists in hip-hop archives is the collection of content that resonates with the local scene. A local scene speaks back to the celebration of commercial-ization and international recognition that resonates less with a community of practitioners and fans and more with mainstream—and often whitewashed—concerns. Case in point is the success of the late DJ Screw in Houston. Screw's "chopped and screwed" DJing technique used on his mixtapes was zealously supported by large portions of the American South, but this local success did not significantly influence how mainstream hip-hop acts produced their music.

A focus on the local community opens the door to an understanding of hip-hop culture as an aesthetic and ethos with paradigms and values that are not solely aligned with a dominant discourse of conspicuous consumption, politi-cal apathy, or indulgence. Black humanity becomes audible and palpable when hip-hop culture attends to local environments and develops signifying practices that furnish a local context with more livable relations—a sense of "peace, love and having fun," in which aesthetic and creative labour dislodge the values of hierarchical ordering that underpin the labour market.

For example, in the Houston context, an understanding of the city's northside and southside divide illuminates how a group like UGK from Port Arthur, Texas, can come to represent Houston on a national stage. So, when UGK member Bun

B uses the word "trill," a term he uses frequently and which that joins "true" and "real" as a signifier of authenticity, he pays homage to his Port Arthur hometown. Houston-based hip-hop's signifying practices speak to a specific social and cultural reality in which an invented word like "trill" holds specific geographic and cultural significance. It is the deciphering of what local slang does, rather than what it means, that reveals the potency of hip-hop's ability to induce notions of belonging in an anti-Black world. A traditional archive, if attentive to vernacular innovations, might seek to make local words more knowable. A deciphering practice is attentive to the work the word can achieve: its circulation, its transformations, and its behaviour-inducing abilities. This is not to simplistically claim that a single word can overcome the real-world politics that animate the northside/southside divide in Houston. Rather, the point is to make more complex the ways in which archiving hip-hop can allow us to move beyond standard concerns and interpretations embedded in the practice of archiving. To move beyond the dominance of established archival methods means taking culture—especially the culture of those being archived—as intimately connected to strategies of decolonizing the archive. The vernacular creations found in hip-hop culture are fluid, and their meanings are contextually bound to time and geography, which challenges the archive's abilities to make static the culture's innovations.

The migration patterns of Caribbean migrants to Toronto from the 1950s onwards similarly deepen our understanding of the emergence of hip-hop music as travelling sound systems from the late 1970s onwards. Sonically, prior to hip-hop's ascendency, calypso and reggae music joined funk and soul to dominate Toronto's Black music scene. Toronto hip-hop often references the Caribbean, for instance the B-side remix to Kardinal Offishal's first single—"Natty Dread," a vinyl-only release, aptly named the "Eglinton West" remix—engages in a geographically signifying act. Like Bun B, Kardinal Offishal honours the community that nourished his artistry; this is a core practice in hip-hop culture called reppin' (Forman 2002). Eglinton West is also known as Little Jamaica, home to a large Caribbean population and the site of Monica's, one of the first stores in Toronto to sell hip-hop records. The hegemony of middle-class taste is upended by hip-hop's focus on the well-being of its local community; the narratives embedded within the dominance of bourgeoise Man only find a home within commercialized music, which is often separated from a specific locale. The representation of a specific area removes a social stigma and affords value to the inhabitants of an area represented in the music. For instance, Kardinal Offishal rhymes in "Kardi's Corner," an unofficial dub remix of Common's "The Corner," "we stopped watching Much when they lost Master T," referencing the Canadian music network's pioneering video jockey whose support of Black music was critical to the growth of the Toronto hip-hop scene. Considering the power of American media's dominance in the Canadian market, Kardinal's decidedly Canadian (and hyper-local) focus as he shouts out areas of significant Black populations—"Rexdale, Scarborough, Regent, Flemington"—speaks to hip-hop's unwavering attachment and support of its local communities and cultures. This reordering of a social hierarchy is audible In frequencies that resonate with hip-hop underground circles and recoverable when we examine signifying practices within their social and cultural environments. "Kardi's Corner" follows a long tradition in Jamaican music of producing versions of a song by recycling a riddim (rhythm) or creating a dubplate (Chamberlain 2010; Veal 2013).

Archiving the work of DJs, particularly in the forms of remixes, dubplates, and mixtapes, presents a challenge to notions of present, past, and future. As Parham reminds us, "the DJ using two turntables to disaggregate songs into discrete soundbites so that they might be used as if they were digital, isolating out samples and breaks so that old texts could be made newly resonant with always present futures" (2019, para 6). Mixtapes as archival items signify more than the recorded content. The impact of mixtapes on their local communities—building a scene or increasing exposure for local artists—combined with the aesthetic innovations of each mix suggest they hold the potential to dislodge audiences from a linear market orientation of new songs, while simultaneously reinforcing an audience's sense of place and belonging by celebrating a specific locale. DJ Screw's practice of chopped-and-screwed mixtapes and his refusal to sign record deals suggests his engagement in hip-hop culture was more than a tactic to gain entry into mainstream society (Walker 2015). Until his untimely death in 2000, DJ Screw produced more than 300 mixtapes, many of them one-of-a-kind, made-to-order cassettes at the request of community members.

Despite hundreds of mixtapes in existence, fewer than two-dozen cassettes were housed at the University of Houston's hip-hop archive in late 2019. While such low numbers might appear to be a specific failure to build the university's collection, far more is signified by this low number of holdings. In making mixtapes on demand and by distributing them hand to hand, DJ Screw refused the commercial music industry and the commodification (through mass production) of his mixtapes. By remaining outside of the industry, DJ Screw cultivated a connection to his art, his locality, and his fans that could not be undermined by the logic of the market. The hundreds of mixtapes not in the archive point to the human connections, built by DJ Screw's fiercely independent ways, that appear incompatible with the ability of an archive to collect, centralize, and make static.

The process of digitizing DJ Screw's mixtapes is currently underway, a long process that presents new opportunities to capture and archive aspects of hip-hop DJ aesthetics. Importantly, the storefront that sells DJ Screw memorabilia is also a site where his cousin, the store owner, digitizes old mixtapes that are brought into the store. Members of the public regularly walk into the shop with old cassettes; the store generates track listings, digitizes the cassettes, and repackages them for sale. This sort of aftermarket commodification of mixtapes, including the sale of drinks, t-shirts, and towels, may somewhat obscure DJ Screw's aesthetic legacy, yet it also demonstrates how far removed from the daily workings of the music industry DJ Screw chose to remain. Further, despite operating as a distribution outlet for DJ Screw memorabilia, the storefront—home to hundreds more mixtapes than the archive—also provides the opportunity for scholars like myself to view more than 340 playlists. In the digitizing and cataloguing of hundreds of mixtapes, in the case of DJ Screw, as well as the Columbia University–based Mixtape Museum, the opportunity exists for the digital humanities to explore how power, geography, gender, and race are situated within mixtape cultures.

Following Risam's (2015) focus on an intersectional framework in the digital humanities, I suggest there are several entry points presented by the digitizing of mixtapes. Risam recommends that user navigation, curation, and visible metatags can amplify the relevance and presence of an item in the archive, especially if intersectional realities are brought into conversation with multiple classification practices. For example, users can organize their viewing preferences to foreground era, geography, race, or gender. User-controlled filters with wide-ranging options provide users control over their navigational experience. In the case of digitally archived mixtapes, metadata could be organized to allow users to filter by regional geographies, DJ mixing techniques, or analogue/digital filters. Making metatags visible and populating these tags with various levels of multiplicity recognizes and promotes an intersectional reading of an archival item. It is worth noting that the metadata categories listed above consciously do not focus on the actual songs recorded in each mix, intentionally moving away from the industry's obsession with copyright to helps users focus on the art of the DJ, the social impact of the mixtape, and the technological possibilities the recording was created within.

Rounding out a trio of sonic innovations found within hip-hop cultures, the remix, often a feature on mixtapes, nicely knits together musical experimentation, illegality, and technological subversion. Remixing is a core aesthetic of contemporary popular music, with sonic lineage in both dub and disco music, and contemporary resonance with hip-hop and dance music. According to Eduardo Navas (2012), remixing is the "activity of taking samples from pre-existing materials to combine them into new forms according to personal taste" (65). Remixes do not often have the same kinds of clear connections to Black geographies as mixtapes, but they have similarities with the act of sampling. Both remixes and the practice of sampling refuse to accept the manufactured obsolesces of "old" music, they both use existing, pre-recorded music to create new music. As they recontextualize and reinvent songs, remixes work through memory, nostalgia, and popular culture. Yet, as David Gunkel (2015) reminds us, remix artists are in clear violation of copyright laws, as a 1991 federal court decision against Biz Markie clearly demonstrated.

The practice of unauthorized remixes urges a re-evaluation of the notion of the author while disturbing intellectual-property concerns of Western economies, signalling generative possibilities for archival sciences. In assigning a data-management system, developing a knowledge graph, and linking archival data across the Web, the remix—with its sonic lineage often extending into multiple musical genres and eras—opens up robust options to link together archival material. Such an archive of deep sonic connections already exists online in sample-focused websites such as whosampled.com, whose academic/pro version boasts more than 400,000 songs.

Several ways to analyze a remix exist, with transformative and regressive remixes accompanying a list Navas developed in 2012, which understands remixes as extended, selective, or reflexive. Important for the archive is how remixes expose the irrelevancy of the concept of the author, illuminating its Western, European origins, and dislodging the authority of an "original." Part of the "good trouble" remixes create is that they provide a clear line of sight to Europe's limited, provincial notions of culture that form the basis of the creative industries. The multiplicity of versions, such as the hundreds of versions of songs using the same riddim as reggae artist Wayne Smith on "Under Mi Sleng Teng," signal other kinds of logic at work—ideals that neither validate nor reproduce the values and tastes of middle-class Europe. For example, the explosion of musical genres, recording-production techniques, and sonic innovations from Jamaican sound-system culture do not reference Europe in its ontology. As legendary sound systems like Stone Love and Bass Odyssey remind us with every dance hall they rock, "nobody owns sound"—an assertion that reflects Dick Hebdige's popular 1987 book, *Cut'n'Mix: Culture, Identity and Caribbean Music*. Adrian Johns's (2009) European values of property ownership do not hold the same weight in the sonic realm; the sonic and the oral continuously reject the notion of individual property rights and even the Enlightenment ideal of autonomy. The antecedents of remix culture, emanating from Jamaican sound-system experiments and American disco clubs, never rested neatly within the existing paradigms of culture as legally enforceable by governments.

The signifying practices of a remix—including the levels of deep listening, and the need to embrace difference, otherness, and the past—are embedded within a copyright regime whose goal is to "promote the progress of useful art" (Logie 2014). The remix highlights a modality of human behaviour that pivots on a paradigm of value unaligned with the mainstream music industry and intellectual-property rights. These aesthetic practices do not lend themselves to archivability, but they speak articulately of hip-hop culture beyond the racial schemas that dominate the genre's reception. If archives collect mixtapes, remixes, and source samples without attending to how these works transform Black cultural life or disrupt mainstream culture, the archive will bring about yet another Black death, replicating the traditional role of colonial institutions in Black life.

NORTHSIDE HIP-HOP

My own digital archival project, Northside Hip Hop Archive, has been a passion project fuelled by the disappearing legacies of Canadian artists from the 1980s and 1990s. Since 2010, digitizing, mapping, exhibiting, and celebrating hip-hop's past in the Canadian context has been my focus. Such efforts began with T-Dot Pioneers, a 2010 public exhibition of art and artifacts combined with live performances, panel discussions, and a collaboratively created timeline. The content of the archive includes oral histories, aspects of material culture such as flyers, and audio clips from radio shows, mixtapes, and public talks. The local plays a central role in how hip-hop in Canada is understood: festivals, radio shows, and DJs continuously remake what we mean by "Canadian." Such a focus ensures no single, grand narrative can overdetermine what hip-hop culture looks like in Canada. With the inclusion of hundreds of videos from Black, Indigenous, francophone, and anglophone artists, as well as content from both mainstream and independent journalism outlets, the various divisions and differences are not buried in an imaginary uniformity.

Rather than foregrounding some of the more traditional aspects of the archival process (such as the reliance on influential people to support and populate the archive), a significant portion of the content housed on nshharchive.ca is organized in personal collections. This is a deliberate move away from the colonial geographies of Canadian cities, which combines with the move away from the language of pioneers and firsts that was employed in the T-Dot Pioneers exhibition. It is an attempt to capture the amplification of our positionality, and stands in stark contrast to what Jarrett Drake refers to as the "delusion of neutrality" (Drake 2016). The collections curated on our site came from personal engagements with and relationships formed through my 17 years on radio with the *Bigger than Hip Hop Show*, and my work with public exhibitions and shows between 2010 and 2020. The site's collections mark the slowness of archival work, the intentional relationship building, the mutual respect required, and a conscious investment in diffusing archival and curatorial power.

Effectively, hip-hop archives as DH projects hold the potential to document the human modality expressed in hip-hop culture, and therefore move beyond the dominance of Western values and middle-class judgments of taste. A deciphering practice in the archiving of hip-hop takes seriously the ways in which hip-hop's intertextual, intersectional aesthetic practices shape social realities. Rather than attempting to archive Black life into a colonial entity—one rife with the residue of Black death—or represent Black life as singular and fixed, hip-hop archives can allow for intersectional understandings of hip-hop cultures, especially its practices that exist beyond the mainstream's view. Focusing on these aesthetic practices dislodges the power of racial schemas that reify colonial logics and reproduce the erasure, power, and so-called objectivity. In the examples offered here, my focus on the sonic innovations of remixing, sampling, and mixtapes—all stressing local expressions of Black life mediated through digital and analogue technologies—underwrites existing cultural entities in the West. Strategies around curation, user navigation, and meta-tagging open the possibilities of another kind of archive—a humanized archive that could emerge out of, and benefit from, a critical Black DH grounding.

REFERENCES

Browne, Simone. 2015. *Dark Matters: On the Surveillance of Blackness*. Durham, N.C.: Duke University Press. https://doi.org/10.1215/9780822375302.

Chamberlain, Joshua. 2010. "So Special, So Special, So Special: The Evolution of the Jamaican Dubplate," *Jamaica Journal* 33, 20–29.

Drake, Jarrett M. 2016. "Liberatory Archives: Towards Belonging and Believing (Part 1)." Keynote address given at Community Archives Forum, Los Angeles.

El-Hadi, Nehal. 2017. "Death Undone." *The New Inquiry* 2.

Forman, Murray. 2002. *The 'hood Comes First: Race, Space, and Place in Rap and Hip-Hop*. Middletown, Conn.: Wesleyan University Press.

Gallon, Kim. 2016. "Making a Case for the Black Digital Humanities." In *Debates in the Digital Humanities 2016*, edited by Matthew K. Gold and Lauren F. Klein, 42–49. Minneapolis: University of Minnesota Press. https://doi.org/10.5749/j. cttcn6thb.7.

Gunkel, David. *Of Remixology: Ethics and Aesthetics After Remix*. Cambridge, Mass.: The MIT Press, 2015. https://doi.org/10.7551/mitpress/10325.001.0001.

Hall, Stuart. 2001. "Constituting an Archive." *Third Text* 15, no. 54: 89–92. https://doi. org/10.1080/09528820108576903.

Hebidge, Dick. 1987. *Cut 'n' Mix: Cultire, Identity and Caribbean Music*. New York: Methuen.

Howard, Philip S. S. 2018 "On the Back of Blackness: Contemporary Canadian Blackface and the Consumptive Production of Post-Racialist, White Canadian Subjects." *Social Identities: Journal for the Study of Race, Nation and Culture* 24, no. 1: 87–103. https://doi.org/10.1080/13504630.2017.1281113.

Johns, Adrian. 2009. *Piracy: The Intellectual Property Wars from Gutenberg to Gates*. Chicago, Ill.: University of Chicago Press. https://doi.org/10.7208/chicago /9780226401201.001.0001.

Logie, John. 2014. "Peeling the Layers of the Onion: Authorship in Mashup and Remix Cultures." In *The Routledge Companion to Remix Studies*, edited by Eduardo Navas, Owen Gallagher, and xtine burrough, 306–319. London, U.K.: Routledge.

McKittrick, Katherine. 2002. "'Their Blood Is There, and They Can't Throw It Out': Honouring Black Canadian Geographies." *TOPIA: Canadian Journal of Cultural Studies* 7: 27–37. https://doi.org/10.3138/topia.7.27.

———. 2021. *Dear Science and Other Stories*. Durham, N.C.: Duke University Press. https://doi.org/10.1215/9781478012573.

———. 2015. "Mathematics Black Life." *The Black Scholar* 44, no. 2: 16–28. https:// doi.org/10.1080/00064246.2014.11413684.

———, ed. 2015. *Sylvia Wynter: On Being Human as Praxis*. Durham, N.C.: Duke University Press.

Navas, Eduardo. 2012. "Remix[ing] Theory." In *Remix Theory*. Vienna, Austria: Springer. https://doi.org/10.1007/978-3-7091-1263-2.

Noble, Safiya. 2019. "Towards a Critical Digital Black Humanities." In *Debates in Digital Humanities 2019*, edited by Matthew K. Gold and Lauren F. Klein. Minneapolis: University of Minnesota Press. https://doi.org/10.5749/j.ctvg251hk.5.

Nowviskie, Bethany. 2020. "A Pledge: Self-examination and Concrete Action in the JMU Libraries." *Bethany Nowviskie* (blog). https://nowviskie.org/2020/a-pledge-self-examination-and-concrete-action-in-the-jmu-libraries/.

Parham, Marisa. 2019. "Sample | Signal | Strobe: Haunting, Social Media, and Black Digitality." In *Debates in Digital Humanities 2019*, edited by Matthew K. Gold and Lauren F. Klein. Minneapolis: University of Minnesota Press. https://doi.org/10.5749/j.ctvg251hk.14.

Risam, Roopika. 2015. "Beyond the Margins: Intersectionality and the Digital Humanities." *Digital Humanities Quarterly* 9, no. 2.

Veal, Michael. 2013. *Dub: Soundscapes and Shattered Songs in Jamaican Reggae*. Middletown, Conn.: Wesleyan University Press.

Walker, Lance Scott. 2015. "DJ Screw: A Fast Life in Slow Motion." Red Bull Music Academy May 20, 2015. https://daily.redbullmusicacademy.com/2015/05/dj-screw-feature.

Wynter, Sylvia. 1992. "Rethinking 'Aesthetics': Notes Towards a Deciphering Practice." *Ex-iles: Essays on Caribbean Cinema*, edited by Mbye B. Cham, 237–279. Trenton, N.J.: Africa World Press.

———. n.d. "Black Metamorphosis: New Natives in a New World." Unpublished manuscript.

Wynter, Sylvia, and Katherine McKittrick. 2015. "Unparalleled Catastrophe for Our Species? Or, to Give Humanness a Different Future: Conversations." In *Sylvia Wynter: On Being Human as Praxis*, edited by Katherine McKittrick, 9–89. Durham, N.C.: Duke University Press. https://doi.org/10.2307/j.ctv11cworj.5.

CHAPTER 17

Sounding Digital Humanities

Katherine McLeod

L istening to sound in the digital humanities involves listening to where and how sound manifests in digital projects. However, although sound has come to play a more audible role, "sound remains perhaps the least utilized, least studied mode within digital humanities" to the extent that "[f]ew projects and fewer tools incite scholars to *listen*" (Lingold, Mueller, and Trettien 2018, 10; original emphasis). While this bias toward the visual is not unique to Canada, Canada is where scholars have launched a digital project that attempts to change the underutilized mode of sound. That digital project is SpokenWeb, a network of researchers studying the sound of literature through interdisciplinary methods and practices of listening. At present, SpokenWeb exists in the form of a seven-year partnership grant funded by the Social Sciences and Humanities Research Council of Canada (SSHRC) and with collaborators from 13 Canadian and American universities, with the number of collaborators from universities and community organizations steadily growing.[1] Given the resistance to sound-based DH projects, what made the formation and funding of a project based on audio recordings possible? Moreover, what makes SpokenWeb—a project that started with analogue recordings—a DH project?

From its circular logo evoking a vinyl record to the audiovisual equipment at labs at various SpokenWeb-affiliated institutions, SpokenWeb is a digital project obsessed with analogue media, but that obsession does not make it any less digital. The digital tool development and research outputs—the making-by-listening—of SpokenWeb are digital, sonic, and literary, all at once. SpokenWeb intervenes within digital humanities in Canada by exemplifying how analogue and digital technologies can inform research methodologies and how the affordances of sound as a medium can be at the forefront of digital development and data management. As a DH project producing new ways of researching sound through sound, SpokenWeb incites scholars to listen.

Initiated and conceptualized by Jason Camlot at Concordia University in Montréal, SpokenWeb began as a single-institution project that received funding to work with one digitized collection of poetry readings, the Sir George

Williams University (SGW) Poetry Series, 1966–1974. Now SpokenWeb consists of numerous collections of literary audio recordings—readings, performances, and interviews, among other literary events—and SpokenWeb-affiliated researchers are designing audio tools to make this content discoverable and usable across different platforms and user interfaces. SpokenWeb uses the term "collection" to refer to recordings belonging to a writer's fonds or recordings from a reading series or produced by a particular festival, institution, or community organization. Collections within SpokenWeb contain their own stories as digital projects, which then become sub-projects under the umbrella project of SpokenWeb—for example, Simon Fraser University's Gerry Gilbert radiofree-rainforest Collection, the University of British Columbia Okanagan Campus (UBCO)'s SoundBox Collection, and Concordia's SGW Poetry Series.[2] By digitizing and describing audio collections, SpokenWeb is making these recordings discoverable, searchable, and adaptable into critical and creative projects. Making literary audio from across Canada more accessible is not only about access to recordings but also about the gathering of metadata about recordings and presenting that information in useful and generative formats; moreover, through this labour of making this spoken *web*, the collaborators in this project are producing another kind of web: a network of scholars conceptually and critically invested in literary audio recordings as objects of study. Their investment in these recordings gives them "stickiness" in the sense of what Sara Ahmed (2010) describes as the process of emotion becoming attached to an object and leaving a residue of that impression upon it. The object itself becomes imbued with emotion—and, in the case of this spoken web, researchers are pulled together by a shared interest in how audio recordings can change the nature of literary studies. The audible medium is the "sticky" web, as explored throughout this chapter in affective-laden descriptions of working with literary audio recordings. Elation in discovering a tape, melancholic nostalgia in listening to a recording, or rage in hearing something unsettling. These feelings, among others, end up attached to the medium of sound. Affect becomes an inevitable part of collaboration: talking about sound with other researchers links the spoken-ness of SpokenWeb to the affective sociality of sound-based research and sonic archives. Affective attachments to and created through sound suggest that sound-based research can *do* something that print-based research cannot. That potentiality also defines SpokenWeb in that audio recordings no longer sit silently on a shelf. Instead, these recordings have the potential to change what literary audio sounds like in Canada. SpokenWeb's audio collections are archives and, like all archives, their contents are shaped by power structures that determine what is "worth" saving; however, within archives there can exist recordings of marginalized voices that have not yet been amplified and, in the case of SpokenWeb, new audio collections are continuing to be acquired through institutions and community partners, thereby destabilizing the traditional gatekeeping of the archival process and expanding the collections to include recordings of literary events that expand the definition of what constitutes a literary audio recording into the realms of performance and sound art, spoken word, and collective happenings.

Of course, the collecting of recordings is not enough to restructure power in the archives, as this process must be accompanied by extensive critical engagement on levels of form and content. But, at the same time, collecting these recordings is an intervention into literary *sound*, particularly within the context of a national literary culture reckoning with whiteness and hetero-patriarchy in its literary culture. What is in SpokenWeb's audio collections will largely determine how radical this change could be in what literary audio recordings and the research produced by them could sound like in Canada. Then, in tandem with dismantling hierarchies within archives through the content of the recordings, what are the digital strategies within SpokenWeb that enable it to make an impact as a research program that scholars pay attention to and that asserts the importance of audio more broadly within digital humanities? This chapter argues that SpokenWeb rethinks digital humanities by combining acts of listening with acts of making, thereby foregrounding the implications for understanding this digital project as a sonic makerspace with its attention tuned in to a making with literary sound.

Within digital humanities, the term "makerspace" connotes collaboration (often across disciplines) and democratization of tools and technologies (albeit while recognizing that the labour of making is not immune to hierarchies): "Makerspaces are community-oriented places in which an ethos of do-it-yourself (DIY) experimentation with new technologies and materials coalesces with the goals of sharing knowledge and collaborating on project design and development" (Elam-Handloff and Rieder, n.d.). A makerspace implies a location where this making is done and, therefore, to suggest that SpokenWeb is a makerspace is a metaphorical application of the term, while, at the same time, the public outputs of SpokenWeb tend to foreground the labour behind the archival materials that inform them.[3] Making sonic objects with digital tools has been recognized as one of the fundamental elements of practising digital sound studies: "That is to say, digital sound studies scholars combine the creative use of sonic technologies with an informed critical inquiry of them, merging the lessons of digital humanities and the 'maker' movement with a thoughtful analysis of digital culture, new media, and the sonic possibilities of technologized learning spaces"

(Clement 2018, 16).[4] So how does SpokenWeb put this into practice—and, specifically, how does it put this into practice in Canada? After first unpacking the story of the recordings that have shaped SpokenWeb's affective attachment to sound and what factors led to this primarily literary project becoming a digital one, this chapter then examines (1) SpokenWeb's processing of audio collections and designing of a metadata schema, a behind-the-scenes making through listening; and, (2) *The SpokenWeb Podcast*, a public making through listening.

MAKING SPOKENWEB

Sounding literature is SpokenWeb. Sound as a medium informs each and every component of SpokenWeb as research. The sound of SpokenWeb starts with audio recorded on analogue media of reel-to-reel tapes; however, despite its roots in analogue media, SpokenWeb has always been a deeply digital project in that what it does with sound inhabits a complex nexus of literary, archival, and sound studies, and digital humanities (see figure 17.1). That nexus is central to SpokenWeb's innovation but it is also what complicates it as a digital project. Listening to SpokenWeb as a researcher who has held and holds "various positions" within it, this section unpacks the story of how SpokenWeb began and situates that story within the broader conditions in which humanities research shifted toward audio-focused methods and materials for analysis.[5] That shift coincided with the building of the partnership that is now SpokenWeb. Led

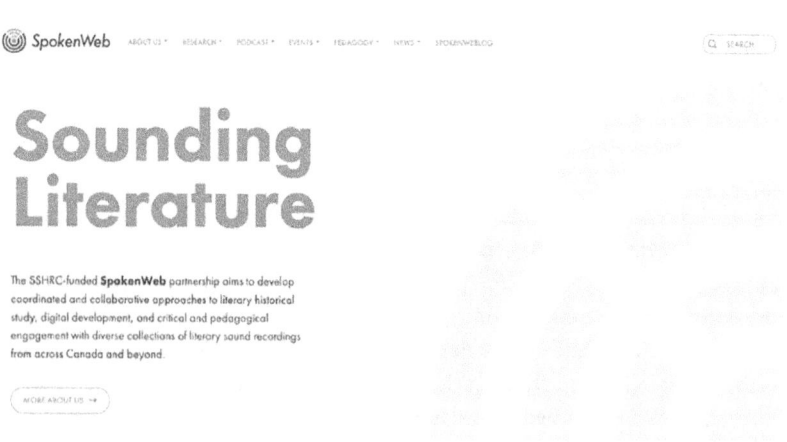

Figure 17.1. SpokenWeb.ca.

Source: SpokenWeb.ca.

by principal investigator Camlot (Concordia) and co-applicants Michelle Levy (SFU), Annie Murray (University of Calgary), Michael O'Driscoll (University of Alberta), Karis Shearer (UBCO), and Tanya Clement (University of Texas at Austin), SpokenWeb as a partnership project is made up of over 50 collaborators, including researchers, graduate students, artists, and community partners. SpokenWeb's multiple axes of inquiry include:

> 1) new forms of historical and critical scholarly engagement; 2) digital preservation and aggregation techniques, asset management and infrastructure to support sustainable access; 3) techniques and tools for searching, visualizing, analyzing, and enhancing critical engagement (for features relevant to humanities research and pedagogy); and 4) innovative ways of mobilizing digitized spoken and literary recordings within pedagogical, performative and public contexts. (SpokenWeb, n.d.-a)

At present, the SpokenWeb task forces focus on sound-signal analysis, rights data management, metadata, pedagogy, podcasting, oral history, and community collections. These multiple lines of inquiry have been part of SpokenWeb since its beginnings at Concordia University, with only one audio collection: the SGW Poetry Series, a poetry series that took place 1966–1974 at Sir George Williams, now part of Concordia University (see figure 17.2). The website for that collection remains online as the digital home for these recordings.

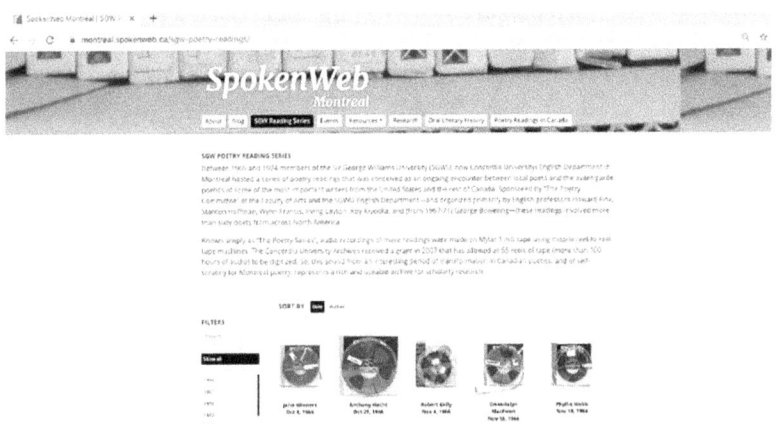

Figure 17.2. Sir George Williams Poetry Series audio collection.

Source: montreal.spokenweb.ca.

The online representation of the SGW Poetry Series is where web developer Max Stein has implemented innovative methods of transcription and time-stamping of extra-poetic speech that allows vertical scrolling during listening, thereby giving listeners control over the tools to navigate the audio through tags linked to time-stamps within the recording. Time-stamping and transcription of extra-poetic speech based on this model have been built into the current meta-data schema. Before turning to its development, more needs to be said about the timing of SpokenWeb.

The evolution of SpokenWeb was not only due to the discovery of the SGW Poetry Series tapes, though, of course, that is part of the story too; rather, as the partnership version of the project formed, it was in dialogue with a coalescing of the digital turn and the acoustic turn—a larger movement also happening at the same time as scholars became interested in new literary methods for studying spoken forms of literature, such as orature, readings, elocution, broadcasts, and interviews. Camlot and Mitchell's (2015) introduction to a special issue of the online journal *Amodern* articulates this moment of convergence:

> This issue of *Amodern* makes an argument for an expanded literary historical critical practice that considers the challenges of migrating literary cultural artifacts and media to digital formats, that registers the specificities of a distinctly audiotextual criticism, that confronts the benefits and risks of recognizing poetics as media poetics and literary histories as media and institutional histories, and that assembles new communities of scholarly practitioners in an effort to understand this manifestation of literary performance, including literary and cultural critics, digital humanists and designers, archivists, librarians, audience members, sound technicians and oral historians.

Articulating what became the goals of SpokenWeb, Camlot and Mitchell's introduction serves as a reminder that this multi-university research program traces back to that one set of recordings. A box of tapes is one place where the story begins: "The SpokenWeb program begins with the preservation and description of sonic artifacts that have captured literary events of the past, and quickly moves into a wide range of approaches and activities that activate these artifacts in the present" (SpokenWeb, n.d.-a); however, that preservation alone could not have sparked a project such as SpokenWeb without the conceptual shifts in humanities and digital humanities around the value of interdisciplinary research that brings sound into the study of literature.

SpokenWeb is at the intersections of digital humanities *and* sound studies *and* literary studies. Amid these intersections, hearing the value of audio recordings as *literary* has been important since the beginning of SpokenWeb, not only due to its connection to English departments and literary recordings but also because SpokenWeb as a research program crystallized at the Literary Audio Symposium, organized by SpokenWeb in Montréal in 2016. An earlier and equally pivotal event that highlighted the convergence of audio and digital humanities in Canada was the student-training session TEMiC (Textual Editing Modernism in Canada), hosted in 2014 by UBCO. With the program organized by UBCO's Shearer, then a newly hired faculty member, this version of TEMiC called "Editing Modernism On and Off the Page" was a foundational event in that it devoted the entire workshop to editing practices for sound-based literature. It built upon previous TEMiC workshops sponsored by the research partnership Editing Modernism in Canada that were text-based workshops on editorial practice and theory.

By 2014, when UBCO's TEMiC applied editorial theories and practices to audio materials, audio-focused sessions were starting to appear on the program for the annual Digital Humanities Summer Institute (DHSI) at the University of Victoria, but the prior and otherwise absence of audio from such a formative DH venue as DHSI was conspicuous. When the Sound and Digital Humanities course was first offered by John Barber at DHSI 2014, it began by justifying the value of sound in DH projects and explaining the need for technical skills to work with sound: "Sound, however, was often overlooked, primarily because participants lacked insight as to how it might be used" (Barber 2016). In that same year of 2014, SpokenWeb researchers Murray and Jared Wiercinski (both of whom were contributors to the first iteration of SpokenWeb and who had already published on the topic of digital sound in 2012 in *First Monday*), published an article in *Digital Humanities Quarterly* about the potential for visualizing, annotating, and "reading" digitized and born-digital audio recordings (see Arnold et al. 2021 for a recent assessment of this same set of potentialities through new case studies). As Murray and Wiercinski explain in expressing an idea that has informed SpokenWeb's mandate: "The primary goal of the SpokenWeb project is to create a sound archive that encourages scholars to engage with the sound recordings in ways that facilitate their research. Whereas for many sound archives the focus is accessibility (i.e., simply making a collection of sound recordings web accessible), SpokenWeb's focus is interactivity and productivity" (2014).

That interactivity and productivity are precisely the qualities that lead me to argue that SpokenWeb is a makerspace for doing things with sound. The digital making of SpokenWeb is about listening practices. In fact, since September

2019, SpokenWeb's Concordia team has been holding a weekly series called Listening Practice, in which a SpokenWeb researcher listeners (rather than delivering a formal talk) through an exploration of different modes of listening. For example, in February 2021, PhD student Julie Funk led a virtual (Zoom) listening practice based upon the technique of "literary machine listening"—a pedagogical approach she is working on under the supervision of SpokenWeb collaborator Jentery Sayers at the Praxis Studio for Comparative Media Studies (University of Victoria), a lab developing techniques for teaching literary audio (Telaro 2021). With its ongoing Listening Practice series, SpokenWeb foregrounds that listening is not an afterthought but rather integral to the research. SpokenWeb's work is making through listening. Moreover, while SpokenWeb's digital tools are designed with literary communities in mind, SpokenWeb pushes at the boundaries of (visually oriented) notions of what constitutes the digital humanities and, more specifically, the sound of the digital humanities.

SpokenWeb as a research project exists at a time when sound has made its way into the digital humanities, though, as noted at the outset of this chapter, debates continue as to the extent to which the digital humanities continues to privilege the visual at the expense of sound, or, for that matter at the expense of other sensory ways of knowing and representing the world. As Mary Caton Lingold, Darren Mueller, and Whitney Trettien outline in their introduction to *Digital Sound Studies*, "scholars have been carving out space for what we call digital sound studies for decades" (2018, 4). In the case of the humanities, sound calls for the humanities "to listen more closely—to attend, that is, not only to *what* but also to *how* we hear—sound studies scholars have productively theorized the sonic technologies that mediate and construct our experiences" (4–5). As for the digital humanities, the same authors go on to suggest that sound studies enters conversation with the digital humanities in two ways, both of which are practised by SpokenWeb: through the use of digital technologies and digital tools and/or through "a more hands-on approach by building digital tools and platforms for humanities research" (8). However, despite this progress, sound has remained on the margins of the digital humanities, thereby impacting any new knowledge created and imagined within these disciplines: "What forms of knowledge— and what embodied experience—are diminished by the humanities' reliance on text and visualist methods? And whose voices are going unheard in the digital turn? Bringing sound studies into meaningful conversation with digital humanities has the power to inspire new questions and foment new methods that are radically different from those of print" (11). Their argument resonates with the significance of SpokenWeb as an intervention at a moment when educational institutions are digitizing and rediscovering archival holdings. Whose voices from the archives will be heard as Canadian universities embrace this "digital turn" and start to digitize their audio collections? Will new voices be heard or will the loudest (i.e., canonized) voices only be heard again, even if in new ways?

It is a timely and necessary moment for developing sound-based methods for preserving and listening to the "unheard" archival audio collections. As SpokenWeb researcher and poet Faith Paré observes on *The SpokenWeb Podcast* ("Talking about Talking," ep. 8), recordings of Lillian Allen, Amiri Baraka, Juliane Okot Bitek, and Esi Edugyan stand out as Black voices in SpokenWeb's audio collections, as compared to, for instance, the predominantly white voices within the first audio collection to be digitized by SpokenWeb (the SGW Poetry Series).[6] The diversity of recordings currently being digitized by SpokenWeb researchers means that the sound-based research produced by SpokenWeb *will* change what literary criticism sounds like in Canada. SpokenWeb's 2021 symposium, with plenary talks by Dylan Robinson, Jonathan Sterne, Mara Mills, and Nina Sun Eidsheim, prioritizes "development of new theories and practices for underrepresented voices in audio archives" (SpokenWeb 2021). In fact, this symposium exemplifies SpokenWeb's evolution as a self-reflexive project that has turned its attention to listening as a relational practice. This practice of listening to listening is in dialogue with the work of a number of sound-studies scholars, with resonances ranging from Brandon Labelle's *Sonic Agency: Sound and Emergent Forms* (2018), where to listen is "to perceive the ever-changing relations in which the self is always embedded," to Dylan Robinson's concept of listening positionality, or how one learns to listen, as explained in *Hungry Listening: Resonant Theory for Indigenous Sound Studies* (2020). By using audio collections as resonant data for unpacking the politics of listening, SpokenWeb is a research program that is listening to how we listen. What, then, are we listening to and how does this shape our listening?

BOXES OF TAPES: ANALOGUE TO DIGITAL

When interviewing founding members of SpokenWeb for *The SpokenWeb Podcast*'s first episode, "Stories of SpokenWeb," the stories about the recordings became soundbites that were spliced together to convey the importance of that "box of tapes" for each researcher. Including transcripts of these clips here takes a cue from "The Pleasure (Is) Principle: *Sounding Out!* And the Digitizing of Community" by Aaron Trammell, Jennifer Lynn Stoever, and Liana Silva (2018), in which they interject screenshots of online conversations where they talk about what *SO!* should be in such a way that captures the tone of the conversations that produced it. With tone often conveying affect, the audio collage transcribed below tells not only the stories of those who initiated the project but also reveals the affect embedded within these stories.

> **Jason Camlot (Concordia):** I remember asking [the department chair] what those boxes contained, and he said to me, "Oh, that's just some poetry reading series that took place here in the '60s."... Then maybe a decade later I thought about those recordings again and I went back to him and asked, "Do you still have those recordings?" (5:41)

> **Annie Murray (University of Calgary):** I think that's an origin story in a lot of people's involvement with SpokenWeb: "Hey! What are those tapes?" (13:21)

> **Deanna Fong (SFU):** [I was] was going to SFU, Simon Fraser University, and just by happenstance came across this box of tapes, as we all do. (13:28)

> **Roma Kail (University of Toronto):** Our research assistant was so excited about the project that she went to our chief librarian with the archivist and they found us this unprocessed box. (13:35)

> **Karis Shearer (UBCO):** He [Warren Tallman] went to get a cardboard box at one point and brought it back to her [Jodey Castricano] and said, "You know, I want to give this to you and someday you're going to know what to do with it." And she said to me, "I think, I think this is it. I think this is what I'm supposed to do with this box." (13:46)

Finding boxes of tapes are the analogue origin stories of SpokenWeb's audio collections; however, SpokenWeb does not exist solely because of these tapes. There must be researchers (notably more than one) compelled to listen and to value the tapes themselves and whatever content that happens to be on them, which is why the timing of Camlot's discovery of the tapes is an important part of this story.

Although poetry reading has always been a form of literary performance, Charles Bernstein's *Close Listening* (1998) was ground-breaking in calling for the poetry reading to be theorized on its own—and not simply treated as a reading aloud of printed text. The increased interest in the poetry reading as a genre coincided with Camlot first noticing a box of tapes when he arrived in 1999 to

Concordia's Department of English.[7] A decade later, he went back to ask about the box, and the chair of the department told him that the tapes had been deposited in the university archives. When Camlot retrieved the tapes from the archive, he couldn't listen to them because they were reel-to-reel tapes. Even once digitized, "[a]s a collection of digitized files on CDs, The [SGW] Poetry Series audio was only slightly less useless for research than it had been when stored on reels of magnetic tape. There were no tape indexes and no contents lists. The only way to find out what was on the tapes was to listen to them, and the next step entailed doing just that" (Camlot and Mitchell 2015). Listening to these tapes involved listening to their content and devising a digital listening strategy for thinking about how the audio itself could be annotated and how the audio could be visualized as sound signals to be "read" and interpreted.

A commonality in all the stories transcribed above is that there is an affective response to finding audio recordings: an affect of excitement about a box of tapes and what they could hold that is even more audible in the vocal inflections while telling these stories in the podcast episode. There is something affective and intimate about the specific materiality of the tapes themselves—physical objects holding sound—an audible glimpse into a moment in time when someone pressed "record," and the added layer of their being stored within a box, which has relegated them to the invisible, absorbed into the storage infrastructure of departments and institutions that have not known what else to do with them because they are "other" to dominant media such as books. The discovery of a box of recordings is exciting for these researchers; even if they did not know what to do with them, the tapes "stuck" with them. I would argue the relation between researcher and audio is established by an affective "stickiness" (to apply Ahmed's term for this affective entanglement between subject and object). Is SpokenWeb, then, a digital project that is shaped by the affective economies of its audio materials? When hearing Shearer speak of her colleague handing her the box with the words—"I think this is what I'm supposed to do with this box"—the this refers to the gifting of the recordings to someone who is excited about them, who will be so pulled to the recordings that she will integrate them into her research program. As such, SpokenWeb becomes a site for theorizing the affective entanglements produced through the exchanges of an object, or rather a "sound object" (Chow and Steintrager 2019); moreover, that affect produces a new making. The object becomes digital audio scholarship, research creation, or some other mode of production. Though SpokenWeb begins with acts of preservation and exchanges of audio recordings, it requires attention given to those media followed by the labour of doing something with them, which would not necessarily happen without the affect that forges attachment between researcher and sonic materials, and audile methodologies. Thus, SpokenWeb brings a complex sociality of affect into the digital realm through the circulation of analogue recordings and the materiality of sound.

MAKING AUDIO COLLECTIONS DISCOVERABLE

Digitization, for SpokenWeb, does not simply mean transferring analogue media to digital formats, but rather compiling a dataset about the media object (form) and the audio/visual content.[8] Metadata is crucial for making recordings discoverable. For instance, when Paré mentioned finding Allen, Baraka, Okot Bitek and Edugyan in SpokenWeb's audio collections, she is referring to discovery via metadata, not by listening to every recording. It is important, therefore, that the metadata does not simply document the asset itself (that is, the physical item being archived). SpokenWeb researchers digitizing the recordings are not just making a copy; rather, they are listening closely to the recording and to the voices on it, listening to how the recordings were made (a record of labour), and even listening to the medium on which they are recorded (including photographs of the physical items). The metadata produced through their listening places recordings in relation to one another, within and across collections. For instance, one could search for an author's name and find recordings of that author from readings across institutions—a visualization of relational listening.

The listening process for digitization is mapped out in the "SpokenWeb Metadata Scheme and Catalogue for Processing," an extensive online and collaboratively written document that thoroughly describes each category of media object that the cataloguer may encounter: "Some have been born via analog media technologies (tape recorders, for example), some have been 'born digital' (recorded as digital files directly onto flash or hard drives), and some may be a mixture of the two. Some have been processed to some extent already using a different kind of cataloguing or metadata schema, and others may never have been organized before" (SpokenWeb 2020). The next step is to transfer the metadata compiled about the audio assets into SWALLOW, "a lean, open-source document-oriented database for ingesting metadata" (SpokenWeb n.d.-b). The current version of SWALLOW is evolving as a metadata management system that must be agile enough to work with collections with varying sets of metadata across collections held by various SpokenWeb-partner institutions. The data of SWALLOW is hosted on a Concordia University server. Developed by Tomasz Neugebauer and Francisco Berrizbeitia, SWALLOW has been designed for directly inputting information about an item as a cataloguer or for uploading data through Islandora or AtoM, among others. Information about recordings is comprised of a range of metadata fields, with differing fields for physical or digital items. As described in the GitHub documentation for SWALLOW:

> The most distinctive characteristic of the Swallow architecture is the complete decoupling of the metadata schema from the database and the system. This is possible by storing the metadata information in no-SQL format and implementing an engine to generate the user interface from a configuration file. As well, in configuration files, there are maps that allow Swallow to batch ingest and export data from and to different systems. These configuration files are defined as JSON objects. (SpokenWeb 2019)

That no-SQL format means that data can be exported in CSV or JSON files, which then gives agency to the user, individual institution, and/or community partner to decide how they will use that data (see figure 17.3).

As an agile metadata management system, SWALLOW continues to adapt—to listen—to the demands of varied sets of data from the different universities and their collections. The year 2023 will include front-end development for SWALLOW in order to present exported data for the user as part of its public release, but the option to export data as JSON will remain and users will be able

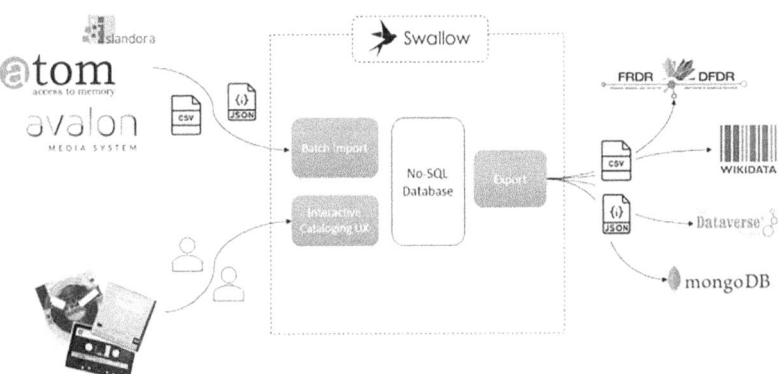

Figure 17.3. Visual representation of SWALLOW.

Source: Camlot, Neugebauer, and Berrizbeitia 2020.

to work with raw data too (see SpokenWeb's GitHub repositories for full documentation of SWALLOW; SpokenWeb 2020).

MAKING *THE SPOKENWEB PODCAST*

While SpokenWeb has been developing its metadata schema for preserving and cataloguing recordings from the past, making new audio content based on archival sound has been the focus of *The SpokenWeb Podcast*. Now in its fourth season, *The SpokenWeb Podcast* has established itself as a sonic space to experiment with DIY making-with-sound. Calling the podcast DIY does not mean that it is not professional—in fact, it is a highly produced podcast, often taking the form of an audio essay—but, rather, the DIY label points to the podcast as an experiment—an experiment both with what you can do with archival literary audio when it is re-presented in a podcast format and an experiment with what scholarly criticism about audio sounds like. In fact, the tagline for the podcast is "Stories about how literature sounds," wherein "stories" could have just as easily be replaced by "audio scholarly criticism" because each podcast makes an argument and provides critical commentary about audio recordings. Moreover, podcasts come out of DIY culture.[9] They are the zines of the airwaves. In contrast to a studio-produced radio show, a podcast can be produced at home with a recording device, editing software, and the ability to upload to the Internet. There can be an affect of excitement around podcasting as a format that is akin to the excitement around audio recordings expressed by SpokenWeb researchers. In fact, the editors of *Podcasting: New Aural Cultures* make this point in the opening pages of their book: "Podcasting imbued in us the enthusiasm of possibility [...]. The medium's hybridity of thought, sound and text perhaps even fosters a reinvigoration of the dialectic, an exchange of ideas beyond what is possible in purely written form—be it in a magazine or academic journal" (Llinares, Fox, and Berry 2018, 1–2). That surpassing of what is possible in written form is exactly what draws SpokenWeb researchers to the medium of the podcast too, with the mandate for the podcast stating that *The SpokenWeb Podcast* presents

> snapshots of literary history and contemporary responses to it, including interviews, panel discussions, lectures, readings, audio essays, and other relevant sound-based forms. The podcast will help share the research discoveries about archival spoken word collections, events, and other topics concerning sound, literature and culture relevant to the SpokenWeb research network, expanding our understanding of the relationship between the fields of literary studies and sound studies. (SpokenWeb 2019)

In season one, sounds of the podcast ranged from audio collages to audio essays, and from broad questions ("How Do Concepts Make Us Feel?," ep. 11) to author-specific ones related to recordings from a particular collection ("Invisible Labour," ep. 3).[10] As the podcast now begins its fourth season, and since each episode is made by different SpokenWeb researchers and/or collaborations among researchers, its sound inevitably changes—and each episode has the potential to surprise listeners with what audio criticism about sound can sound like—and that it need not try to replicate the sound of the printed page.

As of September 2020, *The SpokenWeb Podcast*'s director Hannah McGregor launched the Amplify Podcast Network, a SSHRC Partnership Development project investigating the podcast as academic scholarship: how best to peer review a podcast, how to circulate a podcast to academic communities, and how to evaluate a podcast as academic scholarship for hiring and tenure committees. As an academic podcast, *The SpokenWeb* presents an opportunity to reconsider the format itself as a mode of digital dissemination, and to consider more broadly the ways in which digital scholarship is evaluated as "sounding" scholarly—and how previous frameworks for the digital representation of research findings are challenged by the audio format of the podcast.

The podcast responds to shortcomings identified in formats for digital publications: "Any given audio file in a Scalar book can be annotated through discrete, time-stamped commentary by the author, and this commentary is displayed within the medium's own temporality. While such features are now typical in visual culture (e.g., annotating lexia in Commentpress or tagging an image in Flickr), few such mechanisms exist for the scholarly treatment of sound" (Sayers 2012). But it does not have to be this way, as Jonathan Sterne remarks upon the generative role that audio can play in reshaping DH scholarship:

> The purpose of digital humanities scholarship is analysis and criticism of (or in) forms other than print on a page. This may include "production" but it doesn't have to—it could be as simple as written commentary accompanying sound files on a page [...]. As with audiovisual texts (film, TV, etc.), "born digital" publication allows commentary to be situated alongside or directly within media that unfold over time. Moreover, sound scholarship stands the most to gain from born digital publication. (2011)

If sound-based scholarship has the most to gain, then it necessarily involves a rethinking of the digital formats and methods by which the research is disseminated. Academic criticism could become noisier and—even as researchers develop ways to mark up this noise, such as with Scalar or AudioAnnotate[11]—embedded within digital audio criticism is a resistance to the two-dimensional page.

Stepping fully off the page and into sound, *The SpokenWeb Podcast* is an experiment in scholarly making with sound—one that resonates with its subject matter, in that it places the presence of the voice at the fore. For example, in episodes of the podcast that undertake a feminist recovery of voices in the archives— "The Voice Is Intact: Finding Gwendolyn MacEwen in the Archive" (ep. 7) and "Producing Queer Media" (ep. 9)—stories are told about voice in a way that could not be told on the page. Embodiment and what is at stake in that embodiment become audible.[12] The range of sounds invites a comparison between the complex and even messy sound of podcast scholarship and the diverse sounds of recordings within SpokenWeb's collections—collections that feature not only readings but also sounds of the classroom, interviews, conversations, festivals, audience noise, and even the sound of blank tapes. The sounds of this generative messiness serve as documentation, and they also lend themselves to endless re-makings. What will you do with these sounds next?

LISTENING AS MAKING

The sound of SpokenWeb largely relies upon the content of its audio collections, but, at the same time, its sound will always be shaped by the methods of listening practised by the scholars working on these collections and what is made through this listening—a listening as making. The DIY culture of the digital humanities informs the making of new literary audio recordings in SpokenWeb, often documentation and research creation simultaneously. Making with archival audio has been performed by SpokenWeb researchers in a range of formats that go beyond the aforementioned podcast: live literary events that engage with archival recordings (Kaie Kellough and Catherine Kidd performing at the Words and Music Show at Concordia University in 2018); readings by a poet alongside an archival recording of themselves reading ("Performing the Archive: An Epic Reading" [UBCO, 2019]; "A Poetry Listening" [Concordia, 2019]); and mediated literary events such as the Ghost Reading series, 2018–2019, in which an entire recording of a poetry reading was played and listeners responded by writing, drawing, and crafting, and then talking afterwards. One might ask, what does this have to do with digital humanities? These literary events have relied upon the DH side of SpokenWeb, drawing on SpokenWeb's archival infrastructure and digitization to enact conceptual understandings of both analogue and digital technologies, and to participate in creative and critical acts of making. This making that informs SpokenWeb has been there since the start (as argued by Lee Hannigan, Aurelio Meza, and Alexander Flamenco 2018), with the research activities of SpokenWeb always being a making with archival materials, or rather an "unarchiving" (Camlot and McLeod 2019). Unarchiving, or re-presenting archival materials through critical and creative methods to new publics, activates the archives in such a way that exemplifies why SpokenWeb itself is not an archive but rather a research program—a research program that is collecting new archival materials but that is also fundamentally about the making with, working with, and listening to these materials in order to understand how they transform the methods of literary and digital scholarship. In this way, SpokenWeb bridges the divide that Sayers articulates between "hands-on" and experiential acts of building within digital humanities and critical disciplines that theorize and contextualize the media objects (media and sound studies, for example).[13] SpokenWeb undertakes a making-with-literary-audio-recordings and a theorizing of audio literary recordings that, in fact, overlap within the project's practices of infrastructure building and research dissemination.

As I return to this chapter to revise it in 2022, SpokenWeb has further demonstrated a making-with that builds community among its researchers; for example, leading up to SpokenWeb's 2021 symposium, "Listening, Sound, Agency," Concordia's SpokenWeb team conducted interviews with presenters that were subsequently published on SPOKENWEBLOG; post-symposium projects include the publication *Quotes: Transcriptions On Listening, Sound, Agency* (edited by Klara du Plessis and Emma Telaro and published by SpokenWeb); lathe-cut polycarbonate records of sounds submitted by symposium presenters; a podcast episode about the symposium (produced by Mathieu Aubin and Stephanie Ricci for *The SpokenWeb Podcast*); an online revisiting of the soundwalks held during the symposium (led by Angus Tarnowsky); and a special double issue of *English Studies in Canada*, "New Sonic Approaches in Literary Studies," edited by Camlot and McLeod (forthcoming).[14] All these creations are versions of *listening* to the symposium (let alone the talks at the symposium itself) and they demonstrate how SpokenWeb undertakes a making-with based on recordings and events, including recordings of its own events. While that making-with as a practice is not unique to SpokenWeb, that practice shows the value placed in exploring how one type of research output can lead to another (e.g., a podcast episode can lead to a peer-reviewed article),[15] which mirrors the kind of making-with archival materials that informs SpokenWeb as a project.

SpokenWeb is a DH research program rooted in sound, but what the first years of SpokenWeb as a partnership has also shown is that SpokenWeb is about *how* research is conducted. That research has been undertaken through an ethos of process and the documentation of that process. Moreover, that research is often collaborative in nature—calling for performance, for oral-history interviews about archival recordings, or for a collective listening, among other strategies— which has resulted in the many DH aspects of SpokenWeb being about its *web*, as much as it is about its sound. Although, of course, its sound is what holds its collaborators together. The collaborative making that happens within the "lab" of SpokenWeb is a making that happens out of a desire to bring sound to new listeners, which recalls the stickiness of the project, or rather the pull of the researchers to the audio archives and to imagine a future value in hearing and studying its sounds.

> **Katherine McLeod**: In many ways, the web of SpokenWeb is not only the collaborative network of researchers, but it is also this web of archival recordings held together by a desire to make these available to more listeners. (39:16)

Those are my own words in the first episode of *The SpokenWeb Podcast*, "Stories of SpokenWeb" (SpokenWeb 2019), when we were still imagining what the podcast would sound like as a series. If this chapter is audible—even to the degree that it has made you think about what writing *about* sound sounds like—then this chapter about SpokenWeb will have succeeded in reaching new listeners, and the next step will be to find new ways to continue the work of making with sound.

NOTES

1. The impact of the SSHRC Partnership Grant program on the development of human-ities research projects in Canadian universities could be the subject of its own study. The partnership grant as collaborative research, the emphasis on the digital, and the impact on career paths of graduate students in Canadian literature, for example, could all could be explored through partnership grants of the past 15 years such as Editing Modernism in Canada (EMiC), Canadian Writing Research Collaboratory (CWRC), Implementing New Knowledge Environments (INKE), and Archives/CounterArchive (A/CA).

2. Gerry Gilbert radiofreerainforest Collection, https://digital.lib.sfu.ca/gerry-gilbert-radiofreerainforest-collection; the Soundbox Collection, https://soundbox.ok.ubc.ca/; SGW Poetry Series, https://montreal.spokenweb.ca/sgw-poetry-readings/.

3. As a self-reflexive record of SpokenWeb's events and the labour to organize them, the metadata of "Archive of the Present" (launched April 2020) strives to make visible the labour of organizing literary events; see https://archiveofthepresent.spokenweb.ca/.

4. The reference to the "and the 'maker' movement" has a footnote in the original text that references Jentery Sayers's lab at the University of Victoria. The same lab under the direction of Sayers is now collaborating with SpokenWeb in making digital tools for audio analysis. It is no coincidence that many of SpokenWeb's foundational researchers are also positioned at the intersection of sound and digital humanities, such as Tanya Clement (University of Texas at Austin), Marit MacArthur (University of California, Davis), Jentery Sayers (University of Victoria), Karis Shearer (UBCO), Geoffrey Rockwell (University of Alberta), Chris Mustazza (University of Pennsylvania), Adam Hammond (University of Toronto), who are all leading collaborators in the partnership grant of SpokenWeb.

5. It is worth taking a moment to consider the position from which I listen to SpokenWeb, or rather the position from which I assemble an audio "snapshot" of it for this chapter (which, of course, cannot contain everything, but rather can just present an assessment of the project at this moment in time). I have written this chapter without consulting SpokenWeb team members, with two exceptions: prior to writing the chapter, I said to Karis Shearer, "Imagine if I wrote a written version of the podcast episode, 'Stories of SpokenWeb'"—and, once the chapter was at the page-proof stage, I shared it with Jason Camlot to hear his response to my response to SpokenWeb. Throughout, I have been listening from *inside* SpokenWeb, as a researcher involved starting with planning workshops for the SSHRC partnership application and throughout (working on symposia planning; developing the Ghost Reading series, which led to the current Listening Practice series; serving as managing editor of SPOKENWEBLOG and as a member of the podcast task force; and creating, writing, and producing *ShortCuts* as a monthly feature on the *SpokenWeb Podcast* feed; among other contributions). I have also been listening to SpokenWeb from *outside* of the traditional roles of academia, and that is something that is not as visible in the ways in which academic work is itemized. At the time of revising this chapter, I have been offered a faculty position as a limited-term appointment in the English Department at Concordia University, and the process is underway for my formal affiliation with SpokenWeb as a co-applicant. That affiliation will formalize with SSHRC my role as a contributor to SpokenWeb but, over these past years, what has been most significant is that SpokenWeb recognized that—amid academic precarity—significant research contributions *can* happen with one foot in and one foot out of academic institutions.

6. Paré made those discoveries by searching the metadata system of SWALLOW (discussed later in this chapter) while undertaking research in collaboration with SpokenWeb's artist-curators in residence (2020–2021), jamilah malika and Jessica Karuhanga, whose own projects are (un)making the sound of Black archives: malika's project is an online archive highlighting Black women sound artists across Canada, and Karuhanga is creating "a sanctified Black space in the form of a website that celebrates aural, visual and somatic witnessing" through shared audio recordings of personal stories ("Talking about Talking" 2021).

7. Here is another telling of this story through its artifacts:

The story of The Poetry Series tapes as artifacts for digital presentation thus begins as the story of a hidden collection, for which there are two possible narrative trajectories: stories of discovery or stories of loss. The story of The Poetry Series became one of discovery. Following their initial deposit and basic cataloguing, the next significant phase of the tapes' discernibility was achieved when the University Archives received a grant in 2010 to have them digitized and stored as WAV files on archival quality compact discs. At about the same time, Camlot remembered that the tapes existed and decided to find out what had happened to them. Directed by his colleague, the depositor, to the English fonds, Camlot saw the CDs and started working towards the expanded discoverability of the collection. (Camlot and Mitchell 2015)

8. Exemplary of this process is the work of SpokenWeb research assistant Leah Van Dyk. As she explains: "I am currently preparing our audio archives for public dissemination and am super excited about being able to share and explore the University of Calgary's archival audio recordings.... Our team at Calgary is small (but mighty!), which means that every audio recording is listened to and has data produced by me, and I can't wait to have the opportunity to see the public begin to engage with the recordings I have come to know so well." Van Dyk is digitizing and producing data, but at the same time the audio is informing research questions: "The most surprising—and often entertaining—moments within my work come from what we might call the 'mistakes' within a collection: botched recordings, blank tapes, seemingly unrelated notes and labels" (Aubin 2020).

9. *The SpokenWeb Podcast* "Invisible Labour" (ep. 3) is made by the UBCO SpokenWeb team and produced by Karis Shearer and Nour Sallam; this episode also aired on the UBCO podcast series, *SoundBox Signals*.

10. At ACCUTE 2018, a conversation took place at a panel on transformations in scholarly publishing. Organized and chaired by Hannah McGregor, the panel included Siobhan McMenemy, Ada S. Jaarsma, Amanda Cooper, Stephen W. MacGregor, and myself. A question was asked by Kevin McNeilly about whether, in becoming institutionally recognized as scholarly criticism, podcasts ran the risk of losing the DIY sound that drew listeners to them to begin with. A lively conversation ensued, but it is not about losing that DIY sound but rather recognizing that the labour of making audio-based criticism need to be validated as scholarly work as an academic. It is a conversation that I return to when thinking of what "scholarly" sounds like—and I ask myself of what assumptions about institutionally "valued" sound end up informing how one creates a scholarly podcast, and to remember that whatever one creates is only version of what a scholarly podcast could sound like.

11. AudioAnnotate, developed by Clement and Sara and Ben Brumfield of Brumfield Labs, is an application and a workflow that "will help users to translate their own analyses of audio recordings into media annotations that will be publishable as easy-to-maintain, static, W3C Web Annotations associated with IIIF manifests and hosted in a GitHub

repository that are viewable through presentation software such as Universal Viewer" (HiPSTAs, n.d.).

12. The podcast also becomes a sonic space to return to past conversations, such as in "Revising Feminist Noise, Silence, and Refusal" (ep. 5) when producers Kate Moffatt and Michelle Levy curate the replaying of a series of talks (delivered by Lucia Lorenzi, Milena Droumeva, Brady Marks, and Blake Nemec) at the first SpokenWeb Symposium, and then revisits one of these talks with a conversation between Kate Moffatt and Milena Droumeva about the impact of the talk and its data sonification.

13. Sayers made this argument in 2012, but it continues to have resonance for the extent to which digital humanities enters into conversation with adjacent disciplines and their practices of scholarly publication:

> Informed by claims from experience and anchored in embodied acts of building, digital humanities arguments necessarily become "hands on," and scholarly distance from technologies no longer holds. Meanwhile, media studies investments in cultural criticism and situated knowledge-making are increasingly important to today's digital humanities practitioners, involved such as they are in multimodal communication (e.g., interactive visualizations, geospatial representations, rich exhibits, and gaming). For instance, Alan Liu argues that "digital humanities should enter into fuller dialogue with the adjacent fields of new media studies and media archaeology so as to extend *reflection* on core instrumental technologies in cultural and historical directions" (Liu 2012, 501, emphasis added). This fuller dialogue would enhance the field's awareness of how work with technologies and data intersects with the relevant social, economic, and political issues of our time. (Sayers 2012)

14. This year, after the graduate symposium—The Sound of Literature in Time (May 2022)—a publication about the convergence of data, sound, and affect is underway, tentatively titled *Affective Signals: Literary Sound as Data* (edited by du Plessis and Wiener, with Camlot and McLeod).

15. As an example of episodes of *The SpokenWeb Podcast* transformed into peer-reviewed articles, see "Pandemic Listening: Critical Annotations on a Podcast Made in Social Isolation," by Camlot and McLeod, *Canadian Literature* 245 (2021), based on their episode "How Are We Listening Now? Signal, Noise, Silence" (May 2020); and Michelle Levy, Kate Moffatt, and Kandice Sharren's forthcoming peer-reviewed article in *English Studies in Canada* based on their podcast episode, "Mavis Gallant, Part 2: The 'Paratexts' of 'Grippes and Poche'" (June 2021).

REFERENCES

Ahmed, Sara. 2010. "Happy Objects." In *The Affect Theory Reader*, edited by Melissa Gregg and Gregory J. Seigworth, 29–51. Durham, N.C.: Duke University Press. https://doi.org/10.1215/9780822393047-001.

Arnold, Taylor, Jasmijn van Gorp, Stefania Scagliola, and Lauren Tilton, eds. 2021. "AudioVisual Data in DH." *Digital Humanities Quarterly* 15, no. 1. http://www. digitalhumanities.org/dhq/vol/15/1/index.html.

Aubin, Mathieu. 2020. "'Mistakes' in the Sound Archive: An Interview with Leah Van Dyk." SPOKENWEBLOG. spokenweb.ca/mistakes-in-the-sound-archive-an-interview-with-leah-van-dyk/.

Barber, J. F. 2016. "Sound and Digital Humanities: Reflecting on a DHSI Course." *Digital Humanities Quarterly* 10, no. 1. http://www.digitalhumanities.org/dhq/vol/10/1/000239/000239.html.

Bernstein, Charles, ed. 1998. *Close Listening: Poetry and the Performed Word*. Oxford, U.K.: Oxford University Press.

Camlot, Jason, and Christine Mitchell, eds. 2015. "The Poetry Series." *Amodern* 4. https://amodern.net/issues/amodern-4/.

Camlot, Jason, and Katherine McLeod, eds. 2019. *CanLit Across Media: Unarchiving the Literary Event*. Montréal, Que., and Kingston, Ont.: McGill-Queen's University Press. https://doi.org/10.2307/j.ctvscxtkg.

Camlot, Jason, Tomasz Neugebauer, and Francisco Berrizbeitia. 2020. "Dynamic Systems for Humanities Audio Collections: The Theory and Rationale of Swallow." In *DH2020 Book of Abstracts*, edited by Laura Estill, Jennifer Guiliano, and Constance Crompton. https://dh2020.adho.org/wp-content/uploads/2020/07/730_DynamicSystems forHumanitiesAudio CollectionsTheTheoryandRationaleofSwallow.html.

Chow, Rey, and James A. Steintrager, eds. 2019. *Sound Objects*. Durham, N.C.: Duke University Press. https://doi.org/10.1515/9781478002536.

Clement Tanya. 2018. "Word. Spoken. Articulating the Voice for High Performance Sound Technologies for Access and Scholarship (HiPSTAS)." In Lingold, Mueller, and Trettien, 155–177. https://doi.org/10.1215/9780822371991-008.

Elam-Handloff, Jessica, and David M. Rieder. n.d. "Makerspace." Digital Pedagogy in the Humanities. https://digitalpedagogy.hcommons.org/keyword/Makerspaces/.

Hannigan, Lee, Aurelio Meza, and Alexander Flamenco. 2018. "Reading Series Matter: Performing the SpokenWeb Project." In *Making Things and Drawing Boundaries: Experiments in the Digital Humanities*, edited by Jentery Sayers. Minneapolis: University of Minnesota Press, 2018. https://doi.org/10.5749/j.ctt1pwt6wq.25.

HiPSTAS. n.d. "AudioAnnotate." http://hipstas.org/audiannotate/.

LaBelle, Brandon. 2018. *Sonic Agency: Sound and Emergent Forms of Resistance*. London, U.K.: Goldsmiths Press.

Lingold, Mary Caton, Darren Mueller, and Whitney Trettien. 2018. Introduction to *Digital Sound Studies*, 1–28. Durham, N.C.: Duke University Press. https://doi.org/10.1215/9780822371991-001.

Llinares, Dario, Neil Fox, and Richard Berry. 2018. "Introduction: Podcasting and Podcasts—Parameters of a New Aural Culture." In *Podcasting: New Aural Cultures*. New York: Palgrave Macmillan. https://doi.org/10.1007/978-3-319-90056-8.

Murray, Annie, and Jared Wiercinski. 2014. "A Design Methodology for Web Based Sound Archives." *Digital Humanities Quarterly* 8, no. 2.

Robinson, Dylan. 2020. *Hungry Listening: Resonant Theory for Indigenous Sound Studies*. Minneapolis: University of Minnesota Press. https://doi.org/10.5749/j.ctvzpv6bb.

Sayers, Jentery. 2012. "Abstract: Writing with Sound: Composing Multimodal Long Form Scholarship." Digital Humanities 2012. http://www.dh2012.uni-hamburg.de/conference/programme/abstracts/writing-with-sound-composing-multimodal-long-form-scholarship/.

SpokenWeb. 2019. "About the SpokenWeb Podcast." https://spokenweb.ca/podcast/spokenweb-podcast/.

———. 2020. "SpokenWeb Metadata Scheme and Cataloguing Process." SpokenWeb Team Revision 1b862b8d. https://spokenweb-metadata-scheme.readthedocs.io/en/latest/1-intro.html.

———. 2021. "Listening, Sound, Agency: An International Scholarly Symposium (May 18–23, 2021)." https://spokenweb.ca/symposia/.

———. 2022. "spokenweb/swallow." https://github.com/spokenweb.

———. n.d.-a. "About us." https://spokenweb.ca./about-us/.

———. n.d.-b. "Toolkit—SpokenWeb." https://spokenweb.ca/research/toolkits/.

Sterne, Jonathan. 2011. "Audio in Digital Humanities." Super Bon! https://superbon.net/2011/07/24/audio-in-digital-humanities-authorship-a-roadmap-version-0-5/.

"Stories of SpokenWeb." 2019. *The SpokenWeb Podcast*. Produced by Cheryl Gladu and Katherine McLeod, October 3, 2019. https://spokenweb.ca/podcast/episodes/stories-of-spokenweb/.

"Talking about Talking, ft. jamilah malika, Jessica Karuhanga, and Faith Paré." 2021. Podcast. *The SpokenWeb Podcast*. Produced by Katherine McLeod, May 3, 2021. https://spokenweb.ca/podcast/episodes/talking-about-talking/.

Telaro, Emma. 2021 "On Literary Machine Listening and Pedagogy: The Praxis Studio with Julie Funk, Faith Ryan, and Jentery Sayers." SPOKENWEBLOG, February 26. spokenweb.ca/on-literary-machine-listening-and-pedagogy-the-praxis-studio-with-julie-funk-faith-ryan-and-jentery-sayers/.

Trammell, Aaron, Jennifer Lynn Stoever, and Liana Silva. 2016. "The Pleasure (Is) Principle: Sounding Out! and the Digitizing Community." In *Digital Sound Studies*, edited by Mary Catton Lingold, Darren Mueller, and Whitney Trettien, 83–119. Durham, N.C.: Duke University Press. https://doi.org/10.1215/9780822371991-005.

Linking Out: The Long Now of Digital Humanities Infrastructures

Susan Brown, Kim Martin, and Asen Ivanov

I think an approach focused on institutions and their infrastructures is particularly appropriate. [...] Acting out through the digital humanities about larger social issues is necessary. But such actions must be complemented by creating infrastructures and practices that make their social impact by being what Susan Leigh Star called "boundary objects"—in this case boundary objects situated between the academic institution and other major social institutions.

—Alan Liu (2016)

A great promise of infrastructure is that it will link us all together. Christine Borgman (2007), for example, views the "added value of linking" (117) as one of the greatest benefits of digital research infrastructure (DRI). In this context, linking often means a system's ability to link data and documents. But work in critical infrastructure studies extends the notion of linking further to designate DRI's ability to link not only data and documents but also technologies, communities, and modes of knowledge organization and management (Ribes and Finholt 2009). Like Liu (2016), we believe DH research infrastructure is best understood through Geoffrey C. Bowker and Susan Leigh Star's (1999) foundational concept of a boundary object: an entity that inhabits "several communities of practice and satisf[ies] the informational requirements of each of them... [an entity that is] plastic enough to adapt to the local needs and constraints of the several parties employing them, yet robust enough to maintain a common identity across sites" (297). This chapter analyzes three DRI projects for the humanities as exemplars of general shifts in infrastructure development over the past quarter century as well as indicators of the Canadian context. The analysis traces their increased movement toward linking people, data, and technologies, describing their increasing function as boundary objects that serve diverse modes and communities of cultural inquiry.

The effective use of digital resources for humanities inquiry is hampered by many factors, the foremost being their distribution across many sites and formats and the lack of systematic interlinking. Since calls for cyberinfrastructure, including infrastructure dedicated to cultural heritage (Comité de Sages 2011; Unsworth 2006), emerged in the earlier twenty-first century, regional, national, and international DH projects in Europe and North America have looked to build DRI to address these and other problems by promoting access, collaboration, reuse, interoperability, preservation, and sustainability, as well as supporting more advanced computing methods (Benardou et al. 2018; Borowiecki, Forbes, and Fresa 2016; Meyer and Schroeder 2015). Reflections on infrastructure have intensified in the past decade, alongside the growing use of an ecological metaphor for infrastructure (Linley 2016) that demonstrates changing understandings of DRI generally and in the humanities.

This chapter's case studies are three successive infrastructure projects that build on one another: the Orlando Project, the Canadian Writing Research Collaboratory, and the Linked Infrastructure for Networked Cultural Scholarship.[1] They date from the late 1990s to the present and reflect aspects of DRI generally, as well as the development of DH infrastructure in Canada. Tracing the development of these three projects demonstrates how DH infrastructure differs from cultural-heritage infrastructure and library infrastructure in its need to provide significant support for managing the efforts of distributed teams, collaborators, or members of communities of practice composed of scholars with varying levels of technological expertise, in contrast with trained information professionals.

In the Canadian context, DRI developed under a set of specific conditions emerging from infrastructure funding geared toward STEM fields and an innovation-driven agenda for the digital economy.[2] This is in contrast to Europe, for instance, where, for decades, there have been dual and coordinated efforts toward both cultural-heritage and research infrastructure, driven by EU policy (Comité de Sages 2011). European DRI acts "as an interface through which the different actor groups (researchers, funders, policy makers) rearticulate their mutual relations" (Kaltenbrunner 2017, 301), but has also been criticized for generalizing and standardizing humanities data to the point of making them unusable (Zundert 2012). The EU approach, in our view, produced DRI more integrated with galleries, libraries, archives, and museums (GLAM) and physical cultural heritage (e.g., Pelagios and Trismegistos; see the appendix below for project hyperlinks) in addition to textual and linguistic data (CLARIAH, CATMA). In North America, there is less coordination overall with GLAM initiatives (although Omeka's exhibition platform is an exception), and emphasis has fallen more on DRI for sophisticated text distribution, editing, and publishing (Perseids, TAPAS, Scalar) as well as smaller, topically focused DRI often based in DH centres and funded by private as well as public agencies (Almas 2017;

Flanders and Hamlin 2013; Kaltenbrunner 2017, 293). Canadian digital humanities in particular have been strong in text analysis and nonmaterial history;
much DRI developed here extends that work (Voyant, TAPoR, Canadian Century
Research Infrastructure). A few infrastructural efforts, such as the Text Encoding
Initiative (TEI) community of practice, have bridged North America and Europe.

This chapter is divided into the following sections: a review of the literature
on research infrastructure in several disciplines, including information science,
critical infrastructure studies, and digital humanities; a tripartite account of the
infrastructures; analysis of humanities DRI in terms of people, technology, and
data; and a concluding discussion of the implications of our analysis.

The analysis below advances a view of DRI as an adaptive ecosystem of technologies, communities, and modes of knowledge management. We develop this
conceptualization by drawing on debates in information science and infrastructure studies, as well as contributions to the larger *infrastructural turn* in thinking
about technology from the digital humanities (Rockwell 2010). Across these
fields, consideration of infrastructure has increasingly emphasized care and
repair (Nowviskie 2015; Ramakrishnan, O'Reilly, and Budds 2020) as well as
understanding of infrastructure as a complex system of information, technology, and people (Bowker and Star 1999; Edwards et al. 2009; Ribes and Lee 2010).
We further review ideas associated with visions of digital libraries at the turn of
the twenty-first century as an antecedent to humanities DRI, and as indicative
of a gradual shift happening within libraries from knowledge organization to
knowledge representation.

INFRASTRUCTURE AS ECOSYSTEM: TECHNOLOGIES, PEOPLE, AND DIVERSITY

While a distinct line in the scholarship on research infrastructure has focused
on identifying requirements for networks, systems, and software development
(Borgman, Wallis, and Enyedy 2007; Borgman 2007; Hey and Trefethen 2005),
a parallel literature has examined infrastructure through a sociological and historical lens. This literature is diverse in disciplinary scope, encompassing contributions from science and technology studies, information sciences, information
systems, and computer-supported collaborative work, among other fields. We
review key contributions to define the concept of infrastructure and highlight its
properties and dynamics as a complex adaptive system (Ribes and Finholt 2009).

One of the most influential sociological and historical contributions to the
study of infrastructure is the "Understanding Infrastructure" report by Edwards

et al. (2007), developed for the National Science Foundation in the United States. The report's authors describe infrastructure development as the process of integrating "locally constructed, centrally controlled systems" into networks "governed by distributed control and coordination processes" (7). The report established an understanding of infrastructure in terms of ecology, advocating diverse cyberinfrastructure initiatives. Infrastructure, in this view, is not "built but grown," and develops less from design and engineering efforts than from dynamic "competition among technological systems and standards" that results in their consolidation into a service that fulfills a social need (8, 10, 42). Subsequent definitions buttress this ecosystem view. For example, Ole Hanseth (2010) describes infrastructure as a continuously evolving network consisting of many heterogeneous elements. Likewise, for Borgman (2015), "[i]nfrastructures are not engineered or fully coherent processes. Rather, they are best understood as ecologies or complex adaptive systems" (33). Discussions in the digital humanities have also subscribed to this understanding of infrastructure as a space within which alignments and configurations among elements emerge organically and cannot be fully anticipated, along with some efforts to unpack the implications of the ecological metaphor (Brown and Simpson 2015; Linley 2016).

The process through which infrastructure consolidates into a service has organizational, political, and technological dimensions. For Edwards et al. (2007), systems and technologies are connected to infrastructure through gateways—"plugs and sockets that allow new systems to be joined to an existing framework easily and with minimal constraint" (15). Gateways are equal parts technological solutions and "social choices" made by a community of practice (15). Borgman (2015) similarly notes that infrastructure "consists of many parts that interact through social and technical processes" (33). Charlotte P. Lee, Paul Dourish, and Gloria Mark (2006) suggest that "human infrastructure is shaped by a combination of both new and traditional team and organizational structures," and that "fluid organizational structures should be embraced and encouraged as a strength" of infrastructure projects (491–92). In all these accounts, infrastructure depends as much on links between people and communities as on links between systems and technologies.

Although sometimes buried beneath terms like "organization," people are a crucial and often overlooked aspect of infrastructure, as DH work regularly notes. Geoffrey Rockwell (2010) proposes infrastructure as a "mix of hard visible components, softer services, and professionals that operate and maintain the two." Joris van Zundert (2012) indeed rejects over-engineered infrastructure development as a "dead end for information technology development and application in the humanities" in favour of more agile development that "value[s] humans and interaction over planning and documentation" (Zundert 2012). Addressing social-humanities infrastructure, Elizabeth Grumbach and Laura Mandell (2014) go further to acknowledge users who bring individual research agendas, projects, and needs as a vital component of infrastructure. Agiatis Benardou et al. (2018) note that "contemporary Research Infrastructures have the aspiration of being not merely collections of research resources or tools to conduct research: they are energized by a community of research institutions and individual researchers, and become living environments of evolving, synergistic but also often competing research, education and communication practices" (3).

As Sheila Anderson (2013) shows, this ecosystem view of infrastructure also underpins the influential Atkins et al. (2003) cyberinfrastructure report, which set the agenda for infrastructure development in the sciences and provided a foundation for the humanities-focused *Our Cultural Commonwealth* report (Unsworth 2006). In Anderson's analysis, both reports share an underlying view of infrastructure as a complex adaptive system constituted by "layers of information, expertise, standards, policies, tools, and services that are shared broadly across communities of inquiry but developed for specific scholarly purposes [...]. It is also the more intangible layer of expertise and the best practices, standards, tools, collections, and collaborative environments that can be broadly shared across communities of inquiry" (2013, 15). This notion of infrastructural layers accords with Benjamin Bratton's (2016) ambitious notion of "The Stack" as the governing geopolitical structure of our time, or what Liu characterizes as "our time's fundamental ideological-cum-technological platform paradigm" (Liu 2020, 133). Working from Bratton's framework, Liu argues that the digital humanities requires a complex, multi-layered "diversity stack" that can support complex cultural inquiry. The infrastructure initiatives described below, which emerge from intersectional feminist research, can be understood as contributing to what Liu describes as "a virtuous circle in which research on diversity helps shape technical innovation and, in turn, technical innovation designs new ways to understand and act on diversity" (136).

DATA INFRASTRUCTURE: FROM KNOWLEDGE ORGANIZATION TO KNOWLEDGE REPRESENTATION

Technology and people are crucial to the linking that is characteristic of infrastructural ecosystems. Likewise, the structuring and modelling of data plays an integral role in how that linking can occur. Data-modelling practices from library and information science are particularly pertinent to considerations of DH infrastructure. Not only are libraries justifiably dubbed the labs of the humanities, as stewards of pre-digital knowledge infrastructure, but digital libraries are also an essential component of research infrastructure ecologies for the humanities. Echoing our emphasis here on people, technologies, and data, digital libraries can be defined as "*organizations* (these services have been developed in certain institutional contexts to serve specific work tasks and user groups), *technological infrastructures* (e.g., hardware and database solutions), and *document collections* (contents and their ordering principles)" (Tuominen, Talja, and Savolainen 2003, 562 [emphasis in the original]; see also Borgman 1999; Rowlands and Bawden 2009). One crucial way in which libraries are redefining anew their role as the new laboratories of DH scholarship (Gitelman 2010; Sula 2013) is through participating in a larger shift in data modelling and management practices fundamental to the digital infrastructural turn.

The debate over digital libraries has, from the start, centred on how technology can unlock the knowledge contained in library resources and is resulting in a gradual move from print-legacy practices of knowledge organization (KO) toward knowledge representation (KR) (Sowa 2000). Practically, this shift involves augmenting metadata schemas and records with forms of knowledge representation such as linked data to provide fuller expression of the knowledge within, relationships between, and contexts of library resources (Giunchiglia, Dutta, and Maltese 2014). KR opens library resources to a variety of new forms of inquiry (Unsworth 2001). Consequently, much early thinking about digital libraries revolves around questions of how to more effectively augment and link library resources. Among key early definitions, for example, Clifford Lynch (2002) describes digital library collections as "databases of relatively raw cultural heritage materials...[with added] layers of interpretation and presentation built upon these databases and making reference to objects within them." On similar grounds, Jeffrey Pomerantz and Gary Marchionini (2007) define the future digital library as a "conceptual space" within which new interpretations of humanities materials are constructed "by augmenting representations of the ideas in the [library] materials with new kinds of extensions, hyperlinks, and annotations...while also supporting more personalized interactions between users and digital libraries" (528).

The halting progress toward KR for digital libraries highlights a social rather than strictly technological challenge: KO schemas and authorities remain embedded in organizational structures, relationships, and professional practices that make them slow to change. Even when older forms are poured into new moulds such as KR formats, they seldom reflect the range of concepts associated with the data by diverse researchers, communities, and disciplines. Relatively static KO systems cannot keep pace with the evolving epistemic commitments of infrastructure users, which is to say scholars in the present and future, as is evident in current efforts to decolonize metadata schemes commenced in the heyday of imperialism (Farnel et al. 2018). Andrew Abbott (2011) confirms this challenge by examining the development of library infrastructure. He concludes that over the course of the twentieth century, library infrastructure has developed in relation to two competing visions—the visions of librarians and scholars. Library infrastructure reflects "a vision of universal knowledge, knowledge without specialization, knowledge potentially available, to everyone" (81). Scholars, by contrast, are interested in infrastructure that supports building knowledge around the concerns of "narrow specialism"; that is, knowledge produced within the context of specific communities (82).

Abbott's insights on large library infrastructure (in contrast to more specialized DH tools or infrastructure on which librarian scholars and faculty researchers frequently collaborate) and the slow pace of KR production suggest that DH infrastructure requires a complement to the kinds of knowledge infrastructure provided by digital libraries, produced through more direct involvement of humanities scholars. An example of how researcher-driven, theoretically nuanced representations of content can complement higher-level metadata descriptions of objects is provided by Ioanna Kyvernitou and Antonis Bikakis (2017) in the development of semantic representation of gendered content that cannot be accommodated by but are compatible with the data model of the Europeana Digital Library, which provides only limited "flat text" options for representing artifact content. Initiatives such as this build both on the work on KR from computer science and the movement toward ontologies and other linked-data structures from within library science and aligned initiatives. DH infrastructure driven by scholarly communities coalesced around particular disciplines, domains, or methods thus complements other components, including digital libraries, within the larger linked infrastructural ecosystem.

THREE DIGITAL HUMANITIES INFRASTRUCTURES

The above insights offer an understanding of infrastructure as an evolving combination of organizations of people, hardware, and software, which create various possibilities and constraints, and data organized and structured through various mechanisms for knowledge organization and representation. Importantly, to serve as infrastructure, these socio-technical systems need to support knowledge systems broad enough to allow for interoperability and data exchange but specific enough to meet the demands of focused research specializations and their attendant epistemological commitments. David Ribes and Thomas A. Finholt (2009) see the main challenge in this process as the tension between the present and the future—what they call the *long now* of infrastructure. Building research infrastructure requires a robust approach that meets the "demands of the present" but is flexible enough to allow for future development that may not be anticipated at the outset. The following narrative of three successive infrastructure projects that span a quarter of a century reflects on the *long now* of DH infrastructure in the Canadian context.

THE ORLANDO PROJECT DOCUMENT ARCHIVE

The Orlando Project's experiment in digital-literary history began in 1995 and is one of the longest-running DH projects. It continues to be cited in publications on women's writing and DH methods (e.g., Liu 2018; Battershill et al. 2018; Looser 2015; Hamilton and Spongberg 2020). The project's longevity and ability to thrive—to produce new research published through semi-annual updates to its flagship publication, *Orlando: Women's Writing in the British Isles from the Beginnings to the Present* (Brown, Clements, and Grundy 2021), and adapt to changes in technology to support new approaches (Elford et al. 2010; Holland and Elford 2016; Martin et al. 2019)—is thanks to robust and evolving infrastructure.

Orlando was a distributed project from the outset in terms of both system architecture and organizational structure. The team's leads—principal investigator Patricia Clements, with Susan Brown and Isobel Grundy—at two institutions—the Universities of Alberta and Guelph—being more than 2,000 miles apart meant that asynchronous research by a team of humanities scholars needed to be managed and coordinated over the web, no mean feat considering that the first widely adopted graphical browser, Mosaic (1993), had appeared only a couple of years earlier and that web applications were rudimentary. The team initially used off-the-shelf tools for more structured data production, managing to synchronize commercial databases for bibliographic and timeline information across locations, but this was a stopgap measure since the core requirement was to produce a born-digital literary history whose narrative content could not be accommodated by a database (Folsom 2007). Thanks to the involvement of DH expert Susan Hockey, the team determined to use semi-structured data—Standard Generalized Markup Language (SGML), the precursor to Extensible Markup Language (XML)—and thus needed to link structured and semi-structured data, as well coordinate the work of a team of about a dozen collaborators.

The Orlando Project thus needed infrastructure to provide "distributed control and coordination processes" (Edwards et al. 2007, 8) through the internet to enable asynchronous collaboration using multiple tools. Although the project had no budget line for programming or infrastructure, it was blessed by the involvement of Terry Butler, director of the Arts Resource Centre at the University of Alberta, one of the earliest dedicated research computing centres. Butler oversaw the production of the Orlando Document Archive (see figure 18.1), infrastructure that was for its time both extensive and astonishingly robust: it served the project reliably for two decades.

The Document Archive coordinated a blend of web-based and local tools. For instance, only structured content (bibliographic records and events) could be

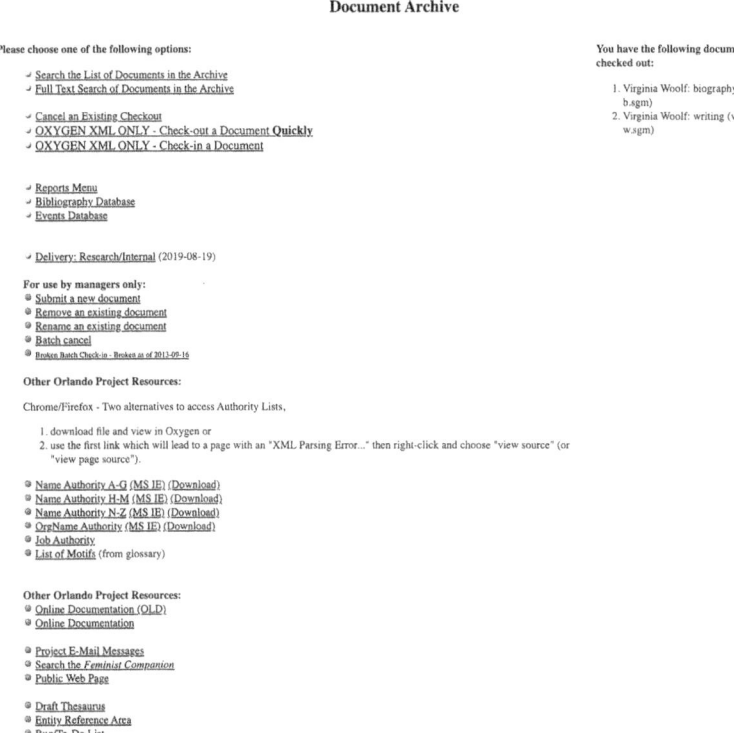

Figure 18.1. Detail of the Orlando Document Archive home page.

Source: Orlando Project.

Create

- SGML/XML timeline entries (form interface or cut-and-paste entry box)
- bibliographic database entries and name authorities (form interface)
- revise timeline and bibliographic entries
- SGML/XML documents (managers only)
- new password

Manage

- authentication with roles
- document checkout/lock; check-in; checkout cancel. -- SGML validation check and error reporting
- responsibility tags saved in database and inserted into objects
- email generation to flag a change in status or request for action to another team member(s).
- remove and rename documents
- batch check-ins or cancellations of checkouts
- reports (see below)

Access

- dropdown lists of documents
- search documents, events, bibliography by name; full text; advanced (SQL) search
- documents for download
- internal publication view (mirror of Cambridge UP)
- same view of in-progress work, regularly updated)
- authority lookups (personal/ organization names, job titles, motifs)
- character entity references
- experimental and testing interfaces (e.g. Degrees of Connection, Breadboard, "Googlesque" search)
- documentation of tags with extensive practice guidelines
- project Document Type Definitions (DTDs)
- searchable Feminist Companion to Literature in English (reference source)
- email archive; commonly used software

In addition, an extensive set of reports including:
- 6 to assist with tag checking, cleanup, tag patterns in single or groups of documents
- 4 workflow tools: a catalogue of documents that could be filtered and sorted, responsibility tag reports, revisions comparison, and a log report
- 6 informational tools to provide various statistics such as tag frequency
- 7 manager-only tools for various forms of batch processing
- pre-publication checks for e.g. invalid dates, broken links and other content patterns that would produce breaking changes in the project's publication environment

Figure 18.2.   Orlando Document Archive functionality.

Source: Orlando Project.

edited directly via the web; semi-structured author profiles had to be downloaded and edited on personal computers using locally installed software. Nevertheless, the system offered a wide array of functionalities, a few of which became obsolete or were added over time (see figure 18.2).

Given the complexity of the system, there were relatively few snags in adoption, and it required little tweaking and maintenance. Butler's attentiveness to how gender matters to IT uptake no doubt informed the design and interface (Butler, Ryan, and Chao 2005). The Document Archive supported and profoundly shaped the project's communication and collaboration. For instance, Orlando weathered changes in managers and managerial styles well, in part because the infrastructure regularized, externalized, and distributed among team members much of the core management activity. One project manager jokingly called herself the "enforcer" (alluding to the intimidating hockey-player role), but, in fact, the system automated and depersonalized much of the quality control.

The Document Archive was crucial to the ability of the Orlando Project to reach initial publication in 2006 and to manage regular corrections and updates to content until 2019, as well as enabling the team to explore and use the data in other contexts. Its tracking of drafting and revision processes, its versioning to support error recovery, and the extent to which it integrated responsibility stamps and research notes that were embedded in the files themselves, and facilitated email communication among team members, all enabled the team to work together efficiently and effectively, reducing errors and confusion. In particular, the Document Archive's sophistication and robustness enabled the involvement of student researchers as collaborators and co-authors to an extent rare in the humanities at that time, creating an extended community of practice that has trained more than 125 new scholars. The assessors for the midterm review of Orlando by the Social Sciences and Humanities Research Council of Canada (SSHRC) recommended that the Document Archive be generalized for use by other projects; however, building a generalized infrastructure was beyond the capacity of the team, which struggled for resources to "finish"—that is, to publish its content (Brown et al. 2009).

Orlando built infrastructure that went beyond a single tool and networked systems to provide distributed processes for coordination and control. But this infrastructure was dedicated to a single project, with Orlando's unique epistemological commitments represented in the infrastructure's encoding structure and the representation of the project data through a custom interface tailored to Orlando's needs. However, when it became necessary to replace the Document Archive—there was still no off-the-shelf equivalent—rebuilding for a wider range of use cases had become feasible because the Canada Foundation for Innovation (CFI) had been established to strengthen Canada's research capacity (Canada Foundation for Innovation, n.d.) by funding infrastructure, including a DH project, the Text Analysis Portal for Research, or TAPoR.[3] The CFI enabled the creation of the generalized, multi-institutional Canadian Writing Research Collaboratory, built from 2011 to 2016.

THE CANADIAN WRITING RESEARCH COLLABORATORY

The Canadian Writing Research Collaboratory (CWRC, pronounced "quirk") is a virtual research environment or science gateway that "brings together researchers working with online technologies to investigate writing and related cultural practices relevant to Canada and to the digital turn" (see https://cwrc.ca/about). Grounded in the experience of building and using Orlando's infrastructure, it aimed to serve a wider humanities user base.

CWRC provides an accessible onramp to digital scholarship for researchers from a wide range of humanities disciplines, particularly for using text as data.

It promotes best practices for metadata and data formats, collaboration, interoperability, and preservation, hosting texts, bibliographic records, and multimedia objects. Scholars can develop, analyze, and publish both research outputs and source content. The free platform offers the most open and flexible scholarly infrastructure in Canada for the production, hosting, management, sharing, and dissemination of humanities research data.

Projects

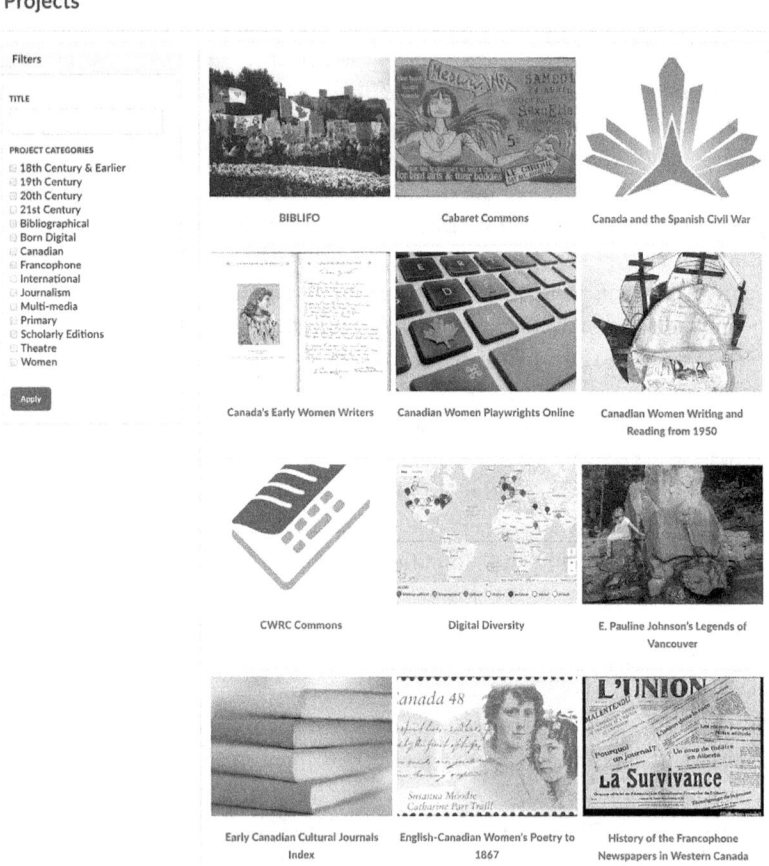

Figure 18.3. Canadian Writing Research Collaboratory project page.

Source: Canadian Writing Research Collaboratory.

Since its 2016 launch, CWRC has steadily gained users, content, and expanded functionality. It has grown organically in response to both researchers and funding opportunities, in ways not planned or fully anticipated. Building on its CFI-funded development and operations, it adapted to new use cases and disciplines thanks to research software development and sustainability grants from Canada's research network provider CANARIE. CWRC houses close to 400,000 digital objects (~1.5 terabytes), a substantial humanities dataset, as it is mostly text, which is efficient to store. The platform hosts 30 research projects (see figure 18.3) such as Canada's Early Women Writers (https://cwrc.ca/project/canadas-early-women-writers), the Orlando Project (http://www.artsrn.ualberta.ca/orlando/), The People and the Text—Indigenous Writing in Northern North America (http://thepeopleandthetext.ca/), and Records of Early English Drama (REED) London (https://cwrc.ca/reed). It provides formal and informal training in digital methods. In 2020, it trained 85 researchers and had more than a hundred active content creators, in addition to thousands of read-only accesses from within and beyond Canada. Based on informal reports from Compute Canada staff, in 2021 it was the third-most-used system on their Cloud.

CWRC supports diverse modes of research. It provides projects with home pages to support credit and reputational practices in the humanities, as well as dashboards (see figure 18.4) for managing members, workflow, reports, and project home pages. CWRC functionality (see figure 18.5) works with a much broader array of media types and activities than did the Orlando Document Archive.

The digital humanities comprise many methodologies, hence CWRC's numerous functionalities. To address the challenge of accommodating multiple methodologies and their respective epistemological and ontological commitments, CWRC concentrated on functionalities general enough to accommodate diverse forms of use. The platform supports all John Unsworth's influential list of "Scholarly Primitives" (2000)—discovering, annotating, comparing, referring, sampling, illustrating, and representing. This enables support for multiple disciplines and sub-disciplines of, for example, literary studies, including literary history, literary criticism, critical editing, bibliography, and Canadian literature; history, including performance history; Indigenous literary studies (under the umbrella of The People and the Text); interdisciplinary and Canadian studies; and a pending project in sociology and anthropology. Pedagogical projects have trained junior highly qualified personnel while creating data on oral history,[4] women trailblazers,[5] and COVID-19.[6] The next iteration of CWRC, based on a rebuild, is forming the basis of an international consortium of institutions using the software, rebranded as the Linked Editing Academic Framework, to establish

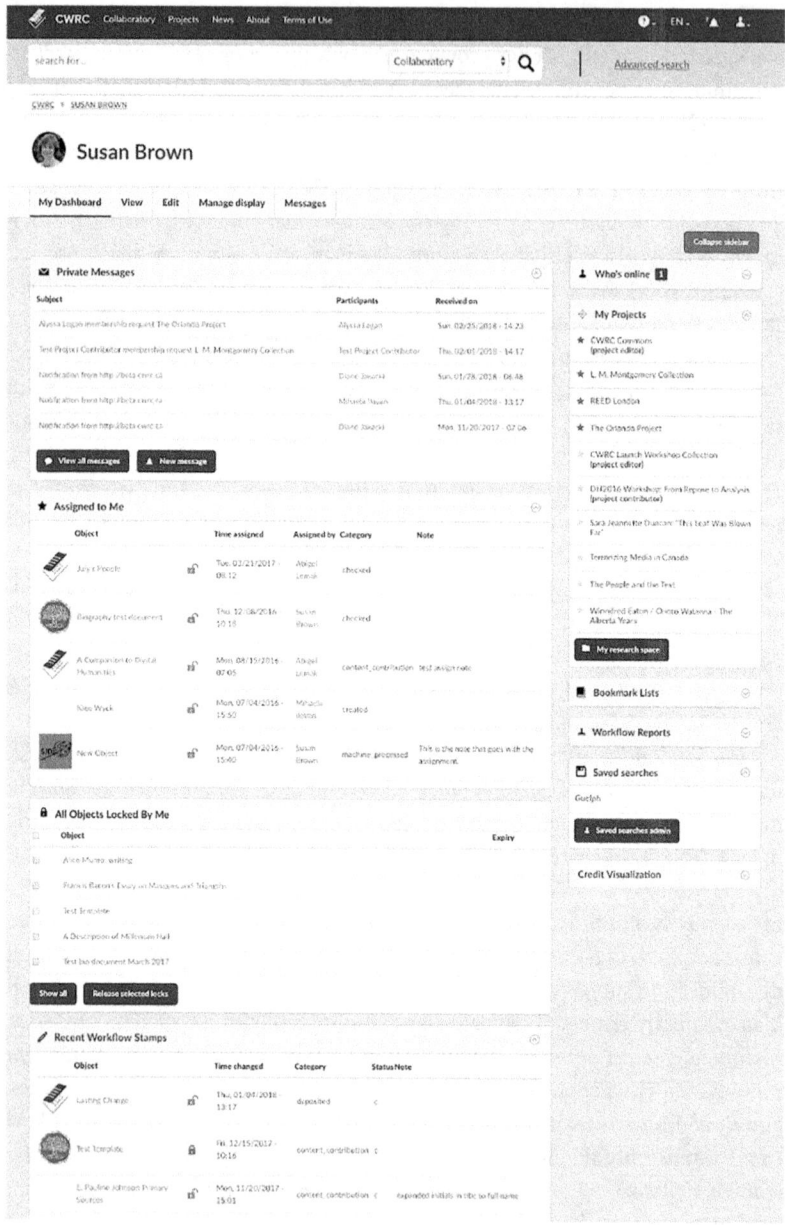

Figure 18.4. Canadian Writing Research Collaboratory user dashboard.

Source: Canadian Writing Research Collaboratory.

Create	Manage	Access
• edit XML content in browser*	• communications	• custom home page
• Linked Data annotation	• membership	• browse
• Named Entity Recognition (NER)	• roles	• Search (Apache Solr)
• digital surrogates of texts	• access*	• Internet Archive Book Viewer
• e-books (scanned or new)*	• sharing*	• CWRC-Reader (XML + CSS)
• object metadata	• workflow reports export/download	• Dynamic Table of Contexts
• bibliographic records	• NER vetting/linking	• PDF viewer
• image, audio, video objects	• backup/archiving	• image viewer
• thematic collections	• object versioning*	• audio & video players
• text from images (OCR)	• project home page	• Voyant visualizations
	• peer review processes	• Timeline/mapping views
	• Traditional knowledge labels	• Entity records
		• Credit visualization

*also openly available via Git-Writer, an open and independent version of CWRC-writer that uses GitHub for storage.

Figure 18.5.   Canadian Writing Research Collaboratory functionality.

Source: Canadian Writing Research Collaboratory.

sites for various purposes, including a cooperative digital publishing platform under the leadership of Diane Jakacki at Bucknell University (Jakacki et al. 2022).

In addition to meeting a wider range of use cases and disciplines, CWRC was also designed to address a number of limitations of the Orlando infrastructure, constrained as it was by project resources and the technologies of the late 1990s. Thanks to advances in web applications and open-source software, CWRC was constructed on different principles than the Document Archive, even as it was informed by the project's understanding of scholarly workflows and the collaboration needs of humanities researchers. A primary aim of the system, along with expanding the data types and use cases it would serve, was to provide an integrated environment to make it feasible for researchers without extensive training to produce digital scholarship entirely from their browser, obviating the need to purchase, download, configure, or manage software or data locally.

The CWRC virtual research environment builds on the open-source Islandora framework. Islandora combines a Fedora Commons repository, a Drupal/Islandora front-end, and a Solr index. The Fedora back-end acts as an object store and defines composite datastreams that hold content (e.g., object metadata, administrative metadata, access policies, and workflow information, as well as datastreams for text, images, audio, or video). Fedora relates objects to one another and to collections. A Drupal web interface supports user, collection, and Fedora object management. A combined Fedora-governed and Drupal rights system allows customized levels of access to collections, objects, and interface components based on user groups or accounts.

Role management and workflow tracking are two key areas where CWRC extended Islandora, which was developed for more static collections, to support collaborative research activities and dynamic content, as well as customizing the Drupal interface as a research space by, for instance, adding dashboards for projects and researchers. Other components enhance content processing, display, and analysis, including a bridge to allow content to be viewed with Voyant. In keeping with the Islandora architecture and more general platform-development practices, CWRC components were developed to be as modular as possible and often set up as external services. Islandora-compatible Drupal modules made the CWRC interface extensions more readily configurable, customizable, updatable, and extensible.

CWRC thus benefited from open-source software becoming more mainstream and flourishing between the late 1990s and early 2010s; this enabled a more extensive and generalized infrastructure for cultural scholarship to be built on free software and on standards that did not exist in the 1990s. Its distributed control and coordination processes (Edwards et al. 2007, 8) relied on the movement toward using application programming interfaces (APIs) both internally and externally, creating a programmable "platform" (Plantin et al. 2018) in a way that the Document Archive was not. Its reliance on open-source software was a result of what Edwards characterizes as "technology transfer" on both a national (Islandora) and international (Fedora Commons, Drupal, TinyMCE editor, etc.) level. Finally, the ability of CWRC to extend Islandora in key ways, through integration of the CWRC-Writer editor, for instance, was a benefit of the shift toward infrastructural gateways that "allow new systems to be joined to an existing framework easily and with minimal constraint" (Edwards et al. 2007, 14).

THE LINKED INFRASTRUCTURE FOR NETWORKED CULTURAL
SCHOLARSHIP

The CFI-funded Linked Infrastructure for Networked Cultural Scholarship (LINCS) cyberinfrastructure project, started in April 2020, uses Semantic-Web technology to mobilize and make interoperable existing humanities datasets from multiple disciplines spanning literary studies; history, including book history, geospatial history, literary and art history; music; Indigenous studies; communication studies; and women's studies. It will also advance digital humanities, information studies, and computer-science research. Canadian universities, research libraries, and memory institutions are collaborating to convert

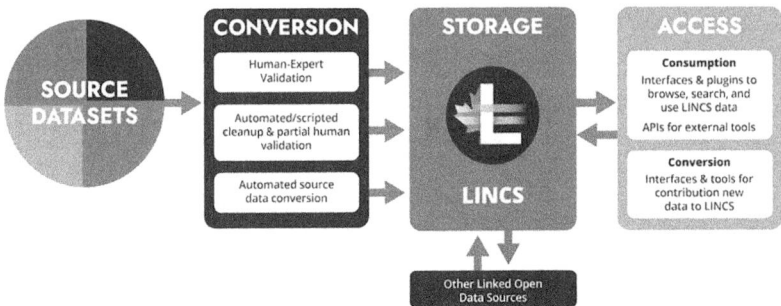

Figure 18.6. Overview of LINCS Linked Infrastructure for Networked Cultural Scholarship research data infrastructure.

Source: Linked Infrastructure for Networked Cultural Scholarship project.

data from 40-plus researchers into linked open data (LOD), provide means of accessing and using linked data effectively, and provide tools for continuing to create and convert data, including via an upgraded CWRC 2.0.

LINCS combines technical sophistication, standardization, and customizability with the promise of more legible data structures and relationships. Despite ramping up during the pandemic the project is making good progress, leveraging open-source software and platforms, creating a large, linked dataset and knowledge graph designed for cultural researchers, and developing a cohort of staff to support its use.

LINCS adheres to an approach increasingly emphasized within DRI contexts, including in Canada, in adopting or adapting existing solutions to build infrastructure platforms, rather than building from scratch (Almas 2017; Plantin et al. 2018), in the LINCS case including ResearchSpace, OpenRefine, and Jupyter Notebooks. Working with Canada's national infrastructure, the project will mobilize large-scale, heterogeneous LOD sets for humanities research (see figure 18.6). LINCS will combining existing and custom-built tools to comprise (see figure 18.7; existing tools marked with an asterisk):

1. a conversion toolkit whereby LINCS will mobilize, enrich, and interlink research data;
2. a linked data storage system; and
3. an access system to filter, query, analyze, visualize, and annotate cultural materials; to modify, evaluate, correct, or reject automated semantic enrichments; and to allow continued use of conversion tools.

Create (Conversion)

Data types:
- Structured
- Semi-structured XML
- Natural language

Entry points
- bilingual overview
- documentation
- APIs

Platforms
- CWRC and CWRC-Writer*
- TEI to RDF workflow based on XTriples
- Named Entity Recognition and Reconciliation (NERVE*) entity and relationship extraction from natural text

Programming notebooks
- Spyral (javascript)*
- Jupyter (python)*

Sandbox
- For users to create and experiment

Manage (Storage)

Managed by:
University of Victoria

Housed by:
National infrastructure (Digital Research Alliance of Canada cloud services)

Mirror by: Scholars Portal research library consortium

Code store:
Gitlab CI/CD environment for code storage and management

Research Space (RS)*
- employing a Blazegraph* Linked Data store
- Entities/authorities
- Ontologies/vocabularies
- Triples

Access

Browsing/Visualization
- Rich Prospect Browser (based on ARC's Big Diva)*
- HuViz (graph explorer)*
- Bridge to text analysis in Voyant Tools*

Search/indexing
- By keyword, entity, predicate
- Advanced queries (RS)*
- SPARQL search

Context Box
- entity overview and property summary

Results view
- List, thumbnail view (RS)*;
- Mapping/spatial view*
- Timeline view;

Feedback form in RS
- Allows suggestions or enhancements for vetting

Figure 18.7.   Projected Linked Infrastructure for Networked Cultural Scholarship functionality.

Source: Linked Infrastructure for Networked Cultural Scholarship project.

LINCS will adhere to standards established by Tim Berners-Lee (see Berners-Lee, Hendler, and Lassila 2001) and other architects of the Semantic Web, drawing on practical solutions implemented for cultural data. In addition to the reuse of tools and platforms, LINCS will use existing ontologies wherever possible and work to adopt, adapt, and extend best practices established by large projects including Europeana, the Digital Public Library of America, and Linked Data for Production (LD4P), as well as learning from enrichment-oriented platforms like the Australian HuNI and European InterEdition projects, and from cultural-heritage providers in Canada who are experimenting with LOD. It will benefit from the recent maturing of some tools for KR for LOD in developing an infrastructure dedicated to advancing diversity.

These three DRI projects reflect changes in the understanding and implementation of infrastructure over the last 20 years as well as some of the specificities of the digital humanities. The changes in technology, organizations (including both the people that run the projects and the researchers that use them), and data evident in the trajectory from Orlando to LINCS indicate that no single infrastructure can support the dynamic, interdisciplinary, iterative work of the digital humanities. LINCS itself is a meta-infrastructure, pulling together a variety of smaller tools and platforms for data conversion, management, organization, linking, sharing, and research to produce an interoperable virtual space that also relies on external infrastructure rather than a single integrated platform like CWRC.

EVOLVING INFRASTRUCTURE TECHNOLOGY:
CARE, REPAIR, AND SUSTAINABILITY

The functionalities and architectures of the three projects understandably reflect general developments in technological tools and strategies. As the challenges of sustaining infrastructure and the costs of care and repair have become evident in data loss, redundant development effort, and the costs of cybersecurity breaches, the earlier stress on innovation has been joined by a recognition that factors such as software reuse, efficient maintenance and upgrades, long-term data management, and security are also essential to DRI. Changes to the Canadian landscape, including the recent reorganization of national DRI, reflect this emerging understanding. Sustainability remains a significant challenge for larger infrastructure systems, and in some specific ways for DH infrastructure, as can be seen with respect to CWRC.

CWRC has sustained itself in the five years since its launch, but its position is precarious. Despite having garnered what is perhaps the best support available to any DH platform in Canada, the project has never had more than a 24-month sustainability horizon. The platform was created with public funds, apart from modest contributions from private partners.[7] The platform is free for users and content is required to be open access where possible, user-fee models being problematic in the humanities for cultural and practical reasons. Within CWRC, projects with bespoke Islandora front-ends (Drupal multi-sites) are responsible for sustaining their sites, and when project leaders apply for grant funding, they are asked to budget modest amounts toward the costs such as onboarding, customizing, or ingestion. However, projects are rarely able to garner such funding for a number of reasons. SSHRC review panels are not used to seeing significant technical expenses in proposals, can have difficulty evaluating them if no DH expert sits on the committee, and have been known to cut technical lines from budgets; applicants are therefore reluctant to include substantial technical costs. This is different from areas of the sciences where grant funding is more predictable and technical costs routine, and where infrastructure maintenance can be subsumed in the other work of technical staff. LINCS, whose development period ends in 2023, faces the same challenges as CWRC but with higher operations costs because the complexity of linked-data technologies demands a larger number of core staff.

MAKING PEOPLE VISIBLE: THE SOCIAL SIDE OF INFRASTRUCTURE

Sustaining human infrastructure—that is, the humans who make infrastructure work—is a major challenge exacerbated for projects that are based in the university system but not formally embedded in any formal unit and indeed spread across institutions; many other researcher-led infrastructure projects in Canada occupy similarly marginal institutional positions, rendering them both invisible and vulnerable. Retaining personnel is essential and a huge challenge when projects are soft-funded. Core developers and project managers provide expertise and tacit knowledge essential to operating research infrastructure: such personnel provide stability while other highly qualified personnel (i.e., students) gain expertise through shorter-term roles. While some view research software development as rather generic, CWRC belies that claim. For instance, it took an experienced Drupal programmer a year to get up to speed with its unusual data structures and software stack, despite the use in Islandora 6/7 of the Drupal framework. CWRC could not survive without seasoned technical staff and project management.

While open-source software has immense advantages, it also increases risk with respect to sustainability. Some CWRC functionality draws on Voyant text analysis and visualization tools. The untimely death of Voyant's brilliant co-founder and lead developer Stéfan Sinclair in 2020, and the jeopardy in which this leaves Voyant, despite its massive user base, underscores the extent to which much DH infrastructure is sustained by individuals through donated time. CWRC's Islandora framework also illustrates the challenges of relying even on a larger open-source project supported by multiple developers at multiple institutions with a coordinating foundation behind it. Specifically, CWRC's ability to upgrade was dependent upon the intermittent progress toward the next release of Islandora, which itself suffered from the same kind of sustainability challenges as did CWRC and was reliant on some of the same funding opportunities.

It is also worth noting that the University of Alberta and University of Guelph libraries showed extraordinary support for CWRC and Orlando in being willing to share personnel positions where project funds could not support full-time ones, and have been highly supportive of LINCS as well. These collaborations attest to a deep understanding on the libraries' part of the value of DH infrastructure. Moreover, the fact that the same skill set was suited to both contexts shows that infrastructures were considerably aligned technically. This alignment provided a basis for collaboration between a much larger, more stable organization and a smaller, more precarious one, but the libraries also benefited from expertise—for instance, in linked data, developed within the research infrastructure contexts. Local organizational contexts can thus have a major impact on DH infrastructural sustainability, particularly when, as is the case in Canada at present, there are no dedicated funding opportunities.

People and their organizations—Orlando, CWRC, LINCS, the Arts Resource Centre at the University of Alberta, libraries, funders, national research infrastructures, scholarly communities, the open-source movement—are thus crucial to both the development and sustainability of DH infrastructure, as well as those that use the infrastructure itself. The details above highlight the exceptional technical work that has gone into each of these projects, from project managers that ensure the work flows smoothly to the students who spend their summers writing documentation, vetting data, or conducting user testing of interfaces. We must remember too that researchers themselves are a vital part of these projects. Everyone who has accessed Orlando to peer into the lives of women writers, every researcher who has uploaded a document, image, or video to their CWRC projects, and every scholar that will work with LINCS to create a national triplestore of humanities data is part of this infrastructure. Orlando and CWRC developed their own communities of practice around their infrastructure. Recognizing this, LINCS knew that it would not succeed without a talented interdisciplinary team and a dedicated group of active researchers for input and testing of tools during development. More than 40 researchers committed to working with LINCS from the outset, which we hope will pave the way for many more.

LINKED OPEN DATA AND THE WEB (AS) INFRASTRUCTURE

The arc from Orlando to LINCS parallels the arc within the library-infrastructure context, from knowledge organization to knowledge representation that prioritizes linking and interoperability. This movement toward interoperable, open data based on web standards also works against the negative "platformizing" impacts of major corporate providers (Plantin et al. 2018).

CWRC tried to address what the Orlando team realized early on as one of the greatest challenges to creating a dynamic, collaborative research environment to support digital scholarship: the silo effect that impeded both discovery and interoperability of related content on the web. While Orlando was creating born-digital content tracking biographical patterns, cultural and other historical influences, reception, and intertextuality networks, and other factors in women's literary history in English, other projects were digitizing and creating corpora, editions, and manuscript/book-history databases: Martha Nell Smith launched the Dickinson Electronic Archives in 1994; the Women Writers Project published online editions from 1995, with current director Julia Flanders already at the helm as managing editor; the Corvey Project driven by professors and research fellows at Sheffield Hallam University and the Victorian Women Writers Project directed by Perry Willett were founded in 1995 alongside Orlando; these were followed in 1997 by others such as the Perdita Project under Rosalind Smith, British Women Romantic Poets founded by Nancy Kushigian, and the Women Writers project led by Suzan van Dijk from the Huygens Institute. All were creating networked texts that begged to be interconnected. However, being online did not mean they were networked: they used different flavours of SGML (often TEI) and employed different naming conventions for entities, which stood in the way of being able systematically to link these related resources. The technological conditions of the time are reflected in a proliferation of websites with separate but related content (or differently structured versions of the same content) and different functionality, even sometimes within larger projects, as happened with the Corvey, both the Huygens and Brown women writers projects, and Orlando. Manual hyperlinks could be created, but there were no APIs or other automated means of supporting access to or aggregation of materials.

Discussion of this challenge with the various recovery projects related to women's writing in English began as early as the summer of 1998 at the joint Association for Literary and Linguistic Computing/Association for Computers and the Humanities conference in Hungary, after which the Brown Women Writers Project and Orlando co-organized a Women's Writing Projects Symposium at the same conference in 2001 that resulted in a nascent project called Naming and Other Metadata for Electronic Networking (NOMEN). However, the project stalled for want of funding and leadership, in part due to other pressures but also perhaps because of technical challenges to achieving the kinds of interlinkages that were desired. Berners-Lee had, just that May, declared with others the potential of a Semantic Web, laying out principles and technical specifications to make the web itself a foundational infrastructure for interlinking and interoperability through what is now called linked open data; however, much work remained to make it feasible. In the meantime, the project that came closest to realizing the vision of an interconnected set of cultural resources was the Networked Infrastructure for Nineteenth-Century Electronic Scholarship (NINES), founded in 2005, which was the brainchild of Jerome McGann brilliantly implemented by Bethany Nowviskie. NINES used the Resource Description Framework (RDF) developed for the Semantic Web to combine aggregated metadata from multiple websites with free-text and keyword search. NINES expanded in 2011 to support community nodes ranging from medieval studies to radicalism under the umbrella of the Advanced Research Consortium (ARC), directed by Mandell (Grumbach and Mandell 2014).

Inspired by NINES and the larger vision of linked open data, CWRC chose the Islandora framework in hopes that its use of RDF and an internal triplestore would make it easier to build a "collaboratory" that could support and promote similar interlinking with related web resources. CWRC's centrepiece tool, the CWRC-Writer editor, was also designed with this in mind. CWRC-Writer allows users to edit texts online using the XML format for the encoding of texts, which is the best practice for scholarly editions of primary text in the humanities, while also creating for key components of the markup (e.g., named entities) annotations using the Web Annotation Data Model. In keeping with best practices for linked data (Berners-Lee 2011; Bizer, Heath, and Berners-Lee 2011), CWRC-Writer incorporates lookups to external authorities for identifiers. Incorporating such lookups in metadata creation forms, CWRC and CWRC-Writer allowed users to create data that could potentially link into the Semantic Web. However, although CWRC, like ARC, used RDF to structure internal relationships in its data, it needed LOD hosting, management, and dissemination infrastructure to mobilize its content as LOD. LINCS will create the more generic LOD infrastructure that the humanities research community requires to explore the capacity of LOD to meet the need for interoperability and diversity in digital scholarship.

CONCLUSION: A WEB OF PEOPLE, TECHNOLOGY, AND DATA

Orlando, CWRC, and LINCS confirm general understandings of research infrastructure in various respects—they operate as coordinated networks, they develop organically in response to social processes and technical change, and they involve layers of information, policy, documentation, and services—and, above all, are responding to changing conditions and opportunities through evolving arrangements of human organization, technology, and data. As case studies in DH infrastructure, they reflect shifts in technology and resulting infrastructural strategies from the later 1990s through to the present. They also demonstrate, in the ways their functionalities overlap and diverge, both changing needs and broadening research communities and stakeholder relationships within digital humanities. They illustrate a high degree of alignment and complementarity with library infrastructure and parallel its shift from KO to forms of KR capable of overcoming institutional and epistemological barriers to interlink related data. The movement of these infrastructural efforts toward deeper collaboration with Canadian research libraries—from Orlando's strong relationship, which saw several project staff trained in Library and Information Science who had gained technical expertise on the project move into library roles, through CWRC collaborations, which involved consultation on technological decisions, publishing, and data-preservation collaboration, to LINCS, which has active involvement from leaders in Canada's library community—provides practical confirmation of the extent to which visions of the digital library and DH infrastructure are aligned and complementary. DH infrastructure manages more diverse, dynamic, and situated KR, and must respond with greater agility to emergent epistemologies and methods in relation to research communities. Its more particular and situated data complements the more stable institutional metadata on which it relies and draws. Such complementarity offers a significant example of the kind of virtuous circle made possible by a multi-sectoral infrastructural ecosystem that continuously recalibrates to accommodate ever-expanding bodies of data and the diverse range of user needs.

Working toward an infrastructure ecosystem that sees a commonality in infrastructural needs among humanities researchers, libraries, and beyond poses exciting possibilities and sobering challenges, given the crucial but tentative past efforts toward collaboration on LOD infrastructure in Canada. With other stakeholders, LINCS has proposed an aligned research partnership to develop a collaborative model for a sustainable national cultural LOD infrastructure. The aim is to build a multi-stakeholder, multi-sectoral partnership comprising researchers and universities; libraries; knowledge-sector non-profits; GLAM institutions; government agencies; university presses; and perhaps eventually also small/ commercial academic presses and corporate arts-sector organizations. These organizations belong to an emergent LOD ecosystem within which shared infrastructure, represented by the cloud at the top of Figure 18.8, would support linkages between parallel content stored and managed across stakeholder groups.

As already noted, the "heavy" LOD technology stack requires specialized expertise to establish and maintain. The LINCS infrastructure funding from CFI has allowed its technical team to extensively research and test candidate technologies (Botha 2020a), to set up a robust development and hosting environment in

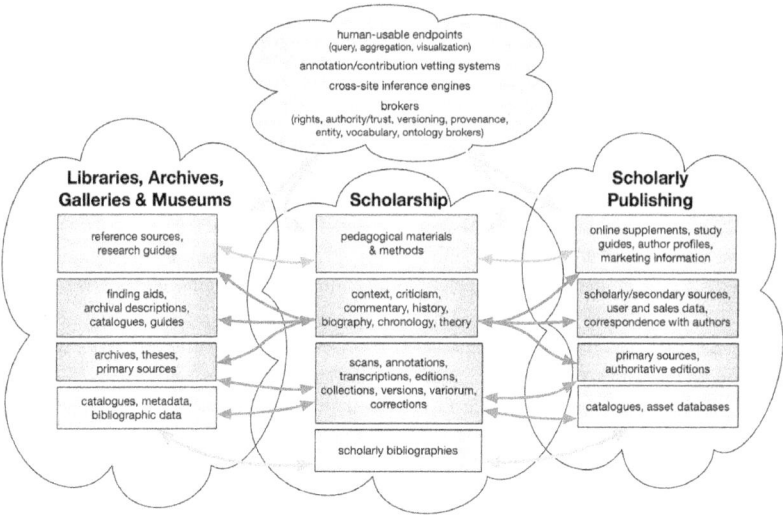

Figure 5: An Ecosystem for Cultural Linked Open Data

Figure 18.8. Multi-stakeholder ecosystem for cultural linked open data.

Source: Linked Infrastructure for Networked Cultural Scholarship project.

Gitlab for managing, testing, and deploying code through a continuous integration/continuous deployment (CI/CD) pipeline. This pipeline integrates with production infrastructure in ways that reduce the costs of provisioning, monitoring, and administering the system but that have also stretched the capacity of the host Digital Research Alliance of Canada (formerly Compute Canada) Cloud (Botha 2020b). The complexity of such infrastructure is a major reason why none of the related sectors in Canada that stand to benefit from it has implemented LOD technology fully. Yet its potential for scholarly, cultural, social, and economic benefits is huge.

A strong network of partners would knit the scholarly community into closer collaboration with publishers, GLAM institutions, and information stakeholders; foster shared expertise and infrastructural costs; and produce a potentially ever-growing community of practice. Current technology enables the production of more agile, flexible, and modular infrastructure that can serve the distinct interests, activities, and needs represented in different areas and groups of people within a large ecosystem, surpassing what any single organization could achieve and constituting a boundary object. Such a shared and interlinked cultural knowledge infrastructure "situated between the academic institution and other major social institutions" would manifest the critical bent of digital humanities in concert with aligned institutions and initiatives, contributing to the realization of a multifaceted diversity stack to the benefit of all (Liu 2016, 2020).

NOTES

1. The authors would like to thank everyone involved in the creation and sustaining of the three infrastructures at the centre of this article. Please see the following links for full lists of contributors: https://orlando.cambridge.org/about/credits, https://cwrc. ca/about/credits-and-acknowledgments, and https://lincsproject.ca/team/.

2. Canada established in 1997 the Canada Foundation for Innovation (CFI), a complement to the research councils charged with advancing both research and partnerships with the private sector through funding scientific infrastructure, later expanded to include humanities and social-science infrastructure. A range of DRI was funded by CFI programs, but the most visible national impact was in high-performance computing managed through regional consortia. As the need to support DRI grew, CFI mandated a new national organization, Compute Canada, to coordinate regional efforts, and CANARIE, which oversees the high-speed network, launched programs to support research software, platform development, and maintenance. Following a series of summits and reports on national DRI led by a council of stakeholders, Canada has restructured the DRI system through the formation of the Digital Research Alliance of Canada with an expanded mandate.

3. Led by Geoffrey Rockwell, the TAPoR project offers another case study in the evolution of Canadian DH infrastructure, having morphed by version 3.0 to a tool assessment and discovery portal, with the Voyant Tools platform created by Stéfan Sinclair and Rockwell operating as a separate entity.

4. "On the Record: A Community History of Guelph," https://cwrc.ca/project/record-community-history-guelph.

5. "Her Story: Women Trailblazers of Guelph and Wellington County," https://cwrc.ca/project/her-story-women-trailblazers-guelph-and-wellington-county.

6. "Contextualizing COVID-19: Plight, Pandemics, and Policy in History," https://news.uoguelph.ca/2020/07/investigating-pandemics-from-a-historical-perpective-focus-of-new-u-of-g-course/.

7. CWRC has been sustained from the following sources: operations and maintenance provided by the original funder, the CFI, which provides 12 percent of the original project budget; operational funds associated with the project lead's Canada Research Chair accompanied by a CFI grant that including some programming; three research software grants from CANARIE, two of which included operations funding; and a partnership with Bucknell University, in Pennsylvania, which received two grants from the Mellon Foundation to build its own instance of CWRC as a digital publishing platform.

APPENDIX: LIST OF DH TOOLS AND PROJECTS

Arts Resource Centre: https://arc.arts.ualberta.ca/research-computing/
Advanced Research Consortium: https://arc.dh.tamu.edu/
Big Diva: https://bigdiva.org/
Blazegraph: https://blazegraph.com/
Canada Foundation for Innovation: https://www.innovation.ca/
Canadian Century Research Infrastructure: http://www.ccri.uottawa.ca/CCRI/ Home.html
CANARIE: https://www.canarie.ca/
CATMA: https://jcmeister.de/projects/catma/
CWRC: https://cwrc.ca/
CWRC-Writer: https://cwrc-writer.cwrc.ca/
DH Dashboard: https://gitlab.dh.tamu.edu/bptarpley/dh_dashboard
Digital Public Library of America: https://dp.la/
Digital Research Alliance of Canada: https://alliancecan.ca/en
Dynamic Table of Context: https://cwrc.ca/DToC_Documentation/
Europeana: https://www.europeana.eu/en
HuViz: http://huviz.cwrc.ca/
HuNI: https://huni.net.au
InterEdition: http://www.interedition.eu/
Islandora: https://www.islandora.ca/
Jupyter Notebooks: https://jupyter.org/
Karma: https://usc-isi-i2.github.io/karma/
LD4P: https://www.ld4l.org/
LINCS: https://lincsproject.ca/
Omeka: https://omeka.org
OpenRefine: https://openrefine.org/
Orlando Project: http://www.artsrn.ualberta.ca/orlando/
Orlando Project publication site: http://orlando.cambridge.org
Pelagios: https://pelagios.org/
Perseids: https://www.perseids.org/perseids-platform
ResearchSpace: https://www.researchspace.org/
Scalar: https://scalar.me/anvc/scalar/
Scholars Portal: https://scholarsportal.info/
SNAC: https://snaccooperative.org/
Spyral Notebooks: https://voyant-tools.org/spyral/
TAPAS: https://tapasproject.org/

TAPoR: http://tapor.ca/
Traditional Knowledge labels: https://mukurtu.org/support/traditional -knowledge-labels-faq/
Trismegistos: https://www.trismegistos.org/index.php
Voyant Tools: https://voyant-tools.org/

REFERENCES

Abbott, Andrew. 2011. "Library Research Infrastructure for Humanistic and Social Science Scholarship." In *Social Knowledge in the Making*, edited by C. Camic, N. Gross, and M. Lamont, 43–88. Chicago, Ill.: University of Chicago Press.

Almas, Bridget. 2017. "Perseids: Experimenting with Infrastructure for Creating and Sharing Research Data in the Digital Humanities." *Data Science Journal* 16, no. 2. https://doi.org/10.5334/dsj-2017-019.

Anderson, Sheila. 2013. "What Are Research Infrastructures?" *International Journal of Humanities and Arts Computing* 7, no. 1–2: 4–23. https://doi.org/10.3366/ijhac.2013.0078.

Atkins, Daniel E., Kelvin K. Droegemeier, Stuart I. Feldman, Hector Garcia-Molina, Michael L. Klein, David G. Messerschmitt, Paul Messina, Jeremiah P. Ostriker, and Margaret H. Wright. 2003. *Revolutionizing Science and Engineering Through Cyberinfrastructure: Report of the National Science Foundation Blue-Ribbon Advisory Panel on Cyberinfrastructure*. https://www.nsf.gov/cise/sci/reports/atkins.pdf.

Battershill, Claire, Alice Staveley, Helen Southworth, and Elizabeth Wilson Gordon. 2018. "Collaborative Modernisms, Digital Humanities, and Feminist Practice." *Modernism/Modernity Print Plus* 3, no. 2. https://doi.org/10.26597/mod.0056.

Benardou, Agiatis, Erik Champion, Costis Dallas, and Lorna M. Hughes. 2018. *Cultural Heritage Infrastructures in Digital Humanities*. London, U.K.: Routledge. https://doi.org/10.4324/9781315575278.

Berners-Lee, Tim. 2011. "5 Star Linked Data." Government Linked Data (GLD) Working Group Wiki. W3. https://www.w3.org/2011/gld/wiki/5_Star_Linked_Data.

Berners-Lee, Tim, James Hendler, and Ora Lassila. 2001. "The Semantic Web." *Scientific American* 284, no. 5: 28–37. https://doi.org/10.1038/scientific-american0501-34.

Bizer, Christian, Tom Heath, and Tim Berners-Lee. 2011. "Linked Data: The Story so Far." In *Semantic Services, Interoperability and Web Applications: Emerging Concepts*, edited by A. Sheth. https://doi.org/10.4018/978-1-60960-593-3.ch008.

Borgman, Christine L. 1999. "What Are Digital Libraries? Competing Visions." *Information Processing and Management* 35, no. 3: 227–43. https://doi.org/10.1016/S0306-4573(98)00059-4.

———. 2015. *Big Data, Little Data, No Data: Scholarship in the Networked World.* Cambridge, Mass.: The MIT Press.

Borgman, Christine L., Jillian C. Wallis, and Noel Enyedy. 2007. "Little Science Confronts the Data Deluge: Habitat Ecology, Embedded Sensor networks, and Digital Libraries." *International Journal on Digital Libraries* 7, no. 1–2: 17–30. https://doi.org/10.1007/s00799-007-0022-9.

Borgman, Christine L., Jillian C. Wallis, Matthew S. Mayernik, and Alberto Pepe. 2007. "Drowning in Data: Digital Library Architecture to Support Scientific Use of Embedded Sensor Networks." In *Proceedings of the 7th ACM/IEEE-CS Joint Conference on Digital libraries.* Vancouver, B.C. https://doi.org/10.1145/1255175.1255228.

Borowiecki, Karol Jan, Neil Forbes, and Antonella Fresa, eds. 2016. *Cultural Heritage in a Changing World.* Cham, Switzerland: Springer. https://doi.org/10.1007/978-3-319-29544-2.

Botha, Pieter. 2020a. "Kubernetes Storage on Compute Canada." *Linked Infrastructure for Networked Cultural Scholarship* (blog). September 24, 2020. https://lincsproject.ca/kubernetes-storage-on-compute-canada/.

———. 2020b. "Open-Source Triplestore Battle." *Linked Infrastructure for Networked Cultural Scholarship* (blog). November 13, 2020. https://lincsproject.ca/open-source-triplestore-battle/.

Bowker, Geoffrey C., and Susan Leigh Star. 1999. *Sorting Things Out: Classification and Its Consequences.* Cambridge, Mass.: The MIT Press. https://doi.org/10.7551/mitpress/6352.001.0001.

Bratton, Benjamin H. 2016. *The Stack: On Software and Sovereignty.* Cambridge, Mass.: The MIT Press. https://doi.org/10.7551/mitpress/9780262029575.001.0001.

Brown, Susan, Patricia Clements, and Isobel Grundy. 2021. *Orlando: Women's Writing in the British Isles from the Beginnings to the Present.* Cambridge, U.K.: Cambridge University Press. http://orlando.cambridge.org/.

Brown, Susan, Patricia Clements, Isobel Grundy, Stan Ruecker, Jeffery Antoniuk, and Sharon Balazs. 2009. "Published Yet Never Done: The Tension Between Projection and Completion in Digital Humanities Research." *Digital Humanities Quarterly* 3, no. 2. http://www.digitalhumanities.org/dhq/vol/3/2/000040/000040.html.

Brown, Susan, and John Simpson. 2015. "An Entity By Any Other Name: Linked Open Data as a Basis for a Decentered, Dynamic Scholarly Publishing Ecology." *Scholarly and Research Communication* 6, no. 2. https://doi.org/10.22230/src.2015v6n2a212.

Butler, Terry, Peter Ryan, and Ining Tracy Chao. 2005. "Gender and Technology in the Liberal Arts: Aptitudes, Attitudes, and Skills Acquisition." *Journal of Information Technology Education: Research* 4: 347–62. https://doi.org/10.28945/281.

Canada Foundation for Innovation. n.d. "Our history." https://www.innovation.ca/about/overview/our-history.

Comité de Sages. 2011. *The New Renaissance: Reflection group on Bringing Europe's Cultural Heritage Online.*

Edwards, Paul N., Steven J. Jackson, Geoffrey C. Bowker, and Cory P. Knobel. 2007. "Understanding Infrastructure: Dynamics Tensions and Design. Report of a Workshop on History & Theory of Infrastructure: Lessons for New Scientific Cyberinfrastructures." Human and Social Dynamics, Computer and Information Science and Engineering, Office of Cyberinfrastructure. https://deepblue.lib.umich.edu/bitstream/handle/2027.42/49353/UnderstandingInfrastructure2007.pdf.

Edwards, Paul. N., Geoffrey C. Bowker, Steven J. Jackson, and Robin Williams. 2009. "Introduction: An Agenda for Infrastructure Studies." *Journal of the Association for Information Systems* 10, no. 5: 6. https://doi.org/10.17705/1jais.00200.

Elford, Jana Smith, Susan Brown, Michael Bauer, Jennifer Berberich, and Jonathan Cable. 2010. "'Elevating Influence': Victorian Literary History by Graphs." *Victorians Institute Journal Annex* 38. https://nines.org/exhibits/Elevating_Influence.

Farnel, Sharon, Danise Koufogiannakis, Sheila Laroque, Ian Bigelow, Anne Carr-Wiggin, Debbie Feisst, and Kayla Lar-Son. 2018. "Rethinking Representation: Indigenous Peoples and Contexts at the University of Alberta Libraries." *International Journal of Information, Diversity, & Inclusion* 2, no. 3. https://doi.org/10.33137/ijidi.v2i3.32190.

Flanders, Julia, and Scott Hamlin. 2013. "TAPAS: Building a TEI Publishing and Repository Service." *Journal of the Text Encoding Initiative* 5. https://doi.org/10.4000/jtei.788.

Folsom, Ed. 2007. "Database as Genre: The Epic Transformation of Archives." *Publications of the Modern Language Association of America* 122, no. 5: 1571–79. https://doi.org/10.1632/pmla.2007.122.5.1571.

Gitelman, Lisa. 2010. "Welcome to the Bubble Chamber: Online in the Humanities Today." *Communication Review* 13, no. 1: 27–36. https://doi.org/10.1080/10714420903558647.

Giunchiglia, Fausto, Biswanath Dutta, and Vincenzo Maltese. 2014. "From Knowledge Organization to Knowledge Representation." *Knowledge Organization* 41, no. 1: 44–56. https://doi.org/10.5771/0943-7444-2014-1-44.

Grumbach, Elizabeth, and Laura Mandell. 2014. "Meeting Scholars Where They Are: The Advanced Research Consortium (ARC) and a Social Humanities Infrastructure." *Scholarly and Research Communication* 5, no. 4. https://src-online.ca/index.php/src/article/view/189/363.

Hamilton, Paula, and Mary Spongberg. 2020. *Feminist Histories and Digital Media*. London, U.K.: Routledge. https://doi.org/10.4324/9780429058073.

Hanseth, Ole. 2010. "From Systems and Tools to Networks and Infrastructures— from Design to Cultivation: Towards a Design Theory of Information Infrastructures." In *Industrial Informatics Design, Use and Innovation: Perspectives and Services*, edited by J. Holmström, M. Wiberg, and A. Lund. Hershey, PA: IGI Global. https://doi.org/10.4018/978-1-61520-692-6.

Hey, Tony, and Anne E. Trefethen. 2005. "Cyberinfrastructure for e-Science." *Science* 308, no. 5723: 817–21. https://doi.org/10.1126/science.1110410.

Holland, Kathryn, and Jana Smith Elford. 2016. "Textbase as Machine: Graphing Feminism and Modernism with OrlandoVision." In *Reading Modernism with Machines: Digital Humanities and Modernist Literature*, edited by S. Ross and J. O'Sullivan, 109–34. London, U.K.: Palgrave Macmillan. https://doi.org/10.1057/978-1-137-59569-0_5.

Jakacki, Diane Katherine, Susan Brown, James Cummings, Mihaela Ilovan, and Carolyn Black. 2022. "The Linked Editing Academic Framework: Creating an Editorial Environment for Collaborative Scholarship and Publication." Talk presented at the Digital Humanities Conference, Tokyo, July 28, 2022. https://www.youtube.com/channel/UClaec96Too6wXWDT_tctYJw.

Kaltenbrunner, Wolfgang. 2017. "Digital Infrastructure for the Humanities in Europe and the US: Governing Scholarship through Coordinated Tool Development." *Computer Supported Cooperative Work* 26, no. 3: 275–308. https://doi.org/10.1007/s10606-017-9272-2.

Kyvernitou, Ioanna, and Antonis Bikakis. 2017. "An Ontology for Gendered Content Representation of Cultural Heritage Artefacts." *Digital Humanities Quarterly* 11, no. 3. http://www.digitalhumanities.org/dhq/vol/11/3/000316/000316.html.

Lee, Charlotte P., Paul Dourish, and Gloria Mark. 2006. "The Human Infrastructure of Cyberinfrastructure." In the *Proceedings of the 2006 20th Anniversary Conference on Computer Supported Cooperative*, 483–492. https://doi.org/10.1145/1180875.1180950.

Linley, Margaret. 2016. "Ecological Entanglements of DH." In *Debates in the Digital Humanities 2016*, edited by Matthew K. Gold and Lauren F. Klein. Minneapolis: University of Minnesota Press. https://dhdebates.gc.cuny.edu/read/untitled/section/47e34952-7b4a-42eb-9b87-d24607a18588#ch34.

Liu, Alan. 2016. "Drafts for *Against the Cultural Singularity* (book in progress)." May 2, 2016. https://doi.org/10.21972/G2B663.

———. 2018. *Friending the Past: The Sense of History in the Digital Age*. Chicago: University of Chicago Press. https://doi.org/10.7208/chicago/9780226452005.001.0001.

———. 2020. "Toward a Diversity Stack: Digital Humanities and Diversity as Technical Problem." *Publications of the Modern Language Association of America* 135, no. 1: 130–151. https://doi.org/10.1632/pmla.2020.135.1.130.

Looser, Devoney. 2015. "British Women Writers, Big Data and Big Biography, 1780–1830." *Women's Writing* 22, no. 2: 165–171. https://doi.org/10.1080/09699082.2015.1011838.

Lynch, Clifford. 2002. "Digital Collections, Digital Libraries and the Digitization of Cultural Heritage Information." *First Monday* 7, no. 5. https://doi.org/10.5210/fm.v7i5.949.

Martin, Kim, Susan Brown, Chelsea Miya, and Shawn Murphy. 2019. *Humanities Centered Design Features: Emergent Serendipity with HuViz.* Digital Humanities Conference 2019. Utrecht, Netherlands. https://dev.clariah.nl/files/dh2019/boa/0817.html.

Meyer, Eric T., and Ralph Schroeder. 2015. *Knowledge Machines: Digital Transformations of the Sciences and Humanities.* Cambridge, Mass.: The MIT Press. https://doi.org/10.7551/mitpress/8816.001.0001.

Nowviskie, Bethany. 2015. "Digital Humanities in the Anthropocene." *Digital Scholarship in the Humanities,* 30 (issue supplement 1): i4–i15. https://doi.org/10.1093/llc/fqv015.

Plantin, Jean-Christophe, Carl Lagoze, Paul N. Edwards, and Christian Sandvig. 2018. "Infrastructure Studies Meet Platform Studies in the Age of Google and Facebook." *New Media & Society* 20, no. 1: 293–310. https://doi.org/10.1177/1461444816661553.

Pomerantz, Jeffrey, and Gary Marchionini. 2007. "The Digital Library as Place." *Journal of Documentation* 63, no. 4: 505–33. https://doi.org/10.1108/00220410710758995.

Ramakrishnan, Kavita, Kathleen O'Reilly, and Jessica Budds. 2020. "The Temporal Fragility of Infrastructure: Theorizing Decay, Maintenance, and Repair." *Environment and Planning E: Nature and Space* 4, no. 3: 1–22. https://doi.org/10.1177/2514848620979712.

Ribes, David, and Thomas A. Finholt. 2009. "The Long Now of Technology Infrastructure: Articulating Tensions in Development." *Journal of the Association for Information Systems* 10, no. 5: 375–98. https://doi.org/10.17705/1jais.00199.

Ribes, David, and Charlotte P. Lee. 2010. "Sociotechnical Studies of Cyberinfrastructure and e-Research: Current Themes and Future Trajectories." *Computer Supported Cooperative Work* 19, no. 3–4: 231–44. https://doi.org/10.1007/s10606-010-9120-0.

Rockwell, Geoffrey. 2010. "As Transparent as Infrastructure: On the Research of Cyberinfrastructure in the Humanities." In the *Proceedings of the Mellon Foundation Online Humanities Conference.* https://cnx.org/contents/PVdHolD@1.3:_USvuzFn@2/As-Transparent-as-Infrastructure-On-the-research-of-cyberinfrastructure-in-the-humanities.

Rowlands, Ian, and David Bawden. 2009. "Digital Libraries: A Conceptual Framework." *Libri* 49, no. 4. https://doi.org/10.1515/libr.1999.49.4.192.

Sowa, John F. 2000. *Knowledge Representation: Logical, Philosophical and Computational Foundations*. London, U.K.: Brooks/Cole Publishing Co.

Sula, Chris Alen. 2013. "Digital Humanities and Libraries: A Conceptual Model." *Journal of Library Administration* 53, no. 1: 10–26. https://doi.org/10.1080/0193 0826.2013.756680.

Tuominen, Kimmo, Sanna Talja, and Reijo Savolainen. 2003. "Multiperspective Digital Libraries: The Implications of Constructionism for the Development of Digital Libraries." *Journal of the American Society for Information Science and Technology* 54, no. 6: 561–569. https://doi.org/10.1002/asi.10243.

Unsworth, John. 2000. "Scholarly Primitives: What Methods Do Humanities Researchers Have in Common, and How Might Our Tools Reflect This?" Symposium on Humanities Computing: Formal Methods, Experimental Practice, King's College London. http://www3.isrl.illinois.edu.subzero.lib. uoguelph.ca/unsworth/Kings.5-00/primitives.html.

———. 2001. "Knowledge Representation in Humanities Computing." Inaugural E-humanities Lecture at the National Endowment for the Humanities. April 3, 2001. https://people.brandeis.edu/~unsworth/KR/KRinHC.html.

———. 2006. *Our Cultural Commonwealth: The Report of the American Council of Learned Societies Commission on Cyberinfrastructure for the Humanities and Social Sciences*. ACLS: New York. http://hdl.handle.net/2142/189.

Zundert, Joris van. 2012. "If You Build It, Will We Come? Large Scale Digital Infrastructures as a Dead End for Digital Humanities." *Historical Social Research*, 37: 165–186. https://doi.org/10.12759/hsr.37.2012.3.165-186.

Unsettling Colonial Mapping: Sonic-Spatial Representations of amiskwaciwâskahikan

Kendra Cowley

A LOW HUM COULD BE HEARD through the universe, rhythmically broken by a consistent lull then the hum would repeat itself over and over
again. No human memory could say when the hum began; only in the oral
tradition of the nêhiyaw and nakawê people has it been told through the
generations that it is foundational in the creation of mother earth.

—Sylvia McAdam (2015)

In the spring of 2018, my co-conspirator, Kateryna Barnes, and I began what
we called our sonic takedown of geographic information systems (GIS).
Unsettling Colonial Mapping: Sonic-Spatial Representations of amiskwaciwâskahikan
began as an installation for *Repurposed: An Exploration of Digital Art and Activism*,
hosted by the University of Alberta's FemLab. This project was designed with
the goal of (re)orienting our relationship to the university as a sounding ecosystem full of vibrations, movement, audible expressions—all part of the sensory
composition of campus. Located as such, *Unsettling Colonial Mapping* asked how
sonic mapping might challenge the visual supremacy of colonial cartographic
technologies to better foreground the fluid, embodied, and complex relationships of spacetime on campus. While Kateryna's energy is now elsewhere, our
collaborations continue to shape-shift, resonating generously and abundantly
for each of us in different ways. This chapter is my elaboration and critical
reflection on *Unsettling Colonial Mapping*: a sonic map of the North Campus of
the University of Alberta, and the collective thinking that went into the project
design.[1] The University of Alberta is located in amiskwaciwâskahikan,[2] Treaty 6
territory and the homeland of the Métis, sitting above the kisiskâciwanisīpiy[3] on
land stolen from the Papaschase Cree only twenty years prior to the university's
founding. This land, and thus the University of Alberta, is governed by Treaty 6,
which, according to Cree legal scholar Sylvia McAdam in *Nationhood Interrupted*
(2015, 24), is a sacred agreement of reciprocity between sovereign Indigenous
nations and the Canadian state based on nêhiyaw[4] laws inextricably connected to
the land. This nêhiyaw interpretation of treaty is a living map of relationality, yet
as Hayden King and Shiri Pasternak suggest in a 2019 paper, practices of settler

cartography have overwritten these relationships of mutuality with colonial boundaries and paternalistic policies, distorting the concept of treaty to allow for multi-scalar practices of dispossession, extraction, and settlement—defining practices of settler colonialism (2019, 17).

To think critically about treaty, cartography, and the university requires an interrogation of space-delineating technologies created and employed by the university to perpetuate settler-colonial claims to land—tools that necessarily flatten the land, representing depth primarily in their desire for extraction. *Unsettling Colonial Mapping* proposes a cartographic practice that might destabilize colonial orientations to the land by tuning into the fluidity of sound that exceeds the silencing, stilling, imperative of settler colonialism. It is my belief that cultivating this type of spatial practice *unsettles* notions of a static landscape and facilitates an attunement to the energetic, material relationships that are, as Sylvia McAdam (2015) and others demonstrate, represented in nêhiyaw legal tradition. By engaging the land through sound, *Unsettling Colonial Mapping* asks that people in Treaty 6 territory listen differently, in order to, as settler scholar and poet Christine Stewart suggests, "locate ourselves within the hum of reciprocations, to locate the relations that bind us" (2015b, 141).

Our map of South Quad, a large outdoor space in the middle of the University of Alberta's main campus, used bioacoustic recordings—sounds of the living in communication with each other and their environment—to engage the sounding relationships of the university. *Unsettling Colonial Mapping* was grounded in the premise that we are enveloped in the sounding world, which hums in/audibly all around us all the time. Sound is vibrational matter, part of the sensory composition of the land that is always participating in a world becoming—an animating force that is both affective and material, part of the earthly composition that shapes the landed projects of the university, Treaty 6 territory, and the nation-state.

South Quad functions as a primary meeting place on campus. In the quad, human and other-than-humans mingle and architecture reveals the time-stamped expansion of the university. From brick, to concrete, to steel and back again, the built landscape of campus reminds us of its continual expansion on Indigenous land. The quad is also punctuated by installation art, including Stewart Steinhauer's *Sweetgrass Bear* sculpture, drawing attention to the university's public reckonings and artistic declarations. According to Steinhauer, the *Sweetgrass Bear* embodies nêhiyaw knowledge and teachings of relationality as engraved on the sculpture: humble kindness, sharing, honesty, and determination (Almond et al. 2018). Steinhauer posits that the use of granite calls into being the Rock Grandfather—a facilitator of communication—"the Rock Grandfather uses a non-linguistic approach to communication, speaking directly, con-sciousness to consciousness" (Steinhauer 2017). The embodiment of the Rock Grandfather, a witness to the treaty-making process (McAdam 2015), and the inscription of *we are all related* on its side, serve as a reminder of the University of Alberta's location within Treaty 6. Our recording process engaged the sculp-ture as a site of repeated listening, where recordings registered that which the *Sweetgrass Bear* sculpture might encounter on any given day.

THE MAP

Unsettling Colonial Mapping is an eight-minute audio assemblage comprising more than 30 different bioacoustic recordings from the area immediately sur-rounding the *Sweetgrass Bear* sculpture. It is accompanied by a 360-degree, slow-motion video of the sculpture, starting in the north and moving through the four directions. Overlaying the 360-degree video is an unrecognizable close-up of the sculpture, calling attention to the atomic materiality of the granite. The video, in its microscopic movements, is meditative, not the focal point of the project, nor in sync with the audio. Functioning as a resting place for busy eyes, the visual asks that listeners drop into the installation (amid a gallery of installations) focused on and open to the sonic movement of the recordings. While a process of bodily attunement that we desire allows for the engagement of all the senses in concert with one another, the colonial elevation of the visual over other sensual experiences, and the unlearning this necessitates, makes disengaging the imper-ative to see difficult. As such, we offer multiple points of invitation and entry into the map that hope to encourage meaningful engagement with the sound space of South Quad.

In addition to the animal, elemental, mechanical sounds listeners hear, the video includes the voices of Kateryna and me reading from Treaty 6 and quot-ing Cree legal scholar Sharon Venne. The track also includes Cree-Métis scholar Trudy Cardinal talking about ceremony on campus. We designed the map as an immersive engagement that foregrounds the relational aspect of listening, operating as a co-constitutive experience that is generated, in part, through the listener's active practice of listening with/and/against the sounds of our compo-sition. Listeners entered the sound space through headphones, where the use of panning facilitates a duo-directional encounter with the recordings.

Unsettling Colonial Mapping used recordings that are both locative and disori-enting: the recorded sounds are heard in relation to one another, always stretch-ing and (re)articulating the bounds of the sound space and the familiar sounds of South Quad, while our voices animate and ground us in treaty obligations. However, these sounds are also disorienting; they have been edited, layered, amplified—thirty-plus recordings resound in under ten minutes, to articulate a fulsome sonic environment rarely experienced as such. Operating as a dis/re/orienting tool that might point toward this opening, *Unsettling Colonial Mapping* foregrounds the University of Alberta as a material, sounding place requiring our attention.

THE UNIVERSITY AS PLACE

> If you are walking around the University of Alberta you are walking on his [Chief Papaschase's] land, Indigenous Land.... What are your treaty rights.... Every non-Indigenous person should know his or her treaty rights. The simple fact is that, without the treaty, no one other than Indigenous people has the right to live on our land.
>
> —Sharon Venne (2007)

The University of Alberta is an important site of convergence for the politics of land, cartography, and treaty. As a site of knowledge production, the university occupies not only the physical land of its campus but contours the intellectual terrain through which disciplines such as cartography (digital and otherwise) are made legible. These cartographic architectures, including the technologies and accompanying academic discourses, both shape and are shaped by settler-colonial ways of knowing the land, affirming the university's role in the ongoing settlement of Indigenous territory. Acutely, in a time governed by the rhetoric of reconciliation, the university is party to the reproduction of a state discourse that celebrates performative moves of atonement and compromise while ensuring the institution remains intact. In the University of Alberta's 2016 strategic plan, *For the Public Good*, it is stated that the institution will develop "a thoughtful, respectful, meaningful, and sustainable response to the report of the Truth and Reconciliation Commission of Canada" (2016, 10), a report whose calls to action require education institutions to honour treaty relationships. Yet, as treaty-feminist Emily Riddle suggests, reconciliation and nêhiyaw interpretations of treaty are incommensurable projects. In fact, as Riddle articulates, adhering to nêhiyaw interpretations of treaty "un-reconciles us" (2019).

The tension between Riddle's decolonial enunciation of treaty feminism as an articulation of Indigenous sovereignty and the University of Alberta's mobilization of treaty rhetoric to foreground reciprocity and "honourable" relationships above the politicized history of dispossession that marks the campus evokes the dissonant irreconcilability between the university and treaty as a nêhiyaw political structure. The University of Alberta—230 acres of forest, river valley, concrete, electrical grid—resounds with life inextinguishable in spite of, though not unmarked by, ongoing displacement and expansion. The persistence of animal and botanical life, undiscriminating weather, and physical processes of growth and decay, not to mention Indigenous activism on/against the university, all call forward material relationships deemed inconsequential to the project of knowledge production at the university. A meshwork, as Tim Ingold writes in *Being Alive* (2011), "of entangled lines of life, growth and movement" (63), the land itself is a sensory body of knowledge that has much to teach about a campus full of life. As such, *Unsettling Colonial Mapping* argues that attending to sounding material relationships is a crucial but often neglected practice that indexes the broader state of (broken) treaty relations in Canada today.

Anishnaabeg scholar, activist, and artist Leanne Betasamosake Simpson writes in "Land as Pedagogy" (2017) that the academy must "make a conscious decision to become a decolonizing force in the intellectual lives of Indigenous peoples by joining [Indigenous peoples] in dismantling settler colonialism and actively protecting the source of our knowledge: Indigenous land." Indigenous scholars have identified what these shifts might entail, for example, non-human research ethics, sustained material support of Indigenous researchers, and land-based pedagogies—all shifts requiring a move away from the settler-colonial infrastructure that continues to define the university. *Unsettling Colonial Mapping* takes seriously the call for the epistemological shifts required to destabilize the way we (come to) know in colonial educational institutions, and suggests that an orientation toward the land as a living, sounding, part of the collective body is needed for the humanities to remain relevant in a world on fire.

SOUNDING PLACE

Rooted in anti-colonial desires, *Unsettling Colonial Mapping* required an approach to place, as articulated by Kathleen Stewart in "Place and Sensory Composition" (2015b), as "a sensory composition [...] not an inert landscape made of dead matter but a composting of bodies, affects and forests, of persons, socialities, and existential ecologies of being in the world" (213). This orientation to place, championed by theorists such as Karen Barad (2007) and Doreen Massey (2005), requires, as Barad writes, that we recognize sensory composition as an apparatus, "a specific material reconfiguration of the world that does not merely emerge in time but iteratively re-configures spacetimematter as part of the ongoing dynamisms of becoming" (Barad 2007, 142). These sensory compositions include sound as a co-constitutive element in the amorphous, forever in motion field of energetic relations that constitute the material world.

Sound is one site of sensory encounter with the material composition of campus. Sound theorist Mickey Vallee writes in a 2018 paper that new theories of sounds are "expanding and evolving explications of the intertwining between sound, body, place, sensation, and, generally, the virtual [...] the virtual, the haptic, the affective—in short, that which vibrates beneath or above the surfaces of perception" (2018b, 50). Sound, differently perceptible, is always present through its vibratory movement—what we hear moves us, what we don't hear *moves* us also. Sound is affective and material, co-generative of the constellation of space, time, and matter that makes up place (Voegelin 2019; Massey 2005; Barad 2007). To contend with place through sound allows for the dynamism of relationships of matter always in flux, perpetually on the move, and full of energetic potential.

Place is political. Sound, as it allows us to attend to the specificities of place, is political too. All living beings are part of the dynamic sound space through which social relations are continually (re)articulated. Place cannot be unwoven from the social, nor can the social be understood as outside of place. As Eve Tuck and Marcia McKenzie suggest in *Place in Research* (2015), place is "a meeting place, not only of human histories, spatial relations, and related social practices, but also of related histories and practices of land and other species" (47). Our bodies, like the land, carry our histories and spatial memories. How we sound *and* how we listen are central to how we are in the world. These embodied practices are contingent on the material and social contours of place, including the settler-colonial re-ordering of the land, which are always already political. Within this frame of reference, a sonic attunement that resonates with the anti-colonial relationships we desire is a practice of embodied presence that not only hears in anticipation of harmony but also listens to all that dissonance divulges about the state of treaty relations.

Kateryna's and my desire for a different way of listening was grounded in a practice of deep listening, as championed by composer, artist, activist, and thinker Pauline Oliveros. Oliveros's life's work was oriented toward deep listening, a practice of "listening in every possible way to everything possible to hear" (2010, 73). Moving through the world attuned to the living resonances of an ecosystem beyond colonial and capitalist containment requires a recognition of the relational and reciprocal nature of sounding—it is in this co-generative becoming that we learn about social relations as they are, and as they might otherwise be.

Bioacoustics are inherently about communication, the sounds of an ecosystem as they communicate with each other and the listener. In *Sounding the Anthropocene*, Vallee states that the power of bioacoustic recording lies in its desire to "repair the damage of aesthetic distanciation as well as corporate extraction, both of which belong to the same colonialist enterprise" (2018a, 206). Deep listening similarly attempts to locate the listener not outside of sound but within, and a part of, the sonic environment. According to Oliveros, listening involves "a reciprocity of energy flow [...] and sympathetic vibration" (2010, 90) between the listener and listened to—an attunement to the interconnectedness of the vital forces that comprise our inner and outer worlds and the spaces between. A practice of deep listening thus requires a particular sonic sensibility, an openness to the possibilities of experiencing the world through our other-than-visual senses.

If we accept Oliveros's assertion that everything is sounding—that is, alive with in/audible vibrations—we understand place as always sonic. Listening, then, informs us about how and where we are. Yet, to better understand our obligations to place, and in particular, place as governed by Treaty 6, requires listening to Indigenous articulations of land-based legal traditions. In *Nationhood Interrupted*, McAdam reinforces the importance of land to treaty: "Indigenous nêhiyaw laws are 'written' in the landscapes of the hills, the rocks, the waters, everything in the land tells of our history and our laws...to follow these laws means to follow a sacred life inextricably connected to the earth: one without the other would die" (2012). According to McAdam, Treaty 6 created a relationship between the nêhiyawak and the nation-state of Canada based on nêhiyaw laws indivisible from the land. Thus, to adhere to treaty means to engage the land as party to the agreement, not a contested property to be allocated to its signatories. To sonically locate oneself on the land and in relation to treaty means to listen for that which vibrates with life and in concert with each other. To hear requires the perceptual body to engage with the sounding environment. To hear not as removed or distanced from the sources of sound requires a practice of embodiment that shifts perception from one that observes to one that participates.

And yet, embodiment is, of course, a composite of perceptive structures shaped by settler colonialism. Embodiment, absent of attention to and action toward Indigenous demands for decolonization and land back, is not inherently radical or subversive. In fact, while the embodiment we argued for in *Unsettling Colonial Mapping* is about connecting with communal life and resistance, embodiment is often taken up as a liberal practice of individuality rooted in spiritual appropriation. To shift one's attunement toward a more fulsome experience of the world is not about proximity to truth or individual enlightenment, but about dismantling colonial structures that reside within the body. To participate through a practice of embodiment means to actively work toward disrupting one's participation in colonial violence. Listening with an intent to experience connection should not only reveal the future potential of interconnection, but the violent and present realities of it.

Exploitation is, in fact, a relationship, even when it seeks to destroy. To listen to Indigenous articulations of treaty and demands for decolonization is, as we mention above, as much about dissonance as it is about harmony. When we begin to tune into the sounds of colonial violence what can we no longer ignore? When we hear the dissonance between articulations of colonial treaty and nêhiyaw treaty, how does our relationship to the land change? When we hear decolonial demands for land back, what material shifts become necessary?

DIGITAL HUMANITIES AND THE MAP

While GIS allows for more multi-dimensional maps (the inclusion of movement, story, manipulation) than traditional cartography, digital maps have been rendered by technologies of visualization premised on commitments to seeing as knowing. As Salomé Voegelin suggests, to see means to occupy a meta-position through which truth is a product of detached objectivity; however, listening means that one is always, everywhere, "simultaneously with the heard" (quoted in Stewart 2015a, 138). Historically, geography positions sight as the measure of truth, observation through seeing as the mechanism for knowing a place. The map not only teaches us about a particular construction of space, but according to Christian Jacob, the map "encompasses many other components of a culture: its conception of the world, physical and metaphysical, its cognitive categories that bring knowledge and truth within reach of the human mind, [and] the social construction and sharing of such knowledge about the world" (1999, 25).

This process of visual world building reifies relations of power as inherent to that which is being represented, allowing for the imposition of colonial ordering practices to be seen as part of the natural order of things. According to Mishuana Goeman, mapping is a physical representation of this authority, a spatial enactment of knowledge/power used to flatten relationships, complexity, and movement, and reify settler-colonial notions of discovery (2013, 16). Visual maps make claims to the stability of land, a resource to be extracted and a foundation upon which to build a nation, not a field of relations always in flux. In mapping the sonic-spatio-temporal composition of the land, this project considers charting sonic relationships and expressions as contributing to a practice of resistance mapping, and in particular, one that challenges the colonial desire to contain and control.

Digital humanities' disciplinary commitment to the design, production, and implementation of tools is not new to anti-colonial critique. As critical DH scholars such as Dorothy Kim, T. L. Cowan, Lisa Nakamura, Michael J. Kramer, and Michelle Moravec suggest, this commitment is rooted in the technocratic dreams of settler-colonial, capitalist desires for accelerated productivity and expansion. The historical emergence of the digital humanities as an academic discipline follows the development and democratization of military and surveillance technologies and techniques such as geospatial mapping and data mining (Drucker 2012, 85). An organizing argument of this chapter, and a commitment that oriented the design of *Unsettling Colonial Mapping*, is that these specific tools of domination shape the way we, as subjects as much produced by our tools as producing them, engage sensorially with the world we live in. Tools condition our movements through space, they modify our feelings, and they change the way we hear the world. Ideological commitments to the state and capital are coded into the tools so celebrated by the digital humanities, even as scholars attempt to use these tools in multiple and anti-disciplinary ways.

While GIS is being increasingly used in participatory, community-driven, and narrativized projects (see, e.g., LaRochelle 2019; Ahmed et al. 2018), its historical and social roots in military and state-management technologies are baked into the core of its technological infrastructure. Many scholars in the fields of feminist, Indigenous, and critical cartography have criticized these technologies as positivist, masculinist, and colonial. In *Re-envisioning GIS as a Method in Feminist Geographic Research* (2002), Mei-Po Kwan situates the optical imperative of GIS in a larger critique of the "decorporealized vision of modern technoscience" (648) articulated by feminist theorists such as Donna Haraway and Gillian Rose. This disembodiment of Western science's gaze positions the patriarchal and colonial researcher outside of the object of its inquiry. In this framework the researcher has power over what they see and survey. Critiquing the supposed scientific objectivity of digital mapping, Liz Bondi and Mona Domosh argue that the "Cartesian space-time grid of GIS implies the existence of an external vantage point" separate from the body and the senses that steer it (1992, 211). As many have argued, the spatial imaginary of Western science that foregrounds the optical (and disregards other embodied knowledges) is one that sees to seize.

While Kwan reminds us that sight—as part of the larger sensorium—is not fundamentally problematic (2002, 649), its valorization over other senses legitimizes colonial articulation of space that erases the energetic relationships and movements that animate it. GIS geocodes this reordered space providing scientific validity to nationalist, industrial, and expansionist interests. Funded, employed, and designed by and for institutions such as the military, resource-development departments, and even universities, GIS is a tool used by the settler-colonial state to seize and maintain control of Indigenous land.

Yet, many projects have used GIS technology to re-story space and address community needs—geolocating memories, providing mutual aid, and challenging settler-colonial land claims. In "Decolonizing Geographies of Power: Indigenous Digital Counter-Mapping Practices on Turtle Island" (2017), Dallas Hunt and Shaun A. Stevenson expose the tensions inherent in Indigenous resistance-mapping reliant on digital tools developed and used by the state to appropriate Indigenous land and exploit it for capitalist accumulation: "[W]e are concerned with how this difficult entanglement can both enter into, and come to bear on, what we might recognize as the most significant Western material ordering practices—that is law, private property and the commodification of land" (379). Maps shore up conceptions of the Canadian nation-state as they act as "instrument(s) of certainty through which the nation-state and ensuing settlers achieve a sense of political, legal, and even sentimental entitlement to the land" (375). Where colonial mapping logics are, according to Gwilym Eades, "inherently cartographic, state-based, and caught up in systems of control [...]. Indigenous mapping practices stress topological and relational aspects of space with much less focus upon precisely defined locations" (2015, 127). While the above conception of Indigenous mapping allows for more fluid movement through space and time, GIS has been politically significant in contemporary Indigenous claims of sovereignty.[5] As Hunt and Stevenson demonstrate, GIS is a product of the state that uses visual technology to delineate, contain, and represent disembodied space, and has been mobilized contradictorily to both reify and confront colonial power (2017, 376). *Unsettling Colonial Mapping* is also bound in this contradiction—a project that works in and against the university and that contributes to a discipline deeply entangled with the technocratic infrastructure we seek to critique.

Of course, sound recording is also a technology of capture and commodification embedded in a history of colonial technological and disciplinary entanglements. Moving between modes of technological capture is not in and of itself enough to destabilize how we come to know in the academy. However, *Unsettling*

Colonial Mapping operates from belief that turning our ears to the ground and listening for the deep reverberations of the land might give the lie to this desire for containment so embedded in our cartographic technologies. What is at stake in this commitment is a practice of encountering land as a sensory body of knowledge in ways that complicate how we engage space and place by amplifying this complexity to a level it can be heard.

With this digital experiment, we hoped to trouble digital humanities' fascination with GIS and its visual focus by detailing spacetime aurally. We suggested that listening to the animate relations that enliven campus—the water, the trees, the birds, the wind—might remind us that the university as an institution does not wholly define the university as place. It was our desire that the map act as a site of embodied connection wherein resonance with the sounds of the environment prompts a (re)orientation to campus—a place sounding with the vibrancies of the land and histories of Treaty 6. Where colonial mapping logics delineate and contain space for the sake of state seizure and control, sonic mapping begins to orchestrate the complex entanglements and fluidity of spacetime that defy borders and static representations of space and place, articulating the important role of sound in resistance cartographies.

ARCHIVING SOUND

In the early stages of *Unsettling Colonial Mapping*, we planned to include historical recordings of campus to highlight the robust sounding of campus across time. However, we were quickly confronted by the absence of archival material that meaningfully accounted for the presence of environmental, ambient sounds in historical recordings. In our research at the University of Alberta Archive, we encountered an anthropocentric focus on archiving *events* that prioritized human-animated sounds at the expense of geological, animal, botanical, elemental, and non-human sounds. Because we know sound to be deeply enmeshed with place, and place deeply enmeshed with history, these absences highlighted significant challenges to the inclusion of historical recordings in our project. Ultimately, we did not end up using any archival recordings in our work, but it is important to note these limitations for the methodological questions they animate for other scholars working at the nexus of sound studies, archiving, and what Foucault has famously called "subjugated knowledge" (1970). While challenges of technological translation and mediation are not new to those doing archival work, our experience of the sonic limitations of the archive has meaningful import for archival work in the fields of sound studies and digital humanities.

In our study at the University of Alberta Archives, we were confronted with three key problems, posed here as questions, that impacted our research: What counts as meaningful archival sonic material? How do material-archival practices of sound recording reproduce the limits of colonial epistemologies? And, how might a practice of critical media archaeology benefit sonic research in the archive?

While important work is already being done in the field of Indigenous studies, digital humanities, library and information studies, and critical media archaeology (see Ernst and Parikka 2013; Ghaddar and Caswell 2019; Christen and Anderson 2019) to address these challenges, the specific points of impasse Kateryna and I faced in the preliminary research for *Unsettling Colonial Mapping* foreground the limits of archival practices to engage sound as a *co-constitutive force* of artifacts in the archive.

The first challenge we encountered is as old as the archive itself: *What counts as meaningful archival material?* Like many official archives, the University of Alberta Archive is structured by an approach to documenting history that is focused on historical events and systematic categorization. For example, the archive's abundance of lectures, commemorative events, and performances. In this framework, the archive acts as a taxonomical resource and record of historical transmission, rather than a site for embodied experience. This type of archival practice poses a particular challenge to those of us interested in sound studies as it adds a layer of institutional interference with the historical matter—sound—that we are interested in accessing. When what counts as archival material is determined by the anticipation of an archival subject or documenter interested in normative historical documentation, many different registers of affective and otherwise information are excluded. For example, "nature" or "natural" sounds become either the focal subject of a recording (as in the case of bird songs) or excess noise to be eliminated. This elimination is not only a question of *what* type of information is given archival import, but also a question of *how* events are recorded in the first place. In the case of sound recordings, the decision to silence the "background noise" occurs not only in the archive, but also in anticipation of an archival model of knowledge designed to transmit information, rather than provoke a practice of deep, historical listening. As a result, environmental sounds as part of the sounding, communicative, design of archivable *events* are rarely found in the archive. They are disappeared from the record of the "event"—silenced as excess noise, their existence and transmutation deemed insignificant in the preservation of artifact. In fact, this "excess" sound that we were interested in serves as a boundary marker or delimiter for the archive itself. The sounding environment becomes that which needs to be quieted so that we can hear productive sounds of campus.

The second challenge is of particular interest to sound-studies scholars committed to anti-colonial methodologies in their work: *How do material-archival practices of sound recording reproduce the limits of colonial epistemologies?* While sound archives do exist (usually as their own collection), there is still work to be done that advances the inclusion of sound as a co-constitutive element of historical encounters present in the archive. When ecological sound recordings exist in the archive, they often take the form of bioacoustic recording that contribute to a particular epistemological taxonomy, one that attempts to name and/or claim the subject of the recordings. For example, research conducted by the Songbird Neuroethology Laboratory at the University of Alberta uses bioacoustic recordings to research the communication patterns of chickadees (Sturdy 2015). These records foreground a particular engagement with sound, one that isolates sounds from their context and selectively listens for information already determined as relevant. Deep listening in the archive thus requires not only paying attention to what is deemed worthy of recording but also how these recording practices reproduce a colonial relation to the world that isolates in order to identify, rather than tarrying with the relational conte/n/xts of archival material. To listen against the grain in archival recording then means to listen for absences, silences, and noise that exceed the technological and authorial preferences that demarcate the archive.

Finally, we questioned *how a practice of critical media archaeology would benefit sonic research in the archive*. As suggested above, bioacoustic recordings are still mediated by and through human-created and -operated technology—to hear (in a recognizable way) environmental sounds, requires a practice of capturing and analyzing—of excavation that is not entirely different from other modes of representation in the archive. "Nature" or "natural" sounds are still recorded and subject to human and technological manipulation. This technological transmutation of data encodes choices of inclusion and exclusion that determine which sounds are worthy of recording. As Kateryna and I recorded, we were reminded of the animating force of wind, often experienced as a nuisance for people interested in recording environmental sounds. Technologies and techniques of recording have long contended with how to minimize its presence in recordings, and thus limit its place in the archive. These attempts teach us about what counts as data and what constitutes interference—wind is understood as noise, yet, if we listen holistically, according to Indigenous knowledge and philosophy, wind is not mere noise, but is an entity that activates a field of relations, and, according to

Gregory Cajete, a vital force that breathes life and movement into the body on all scales (2000; see also McAdam 2015). To erase wind is to erase a relational force of activation that can teach us much about sounding relationships on campus.

I offer these notes on the methodological challenges Kateryna and I faced so that we might contribute to deepening an archival engagement with sonic research. We know that archives are one of the many institutions that naturalize settler-colonialism forms of knowing. In order to disrupt the infrastructure of extractive listening that hears in service of colonial knowledge production, we propose a critical engagement with sound in the archive as a site of further research.

SOUNDING RELATIONS

Unsettling Colonial Mapping sought to pose a multi-scalar challenge to the university: what obligations can we no longer ignore when we hear ourselves as part of a living, sounding landscape, one that is governed by Treaty 6, a continually broken promise of reciprocity?

Constellating sounds to draw attention not only to a particular soniferous location on campus, but also the field of relations in which they sound, and we exist, *Unsettling Colonial Mapping* asked participants to locate themselves within the sounding environment as both a recognition and expression of relationality, one that leads to practices of anti-colonial solidarity and resistance.

This project is hopeful, operating from the belief that listening can draw attention to connections that makes the need for more just relationships undeniable. Our hope is guided by the claim that experiencing campus as a sounding place might (re)invigorate conversations around dis/placement—of people, of animals, of plant life, of obligations; to challenge the static claims of a map in favour of the energized and mobile sonic presence that is co-constitutive of place. And so, in ending, I want to echo Voegelin's sentiment in favour of sonic futurity.

I want to hear the sonic world as possible worlds, as counterfactual positions—that I reciprocate, to investigate its semantic substance, "what it is," through listening beyond the frame of factuality, knowledge, ideology, and aesthetic certainty, and come to understand how I inhabit that substance, how I partake in the construction of its reality, and how I can negotiate its value within the notion of actuality as a plurality, to know "what it is like" and "what it could be like also." (2019, 46)

NOTES

1. Parts of this chapter are taken from conference talks and an accompanying blog post at the Humanities, Arts, Science and Technology Alliance and Collaboratory, SpokenWeb, and the Canadian Society for Digital Humanities in June 2019. Kateryna read this essay prior to publication and enthusiastically supports Kendra's solo byline on a piece that honours our collaboration.
2. Cree for Edmonton, meaning Beaver Hills House. There is no capitalization in Cree.
3. Cree for the North Saskatchewan River.
4. Cree.
5. For examples, see the Ogimaa Mikana Anishinaabemowinplace-names project, "Reclaiming/Renaming and Pasikôw, https://ogimaamikana.tumblr.com/.

REFERENCES

Ahmed, Manan, Maira E. Álvarez, Sylvia A. Fernández, Alex Gil, Merisa Martinez, Moacir P. de Sá Pereira, Linda Rodriguez, and Roopika Risam. 2018. "Torn Apart / Separados." xpmethod, June 25, 2018. https://xpmethod.columbia.edu/torn-apart/volume/2/.

Almond, Amanda, Rob McMahon, Diane Janes, Greg Whistance-Smith, Diane Steinhauer, and Stewart Steinhauer. 2018. "We Are All Related: Using Augmented Reality as a Learning Resource for Indigenous-Settler Relations—Northern Public Affairs." Northern Public Affairs (blog). http://www.northernpublicaffairs.ca/index/volume-6-special-issue-2-connectivity-in-northern-indigenous-communities/we-are-all-related-using-augmented-reality-as-a-learning-resource-for-indigenous-settler-relations/.

Barad, Karen Michelle. 2007. *Meeting the Universe Halfway: Quantum Physics and the Entanglement of Matter and Meaning.* Durham, N.C.: Duke University Press. https://doi.org/10.2307/j.ctv12101zq.

Bondi, Liz, and Mona Domosh. 1992. "Other Figures in Other Places: On Feminism, Postmodernism, and Geography." *Environment and Planning D: Society and Space* 10, no. 2: 199–213. https://doi.org/10.1068/d100199.

Cajete, Gregory. 2000. *Native Science: Natural Laws of Interdependence.* Santa Fe, N.M.: Clear Light Publishers.

Christen, Kimberly, and Jane Anderson. 2019. "Toward Slow Archives." *Archival Science: International Journal on Recorded Information* 19, no. 2: 87–116. https://doi.org/10.1007/s10502-019-09307-x.

Drucker, Johanna. 2012. "Humanist Theory and Digital Humanities." In *Debates in the Digital Humanities*, edited by Matthew K. Gold, 185–99. Minneapolis: University of Minnesota Press. https://doi.org/10.5749/9781452963754.

Eades, Gwilym Lucas. 2015. *Maps and Memes: Redrawing Culture, Place, and Identity in Indigenous Communities*. Montréal, Que., and Kingston, Ont.: McGill-Queen's University Press.

Ernst, Wolfgang, and Jussi Parikka. 2013. *Digital Memory and the Archive*. Electronic Mediations 39. Minneapolis: University of Minnesota Press.

Foucault, Michel. 1970. "The Archaeology of Knowledge." *Social Science Information* 9, no. 1 (February): 175–185. https://doi.org/10.1177/053901847000900108.

Ghaddar, J. J., and Michelle Caswell. 2019. "'To Go Beyond': Towards a Decolonial Archival Praxis." *Archival Science: International Journal on Recorded Information* 19, no. 2: 71–85. https://doi.org/doi:10.1007/s10502-019-09311-1.

Goeman, Mishuana. 2013. *Mark My Words: Native Women Mapping Our Nations*. Minneapolis: University of Minnesota Press. https://doi.org/10.5749/minnesota/9780816677900.001.0001.

Hunt, Dallas, and Shaun A. Stevenson. 2017. "Decolonizing Geographies of Power: Indigenous Digital Counter-Mapping Practices on Turtle Island." *Settler Colonial Studies* 7, no. 3 (July 3): 372–92. https://doi.org/10.1080/2201473X.2016.1186311.

Ingold, Tim. 2011. *Being Alive: Essays on Movement, Knowledge, and Description*. Abingdon, U.K.: Routledge.

Jacob, Christian. 1999. "Mapping in the Mind: The Earth from Ancient Alexandria." In *Mappings*, edited by Dennis Cosgrove, 224–70. London, U.K.: Reaktion Books.

King, Hayden, and Shiri Pasternak. 2019. "Land Back: A Yellowhead Institute Red Paper." Yellowhead Institute. October 2019. https://redpaper.yellowheadinstitute.org/wp-content/uploads/2019/10/red-paper-report-final.pdf.

Kwan, Mei-Po. 2002. "Feminist Visualization: Re-envisioning GIS as a Method in Feminist Geographic Research." *Annals of the Association of American Geographers* 92, no. 4: 645–61. https://doi.org/10.1111/1467-8306.00309.

LaRochelle, Lucas. 2019. Queering the Map. Accessed on October 27, 2019. https://www.queeringthemap.com/.

Massey, Doreen B. 2005. *For Space*. London, U.K.: SAGE.

McAdam, Sylvia. 2012. "Idle No More—I Hear Many People Talk about Treaty, Far Too Many of Us Do." *Net News Ledger*. December 15, 2012. http://www.netnewsledger.com/2012/12/15/idle-no-more-i-hear-many-people-talk-about-treaty-far-too-many-of-us-do/.

———. 2015. *Nationhood Interrupted: Revitalizing Nêhiyaw Legal Systems*. Vancouver, B.C.: Purich Books.

Oliveros, Pauline. 2010. *Sounding the Margins: Collected Writings 1992–2009*. Kingston, N.Y.: Deep Listening Publications.

Riddle, Emily. 2019. "Thoughts from a Traitorous Albertan: Treaty Feminism in 2019." Talk presented at the Parkland Institute 22nd Annual Conference, Edmonton. https://www.youtube.com/watch?v=63ymjDsUOfM.

Simpson, Leanne Betasamosake. 2014. "Land as Pedagogy: Nishnaabeg Intelligence and Rebellious Transformation." *Decolonization: Indigeneity, Education & Society* 3, no. 3, p. 1–25.

Simpson, Leanne Betasamosake. 2017. *As We Have Always Done: Indigenous Freedom through Radical Resistance.* Minneapolis: University of Minnesota Press. https://doi.org/10.5749/j.ctt1pwt77c.

Steinhauer, Stewart. 2017. "Consider This: Stewart Steinhauer on the Sweetgrass Bear in Treaty Six Territory." *The Quad* (blog). July 25, 2017. https://blog.ualberta.ca/consider-this-stewart-steinhauer-on-the-sweetgrass-bear-in-treaty-six-territory-4077941a254c.

Stewart, Christine. 2015a. "Propositions from Under Mill Creek Bridge: A Practice of Reading." In *Sustaining the West: Cultural Responses to Canadian Environments,* edited by Lisa Szabo-Jones and Liza Piper, 241–58. Waterloo, Ont.: Wilfred Laurier University Press.

———. 2015b. "Treaty Six from Under Mill Creek Bridge." In *Toward.Some.Air,* edited by Fred Wah and Amy De'Ath. Banff, Alb.: Banff Centre Press, 2015.

Stewart, Kathleen. 2015. "Place and Sensory Composition." In *The Intelligence of Place: Topographies and Poetics,* edited by Jeff Malpas, 205–20. London, U.K.: Bloomsbury. https://doi.org/10.5040/9781474272872.ch-011.

Sturdy, Christopher B. 2015. "Research in the SNL." Songbird Neuroethology Laboratory. Department of Psychology, University of Alberta. Last modified July 18, 2015. http://www.psych.ualberta.ca/~csturdy/research.htm.

Tuck, Eve, and Marcia McKenzie. 2015. *Place in Research: Theory, Methodology, and Methods.* New York: Routledge.

University of Alberta. 2016. *For the Public Good.* Edmonton. https://www.ualberta.ca/strategic-plan/index.html.

Vallee, Mickey. 2018a. "Sounding the Anthropocene." In *Interrogating the Anthropocene: Ecology, Aesthetics, Pedagogy, and the Future in Question,* edited by jan jagodzinski, 201–16. Cham, Switzerland: Palgrave Macmillan. https://doi.org/10.1007/978-3-319-78747-3_7.

———. 2018b. "The Science of Listening in Bioacoustics Research: Sensing the Animals' Sounds." *Theory, Culture & Society* 35, no. 2: 47–65. https://doi.org/10.1177/0263276417727059.

Venne, Sharon. 2007. "Treaties Made in Good Faith." In *Natives & Settlers, Now & Then: Historical Issues and Current Perspectives on Treaties and Land Claims in Canada,* edited by Paul W. DePasquale, 1–16. Edmonton: University of Alberta Press.

Voegelin, Salomé. 2019. *The Political Possibility of Sound: Fragments of Listening.* New York: Bloomsbury Academic. https://doi.org/10.5040/9781501312199.

Beyond "Mere Digitization": Introducing the Canadian Modernist Magazines Project

Graham H. Jensen

C anadian modernism has a digitization problem. But so do the digital humanities. According to Kathleen Fitzpatrick, the term "digital humanities," which gradually replaced "humanities computing," was coined when Susan Schreibman, Ray Siemens, and John Unsworth rejected their publisher's suggestion that they call their co-edited collection of essays "A Companion to Digitized Humanities"—a title that would have undermined their efforts "to keep the field from appearing to be about mere digitization" (Fitzpatrick 2011).[1] Given the field's ongoing struggles to carve out space for itself—both on campuses and in humanities scholarship—such attempts to move beyond "mere digitization" are wholly understandable. DH scholars do much more than scan documents. And yet this now-instinctual reflex to frame the field as so much more than the sum of its digitized parts has had unintended negative consequences, not only in the digital humanities but in other academic disciplines—including what Zack Lischer-Katz refers to as the "continued casualization and denigration" of digitization work in libraries and across "information institutions" (2019, 242). My own project, the Canadian Modernist Magazines Project (CMMP), relies heavily on this still-undervalued work of digitization. Nevertheless, like Lischer-Katz, I argue that this work is of theoretical as well as practical significance; it involves processes that translate both humanities materials and methodologies into the digital realm. In the case of the CMMP, the digitization of Canadian modernist magazines has meant actively testing and embodying the project's guiding supposition (which, in turn, was informed by recent developments in modern periodical and new modernist studies): that small-scale periodicals or "little magazines" were crucially important not only to the formation of the Canadian modernist canon as it has traditionally been discussed but to the formation of diverse and previously unacknowledged movements in Canadian modernist literature.

Conceptually, the CMMP began first and foremost as a digitization project—which is not to say that I was completely ignorant of the theoretical implications of the work of digitization, or that I had no aspirations to "[go] beyond simply wishing to preserve" a collection of literary "artifacts" (Schreibman,

Siemens, and Unsworth 2004, xxiv). Rather, when I started to develop the project in 2014, my initial goal was "simply" to address a well-known access issue (of which more soon) by establishing an open, digital repository of Canadian modernist little magazines, starting with just two of the most obvious candidates: Montréal-based periodicals *Preview* (1942–1944) and *First Statement* (1942–1945). Quite quickly, however, "mere digitization" became "more digitization" as my collaborators and I dreamed of a more diverse and representative collection of magazines—and then, just as quickly, it became not "mere" but "dear"; it took almost five years of genealogical and literary-critical detective work, letter- and grant-writing, studying copyright law, shameless haggling, cultivating institutional partnerships, and plenty of waiting before I had scanned copies of *Preview* and *First Statement* in my possession. In Canadian modernist studies, at least, it seems that the work of digitization often involves everything but the act of converting physical objects to digital ones; to digitize is to research and collaborate, to analyze, ontologize, incentivize, and prioritize.

Since 2014, the CMMP has moved beyond mere digitization in other ways. In grant proposals, I have reframed the project as a public-facing research and knowledge-mobilization platform for reading, analyzing, and teaching Canadian modernist literature in its many permutations. In other words, the CMMP is now guided by twin objectives: (1) to make canonical texts available to academics and non-academics alike (thereby supporting ongoing research in, and existing literary-critical narratives about, Canadian modernism), but also (2) to expand the canon to include a variety of non-canonical magazines as well as paratextual materials, such as refereed critical introductions, a bibliography, and relevant study guides or syllabi (all of which will wittingly or unwittingly alter those existing narratives). By digitizing and transcribing a diverse selection of magazines, I hope to provide scholars of Canadian literature, modernism, and periodical studies with access both to canonical and to largely unknown and inaccessible modernist texts. These texts will serve as the primary sources for my own analyses of how extant interpretations of Canadian modernism must be revisited in light of recent developments in the interrelated fields of new-media studies, periodical studies, and modernist studies. However, because the CMMP website will include accessible introductions to its digital assets as well as supplementary teaching materials, the project also hopes to encourage new means of public engagement with Canadian literatures and cultural histories of the twentieth century.

Still, the work of digitization continues—and it drives each of the CMMP's research outputs. With the digitization of *Preview* and *First Statement* now completed, as well as several new additions in the works—Tarot (1896), Neith (1903–1904), and Le Nigog (1918)—the project recently launched its primary contribution to scholarship: the CMMP's virtual research hub (modernistmags. ca). In the next phase of the project, the website will be transformed into a full-fledged research portal featuring the supplementary research and teaching materials described above. As the project secures additional funding, we will continue to add to the project's curated collection of pedagogical materials and to its established repository of magazine facsimiles (shared in PDF format, but based on archived TIFF files), plain-text transcriptions of each magazine (using optical character recognition software), and TEI-XML (Text Encoding Initiative–Extensible Markup Language) files with minimal markup of magazine metadata (to encode basic information about each issue's contributors, contributions, and publication details).

This, too, is part of the larger process of digitization; but it also involves what N. Katherine Hayles refers to as "media translation" (2005, 89)—that is, the material and critical remediation of print forms into digital ones. Because we are aware of the potential methodological and literary-critical problems of forcing modernist periodicals into pre-existing bibliographic ontologies and digital frameworks, we have attempted, as much as possible, to encode only metadata essential for navigational and basic research purposes. Even the seemingly straightforward task of assigning genre (is this text a "manifesto," an "editorial," or an "essay"?) involves editorial choices that risk imposing bibliographic interpretations and structures on magazines so as to materially reconstruct them—and therefore undermine the extent to which they function as faithful witnesses to modernist projects of self-fashioning.

However, the CMMP's paratextual materials necessarily craft new narratives of Canadian and modernist literature, and its decision to digitize modernist periodicals, specifically, was also a choice with theoretical implications, not least because it placed the project at the nexus of emergent, intersecting fields. This project was initiated at a moment when modernist studies, modern periodical studies, and Canadian literature were reassessing their definitions of— and methodological approaches to—modernism. In the Canadian context, for example, critics such as Glenn Willmott (2004), Candida Rifkind (2009), Dean Irvine, and Gregory Betts have answered the call of the so-called new modernist studies (now 20 years old) for the expansion of the field in "temporal, spatial, and vertical directions" (Mao and Walkowitz 2008, 737). They have done this by constructing alternative narratives of proto-modernist, modernist, late-modernist, and postmodernist movements and figures—something that the CMMP accomplishes by virtue of its digitization of a curated selection of Canadian periodicals. But the CMMP was also established in response to other critical conversations. In the field of periodical studies more generally, for example, critics such as Robert Scholes, Sean Latham, Susan Smulyan, Jeffrey Drouin, Clifford Wulfman, Mark Gaipa, Hannah McGregor, and Nicholas van Orden have harnessed emergent DH methodologies and technologies to read high-modernist and middlebrow texts in novel ways—and on an entirely new scale.[2] The CMMP intervenes in these ongoing critical discussions through its digitization and analysis of magazines with varying levels of commitment to what might be called modernist aesthetics or attitudes. Because periodical publications contain advertisements, manifestos, letters, editorials, poetry, drama, and fiction, they are well suited to the study of modernism as a heterogeneous, cross-genre phenomenon (see, e.g., Latham and Scholes 2006; Hammill, Hjartarson, and McGregor 2015; McGregor and van Orden 2016).

As I have already suggested, though, the CMMP began as a response to a much more urgent need: the need to digitize Canadian modernist texts. Canadian modernism's digitization problem, as I have termed it, resulted from its lack of a concerted response to this need. As a DH project invested in the study of magazines, the CMMP took its cue from periodical studies initiatives such as the Modernist Journals Project (modjourn.org), the Blue Mountain project (bluemountain.princeton.edu), and the Modernist Magazines Project (modmags.dmu.ac.uk). However, the CMMP is the first project of its kind in Canada. While Latham asserts that "modernism cannot be fully thought or understood apart from the magazines" (2015, 267), even Canada's most canonical modernist magazines have not been digitized, and few of the institutions holding these materials have original or complete print runs. As Irvine remarks in his survey of modernist magazines in Canada: "Access to Canadian 'little magazines' probably represents the major obstacle to the advancement of research in the field. Only a handful of magazines are available in facsimile editions, others on microfilm" (2009, 628). Despite the importance of these texts to a broad network of scholars, to date there is simply no equivalent in Canada of the Modernist Journals, Blue Mountain, or Modernist Magazines projects; digitization efforts have been directed instead at middlebrow periodicals or online editions of individual manuscripts. Consequently, the CMMP proceeds with the understanding that if Canadian modernist magazines are to be read—let alone understood— by a wide readership, they must first be made accessible online. As the modern adage goes, if something doesn't exist online, it doesn't exist.[3]

What this absence suggests, and what the longevity and success of projects such as the Modernist Journals Project make obvious by comparison, is that

Canadian modernist digital humanities —or rather the *digitization* of Canadian modernism, particularly its little magazines—is "belated" in a very real sense, despite the tremendous recent efforts scholars have made to put Canadian modernism on the global modernist map and place Canadian modernist writers in conversation with their Anglo-American contemporaries.[4] In this way, our belatedness is less a literary-critical or literary-theoretical issue than a logistical and material one. Questions of periodization and modernist aesthetics aside, the fact of this digital belatedness in Canadian periodical studies has serious implications regarding the visibility and reception of modernist writers both within and beyond disciplinary, institutional, and national borders. Although Matthew Jockers has issued an urgent appeal for literary historians to follow in the lead of "science," which "has welcomed big data and scaled its methods accordingly" (2013, 2), his utopian call for the upscaling of literary-analytical methods cannot be answered in the field of Canadian modernism until its foundational texts have been rendered into machine-readable forms. Similarly, when J. Stephen Murphy confidently asserts that "we have now reached the stage at which the data and metadata are accessible, reliable, and rich enough that scholars can start digging into them *qua* data itself and not only as texts to be read and interpreted as literary texts are" (2014, vi), he clearly has other corpora or canons in mind. In what follows, I gesture to a few of the factors that have impeded—and in some cases continue to impede—efforts in this direction, including issues related to copyright, the failure of academic institutions to recognize and reward digitization or other DH work, and precarity. At present, suffice it to say that most of the magazines that contributed to the development of a rich, heterogeneous modernist tradition within Canada remain invisible to machines; they are not yet data at all in the sense required either by machines or by scholars such as Tanya Clement (2008) and Adam Hammond, Julian Brooke, and Graeme Hirst (2016), who have productively combined "distant" machine readings of large textual corpora with traditional "close" readings.

While working to supply the data and metadata that make such readings of Canadian modernist magazines possible, the CMMP has attempted to respond to two key questions, corresponding to the twin objectives outlined above: What new or canonical narratives of modernism do we want to construct? What outmoded or problematic narratives might we be unwittingly re-constructing? Below, I outline the steps the CMMP is taking to meet its objectives while remaining responsive to these central questions—questions which foreground how the seemingly neutral, mechanical act of digitizing the modernist archive is inseparable from theoretical considerations of this archive, its digital surrogate, or the process of its remediation. In the third and final section, I identify some of the larger challenges currently faced by the CMMP and other Canadian DH projects.

In 1999, the Canadian modernist poet Margaret Avison mused in *Hi-Lites* (the official newsletter of Toronto's Fellowship Towers senior residence, where Avison lived from 1984 until her death in 2007) about the dangers and benefits of technology in "Putting Computers in Perspective, or A Chip on Our Shoulders." Comparing the pre-digital past, with its rich linguistic history, and the increasingly digital present, with its "plastic" words and alienating, frenetic culture, she nevertheless concludes on a hopeful note: "Quick connections, and gradually resonant overtones: we need them both. Maybe we can reaffirm the essential words that will never pass away, too, to this high-flying worldwide-reaching new generation. Especially if we appreciate how, through them and their Internet, there is in fact no longer any west or east" (Avison 2009, 310). What the CMMP proposes, in essence, is to facilitate both "quick connections" and "gradually resonant overtones": to undertake the often thankless task of "mere digitization" to facilitate "quick connections" or easy digital encounters with modernist texts; to digitally preserve "the essential words that will never pass away" that Avison and other Canadian writers first committed to print; but also, in the process, to challenge monolithic, "plastic" conceptions of Canadian modernism by furnishing material proof of how its many articulations cut across borders, cultures, religions, and neatly defined periods.

COPYRIGHT, COLD CALLS, AND COLLABORATION: DIGITIZING THE CANADIAN MODERNIST CANON

The CMMP's first goal was to digitize a selection of the little magazines most commonly associated with Canadian modernist literature in its canonical forms, starting with *Preview* and *First Statement*. While this choice reifies the Montréal-centric, literary-critical narratives of the 1940s that members of the Editing Modernism in Canada project (EMiC; 2007–2016) or scholars influenced by the new modernist studies have attempted to challenge or augment, it will also render visible the poets, critics, and prose writers still considered central figures in many of these newly articulated narratives of Canadian modernism as a plural, heterogeneous phenomenon.

Even when it comes to *Preview* and *First Statement*, modernist critics have long relied on narratives of these magazines and their authors circulated by others in previous decades, as well as reprinted versions of their most canonical texts, which have circulated largely via anthologies, collected or selected editions of poetry, and criticism. That is to say, the problem of access described by Irvine has shaped our criticism. While Neil Fisher's *First Statement, 1942–1945: An Assessment and Index* (1974) and Don Precosky's *"Preview:* An Introduction and Index" (1981) provided literary scholars with invaluable records of the basic metadata associated with two of Canadian modernism's most canonical magazines, access to the literary texts themselves remains extremely limited. Unsurprisingly, magazine contributions not republished in our major anthologies or excerpted in the criticism have received much less scholarly attention—an effect which has arguably encouraged a piecemeal understanding of these periodicals and an impoverished sense of the ways they function as discrete but also culturally, socially, and politically embedded objects replete with intertextual allusions, juxtapositions, and prosopographical clues that are constitutive of, not incidental to, their critical import.[5]

One of the central issues underlying Canadian modernism's digitization problem has to do with copyright. Many of the contributions to Canadian modernist magazines of the 1940s, for example, are still under copyright, since—at the time of this writing—copyright in Canada typically expires 50 years after the end of the calendar year in which the author expires. In the case of some contributors, such as Denis Giblin, who appears to have been only 16 years old when two of his poems were published in *Preview*'s final issue, the problem of copyright is particularly acute; Canadian modernism was alive and well in the postwar period, as were a great number of the authors who filled the pages of its little magazines. However, in the United States, where copyright laws differ, modernist DH initiatives such as the Modernist Journals Project and the Blue Mountain project most often focus on pre-1922 or -1923 texts.[6] As the Modernist Journals Project website explains:

We end at 1922 for two reasons: first, that year has until recently been the public domain cut-off in the United States; second, most scholars consider modernism to be fully fledged in 1922 with the publication of Virginia Woolf's *Jacob's Room*, James Joyce's *Ulysses*, and T. S. Eliot's *The Waste Land*. We believe the materials in the MJP will show how essential magazines were to the rise and maturation of modernism. (Modernist Journals Project, n.d.)

Similar justifications do not map neatly onto conversations of Canadian modernism and the laws that govern its digitization. Nevertheless, even in the broader Anglo-American modernist tradition, 1922 is not a cut-off date that corresponds to the end of modernism or even of high modernism as these terms have been defined by "most scholars." Nor does it necessarily follow, of course, that if modernism were "fully fledged in 1922" its post-1922 texts were suddenly less worthy of critical consideration. This is only to say that, practically speaking, the Modernist Journals Project's chronological demarcation is wholly justifiable; theoretically, less so. To push pause on modernism in 1922 or 1923—something modernist DH projects have done for perfectly legitimate reasons (legal, financial, and other)—is to exclude many of the novels, poetry collections, and periodicals associated with "fully fledged" and "matur[e]" modernisms, and not only in the Canadian context. The CMMP has elected to take a different route, digitizing Canadian modernist periodicals currently under copyright as well as those in the public domain—with full knowledge that this decision will constrain the project and the narrative that it creates in other ways.

Indeed, the work of securing copyright permissions for the digitization and online publication of periodicals not in the public domain is incredibly time and labour intensive. To secure permissions for *First Statement*, we had to identify and reach out to the literary executives or surviving relatives of over 50 contributors from across Canada, the United States, and Europe. This process involved detective work that sometimes felt uncomfortably invasive. It required familiarity with obituaries; visits to strangers' social-media profile pages and family-reunion websites; trawling online directories for personal addresses and phone numbers; cold calls to lawyers, ship captains, and ailing seniors that felt like a high-stakes version of the "Three Minute Thesis" competition;[7] impromptu literary analysis ("I didn't know grandma wrote poetry! Was it any good?"); and, occasionally, sensitive financial negotiations. To my knowledge, none of the courses offered by the growing number of DH institutes around the world includes preparation for this range of scholarly activity.

In this way and others, the labour that collectively constitutes the digitization of Canadian modernist magazines is fundamentally collaborative. This reality admits both new opportunities and new challenges. For scholars such as Susan Brown: "Collaboration and interdisciplinarity are virtually inevitable in digital humanities work" (2011, 226); for Latham and Scholes, the founders of the Modernist Journals Project, collaboration and interdisciplinarity are not inevitabilities so much as imperatives for scholars interested in periodical cultures: "To be as diverse as the objects it examines," they argue, "periodical studies should be constructed as a collaborative scholarly enterprise that cannot be confined to one scholar or even a single discipline" (2006, 528). Whether it is framed as an inevitability or an imperative, then, collaboration offers the promise of helping projects meaningfully break out of disciplinary "silos"—even if it often requires significant investments of time, energy, and other resources. At each stage of the CMMP's digitization process, at least, collaboration has been vital to the project's operations as I have considered how best to develop a rigorous data-management plan to ensure the project's long-term viability through the secure storage, automatic backup, and responsible management of its digital assets; scan magazines in a high-resolution format; save them for archival purposes; process the raw scans; create PDFs and plain-text files to disseminate via the project's website; and identify the basic metadata most important to the website's end users by encoding each magazine issue in TEI-XML. In order to actually locate and scan material for the project's repository, the CMMP has also collaborated with multiple librarians, archivists, digitization specialists, scholars, and institutional partners.

PERIODICAL INTERVENTIONS: DIGITIZATION AND THE PLURALIZATION OF CANADIAN MODERNISM

After establishing the processes and relationships required to digitize full runs of some of Canadian modernism's most canonical magazines, the CMMP is now focused on its second major objective: the expansion of the CMMP's digital repository to facilitate new narratives of Canadian modernism, as well as new readings of existing narratives and canonical formations. Thanks to critics working under the rubric of the new modernist studies, the pluralistic nature of literary modernism is now generally understood as a given—as are its transnational underpinnings. As Irvine, Vanessa Lent, and Bart Vautour remark in their introduction to *Making Canada New: Editing, Modernism, and New Media*, such developments have encouraged greater receptivity to Canadian modernist writers as well as greater recognition of their contributions to a globally and temporally reconstituted field:

> Instead of periodizations predicated upon notions of cultural belatedness, or narrativizations that identify emergent, marginal, or peripheral modernisms in relation to a dominant, originary centre, the push towards new modernisms has not only opened up the field of transnational modernisms to embrace Canada's early to mid-century cohort but also coincided with the resurgence in Canadian modernist studies over the past two decades. (2017, 4–5)

Yet in both Canadian literary criticism and digital humanities, more can be done not only to correlate Canadian modernisms with the dominant Anglo-American tradition but also to highlight the formation of peripheral and previously unacknowledged modernist movements within the Canadian tradition itself. Accordingly, the CMMP intends to move from "mere" digitization of the dominant modernisms associated with the McGill group and magazines such as *Preview* and *First Statement* to "more" digitization—a pluralistic and digital reimagining of the formation and articulation of Canadian modernisms, along lines already proposed by Irvine, Lent, Vautour, and others. At the same time, though, this reimagining, while carried out under the seemingly niche banner of DH modernist studies in Canada, does not just serve an additive function, complementing existing work within the fields of digital humanities, modernist studies, and Canadian literature; the kind of material and theoretical expansion proposed by the CMMP also invites re-examinations of that work and its underlying terminological or methodological assumptions.

Within modernist studies, for example, this expansion necessitates a reckoning with the field's commitment to truly "global," "transnational," or "late" articulations of modernist culture. By digitizing periodicals that fall outside of modernism's usual purview, the CMMP hopes to supply material evidence of modernism's soft shoulders, of its instantiations across temporal, national, racial, or linguistic lines. To this end, the CMMP is now in the process of digitizing three historically and culturally significant magazines that, until quite recently, have largely been ignored by critics: *Tarot* (1896), a short-lived "bohemian" Toronto monthly featuring the first female editor of a Canadian little magazine, Harriet Ford; *Neith: A Magazine of Literature, Science, Art, Philosophy, Jurisprudence, Criticism, History, Reform, Economics* (1903–1904), a Saint John–based publication featuring the first African Canadian editor of a little magazine, Abraham Beverly Walker; and *Le Nigog* (1918), a controversial francophone inter-arts venture published by Fernand Préfontaine, Robert de Roquebrune, and Léo-Pol Morin.[8] Fortunately, the majority of the content in these magazines is already in the public domain in Canada. Shared via the CMMP website, these magazines will provide a more nuanced picture of peripheral modernisms and proto-modernist aesthetics, and the critical introductions to *Tarot* and *Neith* that I have solicited for publication on this research platform will feature women and Black Canadians—both groups frequently excluded from conversations about prewar twentieth-century Canadian literature. In line with what Faye Hammill concisely refers to as "recent accounts of Canada's modernism as transcultural, global, dislocated" (2017, 29), the CMMP and its various critical components will document how these lesser-known periodicals provide insight into the development of discrete Canadian modernisms as they responded to, but also helped shape, international expressions of modernist culture.

One of the CMMP's driving arguments—and one of the literary-critical motivations for its recent digitization efforts—is that narratives of Canadian modernism predicated on the classification of writers into two distinct, belated generations (in the 1920s and 1940s, respectively),[9] belie the porous nature of the boundaries between myriad iterations of pre-modernist, proto-modernist, modernist, and postmodernist expression. Furthermore, the project's current digitization work and concomitant reimagining of the field proceeds, in part, from an understanding that such narratives have often elided feminist, race-based, and post-colonial critiques of Canada's modernist media ecologies and literatures. Despite the recent renaissance in Canadian modernist studies stimulated by the Canadian Writing and Research Collaboratory, EMiC, and SpokenWeb, one could argue that Colin Hill's observation from 2012 still rings true: "There is nothing approaching a consensus about how Canadian modernisms [...] ought to be situated relative to other foreign and indigenous literary modes" (8). The CMMP will attempt to address this issue by supplying primary materials whose sheer variety provokes not any kind of definitive consensus about what Canadian modernisms are or were, but rather a sense of the unique and sometimes conflicting genealogies of Canadian modernisms. That being said, I have no intention to resurrect the oppositional or conflict-based rhetoric that previously animated much of our mid-century periodical criticism. Nor do I intend to use the CMMP to place undue emphasis on the global at the expense of the local, or (to borrow the terms critics have repeatedly used to pit *Preview* against *First Statement*) on "cosmopolitan" instead of "native" strains of modernist poetics; rather, these tensions will be used to generate new research questions and highlight some of the paradoxes of Canadian literary history.

In the aftermath of the new-modernist explosion of monolithic or strong conceptions of Canadian modernism, however, the CMMP and other projects like it will have to consider how, and to what extent, the labels "Canadian modernism" or "Canadian modernisms" remain meaningful. To repurpose a phrase from Brian Trehearne's 2018 survey of "Canadian Modernism at the Present Time," according to what "*sine-qua-nons* of modernism" (488) should we select magazines for inclusion in our digital repository? This question is particularly salient when it comes to the recovery of popular writers belonging to an ill-defined modernist "project," as Trehearne points out (478). But one must also consider the ethics of what happens when expansion means Indigenous writers who would consider the labels "Canadian" and "modernist" both irrelevant—or, worse yet, as extensions of neocolonial desire—get swept up in the widening gyre of the new modernist studies. Such questions need to be taken seriously, and asked repeatedly, as part of larger conversations about the politics of globalization, Indigenous sovereignty and cultural protocols, decolonization and reconciliation as ongoing imperatives, and the emergence of transatlantic and trans-Pacific literary studies. These are questions that the CMMP tacitly invites, but that its critical introductions and other secondary materials will have to address more directly as the project unfolds.

Canadian digital humanities has the potential to pluralize modernism in ways that address these concerns, particularly as it recreates traditional scholarship in the digital sphere, for new and wider audiences, engendering new forms of critical engagement, new critical outputs, and new ways of defining and delimiting Canadian modernism. As new-media theorists, modern periodical scholars, and modernists have argued, the digital remediation of modernist or other cultural materials introduces transformations that make such shifts in literary theory and praxis unavoidable. As McGregor and van Orden note, "digitizing periodicals leads to an opportunity to rethink print differently" (2016, 140)—and they go on to quote James Mussell, for whom digital remediation similarly foregrounds not "a deficit, a misrepresentation, in digital resources" but a generative sense of "difference, introduced through transformation" (Mussell 2015, 355). Yet for certain critics, including Trehearne, it appears that Canadian modernist digital humanities has been valuable largely to the extent that, through "exhaustive replication" (2018, 481), it has helped produce digital facsimiles of traditional print media and scholarly products, such as critical and genetic editions ("actual scholarly editions," as he refers to them [481]). To be fair, Trehearne gestures to the potential of digital projects "to expand Canadian modernism's readership at this crucial point in its afterlife" (481). Even so, he ultimately concludes that, to date, Canadian DH projects have failed to deliver on this promise to broaden modernism's audience. While he writes of his "conviction that modernism has meaning *now* for the way Canadians live their lives, and for all possible readerships, hardbound or digital" (483), he also appears largely unconvinced that the "new modes of dissemination to a wider public" he champions—by invoking EMiC and the "exemplary editorial work" of Irvine and Zailig Pollock (481)—have, in fact, made Canadian modernism meaningful to anyone beyond the walls of our universities. Accordingly, Canadian digital humanities in this view is tacitly reduced to a *vehicle* for critical inquiry. This is not simply a version of "mere digitization," but "mere DH."

Trehearne praises EMiC as an admirable experiment in public-facing DH work that refreshingly "took for granted...the legitimacy and breadth of the Canadian modernist movement and its relevance to contemporary national life" even as he bemoans its lack of sustained engagement with the same (2018, 482). In this sense, Trehearne's assessment of DH activity in Canada recreates in miniature an unresolved conflict in his larger argument—that is, between his own utopian desire to make Canadian modernism newly relevant to everyday Canadians and their contemporary concerns and his stated ambition to shore up modernism's walls as a discrete, autonomous field.[10] To this end, he "suggests some ways forward for Canadian modernist criticism that seem [...] likely to provoke a positive, durable response in the new and younger audiences modernism must now find if it is to retain cultural meaning and prestige" (466). On the one hand, this call for critics to help an ailing and besieged modernism "retain cultural meaning and prestige" appears decidedly out of step with the new modernists who have toiled to interrogate the field's hierarchies of prestige—what David James refers to as "modernism's recognizability as a wellspring of creative audacity and resounding prestige" (2018); on the other, Trehearne's concerns need to be taken seriously and addressed concretely. Even those critics not at all interested in recuperating Canadian modernism under the besmirched banner of prestige would do well to consider whether it is possible to expand the canon and embrace the tenet of new-modernist expansion while also clarifying modernism's legibility.

To put this another way, the kind of critical and digital reconceptualization of Canadian modernism that the CMMP encourages is not likely to occur if digitization efforts are not supplemented by traditional as well as DH analysis and theorization. Like Hammill, Paul Hjartarson, and McGregor in the introduction to their "Magazines and/as Media" special issue of *English Studies in Canada*, the CMMP "insist[s] on a reorientation of periodical studies that moves past the accepted canons and bibliographies and shifts beyond the familiar cosmopolitan centres, while also demonstrating the exciting juxtapositions that emerge not only between diverse items on the periodical page but also between studies of diverse periodicals" (2015, 12). To accomplish these goals, the CMMP will digitize modernist periodicals and disseminate them through its website; however, it will also mediate visitors' experiences of these texts overtly, through its critical introductions, teaching guides, and crowdsourced syllabi on various aspects of Canadian literature, modernism, and periodical culture. These materials will address Trehearne's concern by remediating Canadian modernist texts into digital forms (PDF, plain-text files, TEI) that are freely available, accessible, and responsive to the needs of "new and younger audiences" (Trehearne 2018, 466)—including new generations of critics and students. After all, it seems unlikely that narratives of Canadian modernist expression will change any time soon if the primary sources available to us, and our ways of teaching them, do not. Although, unlike some DH projects, the CMMP's current focus is not on providing digital tools, let alone an integrated digital platform for analysis, editing, and publishing, it aims to make Canadian periodicals amenable to a range of critical and pedagogical tools or activities.

FUTURE OPPORTUNITIES AND CHALLENGES

In "Modernism Meets Digital Humanities," Stephen Ross and Jentery Sayers open with a vision of "digital humanities frameworks for modernism" that is at once guarded and hopeful: "To be sure," they write, "this fusion [of digital humanities and modernist studies] is not without its shortcomings, and it is in a nascent state. However, we believe it promises exciting research, which was not possible—at least technically speaking—until somewhat recently" (2014, 625). Despite its "nascent state," the fused field which the CMMP occupies has already begun to identify and address some of the difficulties facing other DH researchers in Canada, as elsewhere. These include the marginalization of digitization and other DH work, despite recent efforts to recognize DH work as scholarly work that "counts" (toward tenure, as part of tenure-track job applications, etc.), and the failure of universities and other academic institutions to acknowledge the various resources and forms of labour required to produce "public" or "open" digital scholarship.

To begin with, the work of digitization upon which Canadian DH projects such as CMMP depends is not only subject to, but might also be cited as part of further attempts to reform, institutional policies and structures. As Lischer-Katz suggests: "Rethinking the invisibility of this type of work will pave the way for material and economic improvements to the working conditions of digital laborers" (2019, 214). Understood as a form of intellectual as well as technical labour, digitization poses a problem to institutions that have been quick to stress the importance of collaborative, public-facing, open, and non-traditional forms of scholarly production, but slow to acknowledge it in ways that matter to everyone engaged in this work. For example, is digitization "service" or "scholarship," or can it be both? Is a monograph more valuable than a digitized archive or digital research platform? For many precariously employed scholars and graduate students, these are not trivial questions to be passed across the breakfast table like half-finished sudokus; they are questions with real, immediate consequences (intellectual and material), and our answers to them structure the way we work, the things we choose to work on, and how we frame that work to ourselves and others. Even for more established scholars, to invest in digital humanities (and perhaps particularly digitization) projects is to commit to something that, in many disciplines, may still be viewed as a mere appendage or stepping stone to more traditional, "intellectual" outputs or forms of academic labour.

Fortunately, some concerted efforts have been made to recognize digital and other non-traditional forms of scholarship. Starting in 2000, the Modern Language Association has maintained "Guidelines for Evaluating Work in Digital Humanities and Digital Media." This evolving document urges "institutions and departments [to] develop written guidelines so that faculty members who create, study, and teach with digital objects; engage in collaborative work; or use technology for pedagogy can be adequately and fairly evaluated and rewarded" (2012). In modernist studies, too, organizations such as ModNets have provided multiple forms of support to DH scholars, including a peer-review service for finished or unfinished projects. And on November 13, 2019, the Social Sciences and Humanities Research Council of Canada (SSHRC) and four other Canadian funding bodies announced that they had formally agreed to the San Francisco Declaration on Research Assessment (DORA). As explained on the NSERC website, "DORA recognizes that scholarly outputs are not limited to published journal articles but can also include article preprints, datasets, software, protocols, well-trained researchers, societal outcomes and policy changes resulting from research" (2019). These developments are all to the good, as far as the CMMP is concerned, though it remains to be seen whether such initiatives will collectively trigger widespread and profound changes in the humanities. While we wait to find out, more can and should be done to reimagine the institutional metrics according to which DH projects receive funding and are granted intellectual legitimacy—both as a retrospective, reparative gesture for past labour and as a precedent for future work in the still-growing field of digital scholarship. The process of recognizing DH work as legitimate, necessary scholarship is implicit in the establishment of digital scholarship centres in recent years.[11] For the CMMP, this ongoing process is informed by larger narratives about what Canadian literary criticism can and should be, or about the role of the digital humanities in the formation and critique of institutional and national identities.[12]

In the coming years, however, such conversations will be dramatically shaped by another, increasingly pressing issue: precarity. Is it irresponsible, as Alix Beeston (2019) has suggested, to even talk about the future of modernist studies—or digital humanities, or any other academic field, for that matter—without also acknowledging the realities of academic precarity? To whom are digital humanities' and modernism's shiny futures foreclosed? In what ways are such scholars forced to remain DH-adjacent or "modernism-adjacent" (to use a recent coinage by Naomi Milthorpe, Robbie Moore, and Eliza Murphy [2019])? For modernist writers and the people who study them for a living, "making it new" is an ongoing imperative; we must continually be up to the task of pursuing productive new directions in the field—and not just because we need to appear relevant to granting agencies and employers, who are similarly dependent on the hyperbolic rhetoric of innovation. Academic cynicism and internecine conflict aside, there really are new claims to be made, new research questions to be asked and answered, and new critical, cultural, and socio-political contexts in which to situate them. But making Canadian digital humanities new, or making Canadian modernist studies new, is a process that must involve paradigm-shifting considerations of the ways pioneering work by non-tenure track scholars, or even tenure-track scholars with limited funding or institutional support for DH work, is being sidelined by the very institutions that have fetishized—and profit from—this kind of "innovative" and "public" scholarship.

In *Generous Thinking: A Radical Approach to Saving the University*, Fitzpatrick posits generosity and openness (including open-access scholarship) as one possible antidote to the institutional problems I have briefly outlined: "Enabling access to scholarly work does not just serve the goal of undoing its commercialization or removing it from a market-driven, competition-based economy, but rather is a first step in facilitating public engagement with the knowledge that universities produce" (2019, 148). But to what extent can precariously employed scholars afford to work against the institutional rubrics of reward and promotion in which they are enmeshed? Can DH projects like the CMMP afford the hidden costs of collaborative DH work, the hidden labour and potential risks of "facilitating public engagement," or the equitable inclusion of graduate and undergraduate students in research while also continuing to produce the forms of traditional scholarship that actually "count"? Or does this line of questioning what academics and academic projects can "afford" miss the point altogether, since true "generosity" confounds the logic behind the university's "market-driven, competition-based economy"? After all, by definition, being "generous" means "giv[ing] more of something...than is necessary or expected."[13] In reality, much DH work—and much scholarship produced by precarious academics and graduate students—is premised on this kind of generous approach to scholarship, whether intentionally or no: to play the "game," people who do this work are forced to forecast long-term outputs and plans for expansion, knowing full well that whether they actually succeed may be outside of their control, and that much of the labour required along the way will not be compensated labour. In the case of the CMMP, I have been struck by the irony of attempting to build a project and plan for its long-term data management in such a way that it may very well survive in the academic ecosphere longer than I do. (I hope it does, whatever and whenever that means.)

In the meantime, the work continues: the work of digitizing our modernist magazines to make Canadian literatures and periodical cultures visible to new audiences, both in Canada and abroad; of expanding the canon and interrogating the implications of that expansion; and of actively fostering practices that are open, sustainable, and resistant to the conditions of precarity under which the CMMP and countless other projects operate. Whatever the result, the CMMP is hopeful that it "can reaffirm the essential words that will never pass away, too, to this high-flying worldwide-reaching new generation" (Avison 2009, 310) by recovering and remediating Canada's modernist magazines—magazines that open new avenues of critical inquiry into Canadian modernism and its divergent material legacies.

NOTES

1. When published, the book was titled A Companion to Digital Humanities (2004).

2. The collaborative work of Scholes, Latham, Smulyan, Drouin, Wulfman, and Gaipa on the Modernist Journals Project, for instance, has played a crucial role in the revitalization of modernist and modern periodical studies, and its "MJP Lab" has invited increased attention to the possible applications of DH methodologies and technologies. Wulfman's Blue Mountain project and excellent "The Rise and Fall of Periodical Studies?" (2017) are also worth noting here, as is McGregor and van Orden's (2016) computational analysis of middlebrow periodicals for Modern Magazines Project Canada (now defunct).

3. One early iteration of this saying appears in Sarah Stevens-Rayburn and Ellen N. Bouton's "'If It's Not on the Web, It Doesn't Exist at All': Electronic Information Resources—Myth and Reality" (1998).

4. For discussions of the apparent belatedness of Canadian modernism, see, for example, Kamboureli (2007, xviii–xix) and Irvine (2005, 4–7).

5. I invoke prosopography here with some misgivings, given the brief history of canonical Canadian modernism and its circulation that I have just traced. Especially prior to the recent push by EMiC scholars and publishers to digitally recover and reconstitute Canadian modernism, the lack of access to Canadian modernist texts I have been lamenting—coupled with the frequent, albeit frequently inevitable, lack of critical distance between Canadian modernist authors and their academic interlocutors that Trehearne laments in his seminal The Montreal Forties: Modernist Poetry in Transition (1999)—also led to a proliferation of reductive, sometimes gossipy, and often self-interested prosopographical accounts of Canadian coteries and modernist movements.

6. For a more detailed discussion of the significance of copyright laws for modernist DH projects in the United States and the United Kingdom, see Ross and Sayers (2014, 626–27).

7. See "The Three Minute Thesis (3MT) in Canada," Canadian Association for Graduate Studies, https://cags.ca/3mt-2.

8. For some notable exceptions to this claim of critical neglect, see Betts 2013; Clarke 2012; Irvine 2009; Johnson 2019; Williams 2005.

9. This narrative appears, for example, in Ken Norris's *The Little Magazine in Canada 1925–80* (1984).

10. See, for example, 465, 466, 473, and Trehearne's complaint about his edited *Complete Poems of A.J.M. Smith* (Smith 2007) failing to stimulate critical—let alone widespread public—interest in one of Canadian modernism's central figures (Trehearne 2018, 485).

11. Three examples of newly established digital scholarship initiatives in Canada include the University of British Columbia Okanagan's AMP Lab (2018), and the University of Victoria's Digital Scholarship Commons (2018).

12. Questions of national identity are particularly timely, given the spate of recent publications and conference panels on Canadian modernism or Canadian literature as a whole. These include, for example, Tanti et al 2017; Mount 2017; McGregor, Rak, and Wunker 2018; Betts and Bök 2019; a 2019 roundtable entitled "Canadian Literature: What Now," Modern Language Association Annual Convention, Chicago, January 3–6, 2019; and the Modernist Studies Association's 2019 conference, which actively sought out and featured work on Canadian modernists as part of a special "Making Modernism in/out of Canada" theme.

13. *OED Online*, s.v. "generosity, n.," last modified March 2022, https://www.oed.com/.

REFERENCES

Avison, Margaret. 2009. "Putting Computers in Perspective, or A Chip on Our Shoulders." In *I Am Here and Not Not-There: An Autobiography*, 308–10. Erin, Ont.: Porcupine's Quill.

Beeston, Alix (@alixbeeston). 2019. "I'm both extremely excited by & proud of the great work emerging scholars are doing in impossible conditions—& absolutely livid about those conditions. We must include them when we talk about the state of the field, otherwise speaking of its future is dangerously whimsical." Twitter, June 23, 2019, 2:22 a.m. https://twitter.com/alixbeeston/status/1142724569087848449.

Betts, Gregory. 2013. *Avant-Garde Canadian Literature: The Early Manifestations*. Toronto, Ont.: University of Toronto Press. https://doi.org/10.3138/9781442696907.

Betts, Gregory, and Christian Bök, eds. 2019. *Avant Canada: Poets, Prophets, Revolutionaries*. Waterloo, Ont.: Wilfrid Laurier University Press.

Brown, Susan. 2011. "Don't Mind the Gap: Evolving Digital Modes of Scholarly Production Across the Digital-Humanities Divide." In *Retooling the Humanities: The Culture of Research in Canadian Universities*, edited by Daniel Coleman and Smaro Kamboureli, 211–53. Edmonton: University of Alberta Press.

Clarke, George Elliott. 2012. *Directions Home: Approaches to African-Canadian Literature*. Toronto, Ont.: University of Toronto Press. https://doi.org/10.3138/9781442666511.

Clement, Tanya E. 2008. "'A Thing Not Beginning and Not Ending': Using Digital Tools to Distant-Read Gertrude Stein's *The Making of Americans*." *Literary and Linguistic Computing* 23, no. 3: 361–81. https://doi.org/10.1093/llc/fqn020.

Fisher, Neil H. 1974. *First Statement, 1942–1945: An Assessment and an Index*. Ottawa, Ont.: Golden Dog.

Fitzpatrick, Kathleen. 2011. "The Humanities, Done Digitally." *Chronicle of Higher Education*, May 8, 2011. https://www.chronicle.com/article/the-humanities-done-digitally/.

———. 2019. *Generous Thinking: A Radical Approach to Saving the University*. Baltimore, Md.: Johns Hopkins University Press.

Hammill, Faye. 2017. "No Canada: How A Country's Modernism Went Global." *Times Literary Supplement*, no. 5956 (May): 29–30.

Hammill, Faye, Paul Hjartarson, and Hannah McGregor. 2015. Introduction to "Introducing Magazines And/As Media: The Aesthetics and Politics of Serial Form," edited by Faye Hammill, Paul Hjartarson, and Hannah McGregor, special issue, *English Studies in Canada* 41, no. 1: 1–18. https://doi.org/10.1353/esc.2015.0006.

Hammond, Adam, Julian Brooke, and Graeme Hirst. 2016. "Modeling Modernist Dialogism: Close Reading with Big Data." In *Reading Modernism with Machines: Digital Humanities and Modern Literature*, edited by Shawna Ross and James O'Sullivan, 49–77. London, U.K.: Palgrave McMillan. https://doi.org/10.1057/978-1-137-59569-0_3.

Hayles, N. Katherine. 2005. *My Mother Was a Computer: Digital Subjects and Literary Texts*. Chicago: University of Chicago Press. https://doi.org/10.7208/chicago/9780226321493.001.0001.

Hill, Colin. 2012. *Modern Realism in English-Canadian Fiction*. Toronto, Ont.: University of Toronto Press. https://doi.org/10.3138/9781442685772.

Irvine, Dean. 2005. "Introduction" In *The Canadian Modernists Meet*, edited by Dean Irvine, 1–13. Ottawa, Ont.: University of Ottawa Press.

———. 2009. "'Little Magazines' in English Canada." In *The Oxford Critical and Cultural History of Modernist Magazines: Vol. II: North America 1894–1960*, edited by Peter Brooker and Andrew Thacker, 602–28. Oxford, U.K.: Oxford University Press.

Irvine, Dean, Vanessa Lent, and Bart Vautour, eds. 2017. *Making Canada New: Editing, Modernism, and New Media*. Toronto, Ont.: University of Toronto Press. https://doi.org/10.3138/9781487511357.

James, David. 2018. "Modernist Affects and Contemporary Literature." *Modernism/modernity Print Plus* 3, no. 4. https://doi.org/10.26597/mod.0074.

Jockers, Matthew L. 2013. *Macroanalysis: Digital Methods and Literary History*. Urbana: University of Illinois Press. https://doi.org/10.5406/illinois/9780252037528.001.0001.

Johnson, Billy. 2019. "Neith." Paper presented at the Modernist Studies Association Annual Convention, Toronto, October 17–20, 2019.

Kamboureli, Smaro. 2007. "Introduction to the First Edition." In *Making a Difference: Canadian Multicultural Literatures in English*, edited by Smaro Kamboureli, xviii–xxxiii. 2nd ed. Toronto, Ont.: Oxford University Press.

Latham, Sean. 2015. "Serial Modernism." In *A History of the Modernist Novel*, edited by Gregory Castle, 254–69. Cambridge, U.K.: Cambridge University Press. https://doi.org/10.1017/CBO9781139542395.012.

Latham, Sean, and Robert Scholes. 2006. "The Rise of Periodical Studies." *Publications of the Modern Language Association of America* 121, no. 2: 517–31. https://doi.org/10.1632/003081206X129693.

Lischer-Katz, Zack. 2019. "Reconsidering Technical Labor in Information Institutions: The Case of Analog Video Digitization." *Library Trends* 68, no. 2: 213–51. https://doi.org/10.1353/lib.2019.0037.

Mao, Douglas, and Rebecca Walkowitz. 2008. "The New Modernist Studies." *Publications of the Modern Language Association of America* 123, no. 3: 737–48. https://doi.org/10.1632/pmla.2008.123.3.737.

McGregor, Hannah, Julie Rak, and Erin Wunker, eds. 2018. *Refuse: CanLit in Ruins*. Toronto, Ont.: Book*hug.

McGregor, Hannah, and Nicholas van Orden. 2016. "Remediation and the Development of Modernist Forms in *The Western Home Monthly*." In *Reading Modernism with Machines: Digital Humanities and Modern Literature*, edited by Shawna Ross and James O'Sullivan, 135–63. London, U.K.: Palgrave Macmillan. https://doi.org/10.1057/978-1-137-59569-0_6.

Milthorpe, Naomi, Robbie Moore, and Eliza Murphy. 2019. "Modernism-Adjacent." *Modernist Review*, September 30, 2019. https://modernistreviewcouk.wordpress.com/2019/09/30/modernism-adjacent/.

Modern Language Association. 2012. "Guidelines for Evaluating Work in Digital Humanities and Digital Media." Last modified January 2012. https://www.mla.org/About-Us/Governance/Committees/Committee-Listings/Professional-Issues/Committee-on-Information-Technology/Guidelines-for-Evaluating-Work-in-Digital-Humanities-and-Digital-Media.

Modernist Journals Project. n.d. "About Us." https://modjourn.org/about.

Mount, Nick. 2017. *Arrival: The Story of CanLit.* Toronto, Ont.: House of Anansi.

Murphy, J. Stephen. 2014. "Introduction: 'Visualizing Periodical Networks.'" *Journal of Modern Periodical Studies* 5, no. 1: iii–xv.

Mussell, James. 2015. "Repetition: Or, 'In Our Last.'" *Victorian Periodicals Review* 48, no 3: 343–58. https://doi.org/10.1353/vpr.2015.0043.

Natural Sciences and Engineering Research Council of Canada. 2019. "San Francisco Declaration on Research Assessment (DORA)." Government of Canada. Accessed November 13, 2019. https://www.nserc-crsng.gc.ca/NSERC-CRSNG/policies-politiques/DORA-DORA_eng.asp.

Norris, Ken. 1984. *The Little Magazine in Canada 1925–80: Its Role in the Development of Modernism and Post-Modernism in Canadian Poetry.* Toronto, Ont.: ECW Press.

Precosky, Don. 1981. "*Preview:* An Introduction and an Index." *Canadian Poetry: Studies, Documents, Reviews,* no. 8: 74–89.

Rifkind, Candida. 2009. *Comrades and Critics: Women, Literature, and the Left in 1930s Canada.* Toronto, Ont.: University of Toronto Press. https://doi.org/10.3138/9781442687707.

Ross, Stephen, and Jentery Sayers. 2014. "Modernism Meets Digital Humanities." *Literature Compass* 11, no. 9: 625–33. https://doi.org/10.1111/lic3.12174.

Schreibman, Susan, Ray Siemens, and John Unsworth. 2004. "The Digital Humanities and Humanities Computing: An Introduction." In *A Companion to Digital Humanities,* xxiii–xxvii. Oxford, U.K.: Blackwell. http://digitalhumanities.org/companion/.

Smith, A. J. M. 2007. *Complete Poems of A.J.M. Smith.* Edited by Brian Trehearne. London, Ont.: Canadian Poetry.

Stevens-Rayburn, Sarah, and Ellen N. Bouton. 1998. "'If It's Not on the Web, It Doesn't Exist at All': Electronic Information Resources—Myth and Reality." *ASP Conference Series,* no. 153. https://www.stsci.edu/stsci/meetings/lisa3/stevens-rayburns.html.

Tanti, Melissa, Jeremy Haynes, Daniel Coleman, and Lorraine York, eds. 2017. *Beyond "Understanding Canada": Transnational Perspectives on Canadian Literature.* Edmonton: University of Alberta Press.

Trehearne, Brian. 1999. *The Montreal Forties: Modernist Poetry in Transition.* Toronto, Ont.: University of Toronto Press. https://doi.org/10.3138/9781442681729.

———. 2018. "Canadian Modernism at the Present Time." *Modernist Cultures* 13, no. 4: 465–95. https://doi.org/10.3366/mod.2018.0226.

Williams, Dorothy L. 2005. "Print and Black Canadian Culture." In *History of the Book in Canada. Vol. II, 1840–1918,* edited by Yvan Lamonde, Patricia Fleming, and Gilles Gallichan, 40–43. Toronto, Ont.: University of Toronto Press.

Willmott, Glenn. 2004. "Modernism and Aboriginal Modernity: The Appropriation of Products of West Coast Native Heritage as National Goods." *Essays on Canadian Writing*, no. 83: 75–139.

Wulfman, Clifford. 2017. "The Rise and Fall of Periodical Studies?" *Journal of Modern Periodical Studies* 8, no. 2: 226–41. https://doi.org/10.5325/jmodeperistud.8.2.0226.

CHAPTER 21

A Legacy of Race and Data:
Mining the History of Exclusion

Allan Cho and Sarah Zhang

Between 1885 and 1923, the Canadian government imposed a head tax, recorded in print registries, on Chinese immigrants entering Canada in order to restrict immigration (see figure 21.1). Although these print registers were used to keep track of the influx of migrants, these detailed records have provided historians with years of demographic information about the immigrants and remain a rich source of data for researchers. Painstakingly transformed into a Microsoft Excel spreadsheet manually transcribed by researchers at the University of British Columbia, these data represent original records that include 97,123 registrants whose arrival time in Canada spreads out over half a century, from 1886 to 1949 (Ward and Yu 2012). As impressive as the large-scale dataset is, the records are largely incoherent as they were captured in the idiosyncratic dialects of the immigrants and resulted in English variations of place names and titles. Figure 21.2, for example, illustrates the various transliterations of a village in Taishan county. The inconsistencies in place names unfortunately lead to difficulties for anyone seeking to conduct deeper analysis into the origins of these individuals. In other words, while there is a treasure trove of data to interpret, it may be unusable unless the data can be manipulated to unlock a better understanding of the gaps.

THE CONTEXT OF LANGUAGE AND DATA

To address these inconsistencies, some work has already been initiated at UBC to normalize various transliterations of the immigrants' origins in order to lay the groundwork for more in-depth research. The immigrants' origins are represented at two hierarchical levels: counties and villages/towns; there are eight counties and numerous villages in the registry. The normalization work was arduous—the project required the work of a large team of researchers, and the UBC Asian Library hosted 20 rounds of community-based meetings to which seniors who spoke distinct dialects were invited to identify the origin names on the records.[1] They ended up mapping the names of villages/towns in three of the eight counties: Sun Woy (now known as Xinhui), Zhongshan, and Taishan.

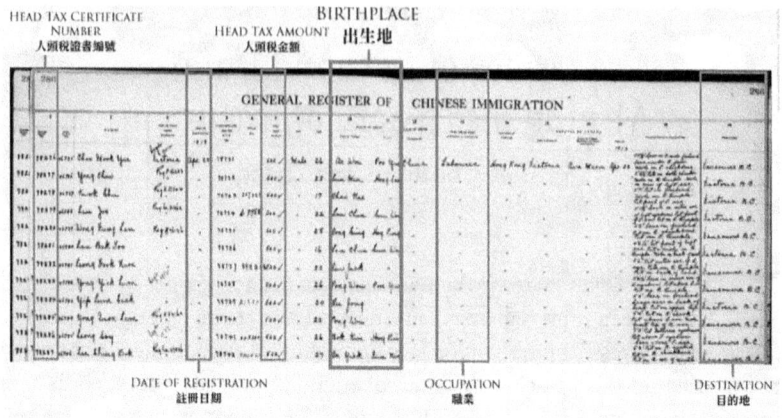

Figure 21.1. Original copies of the head-tax register.

Source: Zhang and Cho 2019.

Although just a snippet of the records, these normalized data offer a glimpse into what is available in the research.

SCHOLARSHIP BORNE OUT OF THE DATASET

Since the completion of the digitization work, scholars have begun to draw on the digital records from the project, manifesting novel methods and research findings. Rudy Chiang, a genealogical researcher, has conducted in-depth research regarding the demographic characteristics of migrants from this county. In his research, Chiang discovered that over 48 percent of head-tax migrants from Sun Woy were young men from a small island within Sun Woy, revealing that certain villages sent considerably more migrants than others (Chiang 2012).

Peter Ward's 2013 publication focused on the changes on the well-being of Chinese head-tax immigrants, particularly analyzing the immigrants' stature, a statistical indicator for well-being. He contrasted mean height by age of different age cohorts (ten years apart), and found a rising trend in stature over time (see figure 21.1): "[A] slow but significant increase in stature within the immigrant population from the middle of the 19th century to the early years of the Sino-Japanese War" (Ward 2013, 494). Additionally, he found there was an adolescent spurt for almost all the cohorts. These increases, he hypothesizes, were directly tied to the improvements in diet that were themselves the result of

Variant Forms of Romanization
白沙
Taishan (台山)

Bak Sa	Back Shor	Pack Shan
Back Saer	Back Shr	Pack Shar
Back San	Bak Shuk	Packshar
Back Sar	Back Shut	Pack Shu
Back Ser	Back Suck	Pah Sha
Back Shar	Back Sun	Pak Cha
Backshar	Back Thut	Pak Sah
Back La	Bark Sar	Pak San
Bak Lai	Bark Sun	Pak Sar
Bak Sa	Bat Shar	Pak Sha
Bak Sak	Bok Sha	Pak Sha Une
Bak Sah	Bok Shar	Pak Shar
Bak Sahr	Buck Sa	Pak Sue
Bak Sai	Buck Sah	Pang Sa
Bak Sam	Buk Sah	Pang Sar Hang
Bak San	Buk Sai	Pang Sha
Bak Sar	Buk Shar	Pang Shar
Bak Sha	Packhar	Par Sar Long
Bak Shai	Pack Sam	Par Shar
Bak Shar	Pack San	Park San
Bak Shat	Packsar	Park Sar
Back Shen	Pack Sha	
Bak Sheuk		
Bak Sar (or Nam Long)		

Figure 21.2. The various transliterations of a village as a result of idiosyncratic dialects of the immigrants.[1]

Source: Zhang and Cho 2019.

financial remittances from their relatives and family who had entered Canada earlier (225–37).

Subsequent to Peter Ward's study, other researchers have continued pursuing historical inquiries using DH tools, specifically GIS (geographic information systems) to investigate the spatial distribution of migration. Sally Hermansen and Henry Yu (2014) examined the spatial elements in the head-tax registry and visualized the distribution of the Canadian destinations of Chinese immigrants. This first attempt at examining the spatial distribution of the immigrants' destinations in Canada, however, depicted only a generalized view rather than zooming in on nuanced patterns of migration such as kinship networks.

As a further step, as they describe in a 2017 article, Yu and Chen used Gephi, a network analysis tool, to reveal patterns of family-chain migration, and found that the "four major Cantonese clans in Saskatchewan all showed a high geographic distribution in the choice of destinations" (Yu and Chan 2017, 46). Combining their Gephi network visualization with oral history, Henry Yu and Stephanie Chan explained the reasons for the highly dispersed nature of the migration network: the dispersion of the migration network happened over recurring generations. After the first generation of migrants that came to build the Canadian Pacific Railway, the subsequent generations who shared kinship ties with the earlier generation, driven by a mythic "golden mountain" and aspiration for economic mobility, often choose to branch out from the pre-existing migration network and settle down in small towns to avoid competition with existing businesses (46–47).

CONTEXT OF THE HEAD TAX

The history of Chinese people in Canada begins as far back as 1788, before the country's founding, when the first groups of Chinese were brought to Nootka Sound by John Meares to build forts and large sailing ships. The head tax worked to limit the "unassimilable populations" influx into Canada, therefore protecting the homogeneity of a Euro-British populace. The spatial segregation of Chinatown, on the other hand, was the physical manifestation of the idea that the Chinese were unassimilable (Der 2018, 41).

The head-tax system was a complex bureaucratic process of categorization. Lily Cho has argued that the head tax constituted the first mass use of identification photography in Canada (Cho 2018, 381–84). Unlike any prior moment in its history, the Canadian government isolated one specific group of people, purely on the basis of race and ethnicity, in order to identify each individual member of that group through the creation of a massive apparatus that would systematically correlate their identity with a photograph. This system was developed so that the Canadian government could contain and police these migrants in order to continue to exercise the processes of their exclusion from citizenship (Ward 2013, 488–501). Ironically, this very systematic recording of exclusion has provided a rich historical detail of the migrants that were intended to be marginalized.

THE MOTIVATION FOR AND METHODOLOGY OF THE STUDY

Looking back at what has been done up to this point—the digitization of the original registry, normalization of origins, and scholarly studies that have arisen from the digital records—the authors of this chapter, both academic librarians, became curious about furthering the research into the historical data that has been used. We are primarily interested in exploring whether previous work on the normalization of place name origins have been undertaken in research.

In terms of methodology, we used R, an open-source statistical computational language, for the sake of research reproducibility, and Palladio, an open-source network analysis tool developed by the Humanities + Design Lab at Stanford University. The intentions of this study are twofold. First, it will demonstrate the untapped potential in the head-tax data. Second, it will provide testimony for new approaches that librarians can help shape digital scholarship and create promising new research questions.

MINING THE DATA OF SUN WOY COUNTY: ANALYSIS OF THE WELL-BEING OF CHINESE MIGRANTS

The normalization of origins, a project by UBC Asian Library, as mentioned above, opened up the possibility of extracting more subtle information about the immigrants who came from the same county and same village. Hoping to test whether Ward's conclusions about the rising height of immigrants over time is true at one finer level—the county—we used the same dataset from the head-tax register and replicated Ward's study. We extended Ward's analysis by subdividing this group of migrants by their county of origin to examine this homogenous group of migrants and narrowed them to even smaller sample sizes, namely the county from which they had arrived.

When we used the same dataset of the 97,123 migrants that Ward had used, we found the same trend of increasing stature. But when we restricted the data to only the immigrants from Sun Woy county, however, different patterns appear,

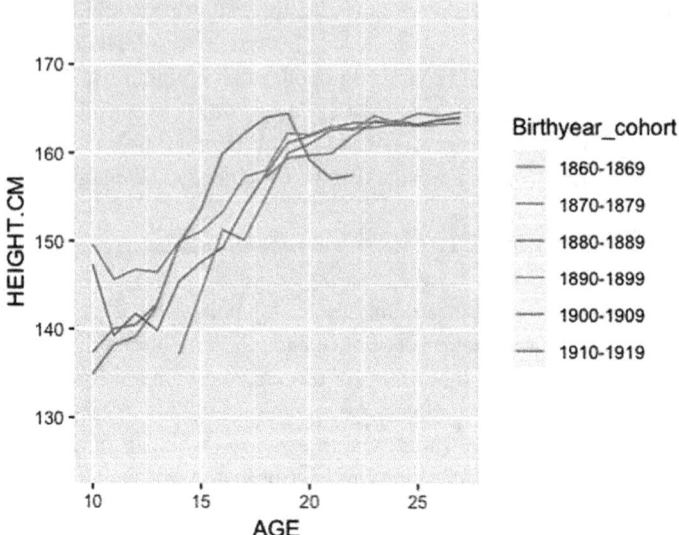

Figure 21.3. Mean height by age, immigrants from Sun Woy County.

Source: Zhang and Cho 2019.

which adds more context to Ward's previous research and may, in fact, counter his hypothesis. The analysis at the level of county, namely Sun Woy, in this case, uncovered more *nuanced* trends in the immigrants' heights, which is an indicator of their well-being before they landed in Canada.

We produced figure 21.2 by filtering the data to Sun Woy county using R. Following Ward's method, we grouped all the immigrants from the county into birth cohorts of ten years apart, and visualized for each cohort how height changed from the preadolescent ages to adult years (around 25 years old). Contrasting this line graph with Ward's (see figure 21.3), different or more nuanced patterns in the immigrants' heights are uncovered.

For example, the birth-year cohort of 1900–1909 and 1910–1919 present surprising and unexpected trends for immigrants originated from Sun Woy:

1. The mean height for most ages for the 1910–1919 cohort is the lowest, instead of highest, among all the cohorts. This goes against Ward's conclusion that well-being of the latter years of migrants correlates to overall taller heights.

2. For the 1900–1909 cohort, the mean height for adolescent years grew uncommonly fast, while plummeting in the subsequent age group. As height is an

indicator for the immigrants' well-being before they landed in Canada, this drastic change in height for this cohort might have been caused by distinct socio-economic circumstances in Sun Woy and the specific purposes emigration served in the county. Our intention, however, is not to try to investigate any of those causal factors in Sun Woy at that historical time, but simply to expose more complex patterns which could serve as gateways to more in-depth historical studies on the migrants from Guangdong to Canada in the late nineteenth and early twentieth century.

DOWN TO THE VILLAGES

In order to test the potential of the dataset of a migrant's origins beyond the county and down to the village level, we used social-network analysis to find another avenue to explore migration patterns. A network is "a pattern of inter-connections among a set of things" (Easley and Kleinberg 2010, 1). Network analysis, as an increasingly popular area in digital humanities, studies the things and relationships in networks (Weingart 2011). There has been previous work on this dataset using Gephi, a social-network-analysis tool by Yu and Chan, as described in the extant scholarship section.

Though the authors make a point of a possible correlation that migrants who share a surname and who came from the same village are "family" from the same clan, as indicated in their study, there are limitations to this assumption. It is a reasonable speculation that people who share the surname *Mah* may not necessarily be related, while, on the other hand, a person whose surname is different may not necessarily be unrelated. Hence, we decided to treat a village, which usually comprises a closely knit social unit, as a clan in order to study patterns in chain migration. As Gephi is not necessarily an easy tool to understand or master, we were motivated to use Palladio, an easily accessible network-analysis tool, to further explore patterns of migration visually that might not otherwise have been captured by previous studies.

NETWORK ANALYSIS BY PALLADIO

Palladio is a web-based visualization tool for complex humanities data. Miriam Posner has likened Palladio to "a sort of Swiss Army knife for humanities data" as it is a tool that includes a number of other tools, each of which allows one to get a different angle on the same data. Palladio offers various ways to visualize data, including maps, networks, and tables, and the ability to filter the data based on facets, allowing us to visualize data based on different criteria or at

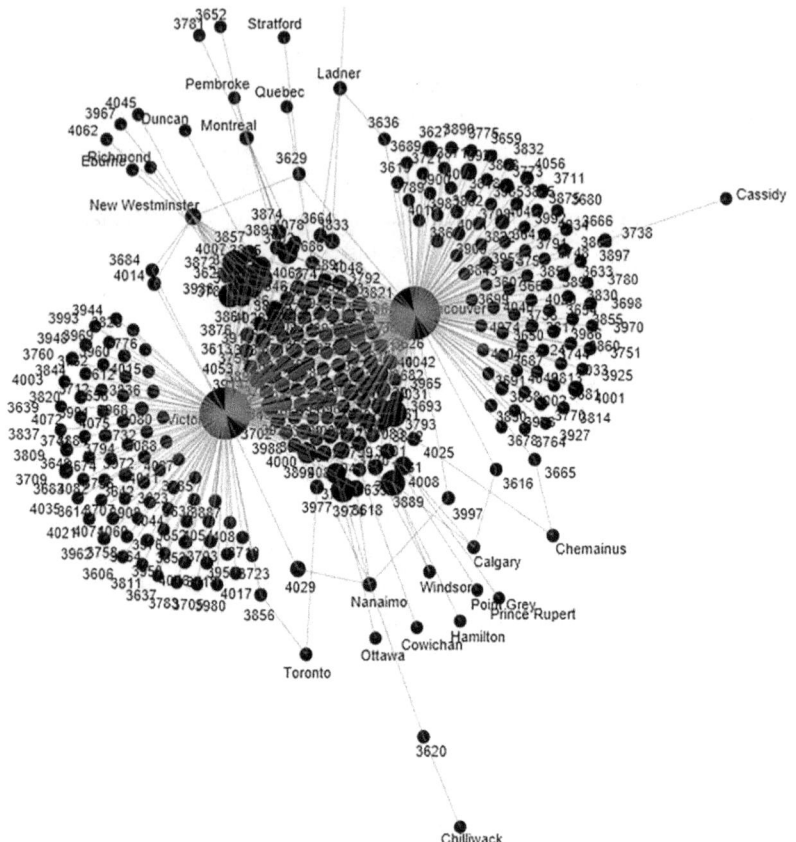

Figure 21.4. Palladio's network visualization diagram (Zhongshan County).

Source: Zhang and Cho 2019.

different points in time. For the purposes of our research, Palladio enabled us to more easily create network graphs to illustrate relationships between two dimensions. We filtered the data for Zhongshan county and uploaded them to Palladio to visualize the connections between the immigrants' origins and their destinations. Figure 21.4 is the Palladio visualization for the migration network for Zhongshan county.

Even though the workflow of using Palladio is simple, it takes deeper analysis to truly understand the nuances of the visualization in order to unlock migration patterns. This graph produced in Palladio enabled two elements for investigation:

connecting lines (in network terms, these are called edges) and node sizes. First, each number represents an origin village in China where immigrants were from, while the immigrants' destinations are spelled. Wherever a village node is connected to a destination node, it means there were immigrants originating from the village in China who chose that destination in Canada.

The other feature—node size—appears deceptively obvious, but it takes probing into the algorithm to understand that it actually corresponds to the number of immigrants a village sent or a destination received. In network terms, this is called "weighted degree," as opposed to "degree" (which simply measures how many edges a node is connected to). The node size equals weighted degree in this graph because each village sent more than one immigrant and hence the edges are weighted. Or, in statistical language, a node size equals the frequency of a destination or a village.

To summarize, the Palladio graph enables a visual representation of two layers of information: which villages are linked to which destinations, and the frequencies of each village and destination (or weighted degree), all in one glance.

The third characteristic that stands out from the graph is the proximal arrangement of the nodes: nodes that are heavily connected to other nodes are placed in the middle. Vancouver and Victoria are connected to the biggest numbers of villages and so they are at the centre; by contrast, the peripheral destinations are connected to fewer villages. Likewise, the villages in the middle have more connecting lines to destinations; while the two "wings" have fewer connecting lines to destinations, which means for these two big chunks of villages, most of them are only connected to one destination.

One of the most intriguing findings revealed by node size and proximity of nodes in the migration network visualization are the "constellations" of village nodes. The proximal villages were gathered like clusters or constellations on two overlapping and yet somewhat different parameters: what destinations they were connected to, and the number of destinations that a village was connected to. For example, the two biggest constellations of villages are connected to Vancouver and Victoria respectively. The node sizes in these two constellations tend to be small to medium. In contrast, a small constellation of village nodes near the upper middle of the graph stands out from the rest (see figure 21.5). The nodes in this cluster demonstrate a number of common characteristics: they are connected to Vancouver and Victoria but also to other destinations; and because of their larger node size, these villages sent a relatively higher number of immigrants. Interestingly, these observations lead to a subsequent research question: Were villages with similar migration characteristics geographically

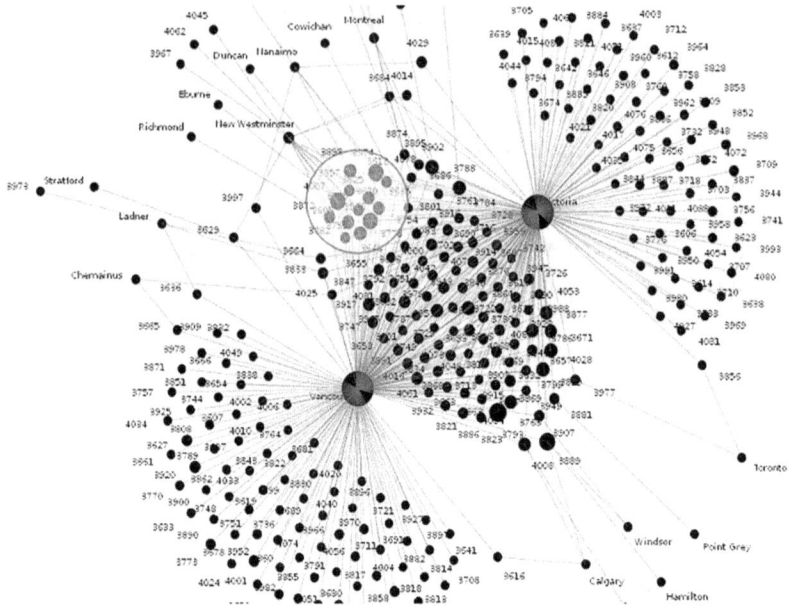

Figure 21.5. The village cluster of interest emerged from the Palladio visualization (Zhongshan County).

Source: Zhang and Cho 2019.

close? Although this may be a question that can be better examined using GIS technology, it may not be an easy task considering many villages had disappeared or their names changed. It remains an exciting research opportunity, regardless, for researchers who can piece together the clues of the past.

Further Analysis of the Diversity of Destinations

The Palladio network visualization is useful in revealing the clusters of villages which showed similar migration patterns, but is unable to shed light on a critical query regarding chain migration that Yu and Chan attempt to address: the diversity of choices of the immigrants' destinations. As Yu and Chan assert, four major Cantonese family clans in Saskatchewan all showed a high geographic distribution in choice of destination. However, it is fair to say that this is a micro-level perspective that highlights only the four clans. We were intrigued by this question: What would a more complete picture look like, and would the diversity of choices of the destinations (how many destinations the migrants ended up in) be

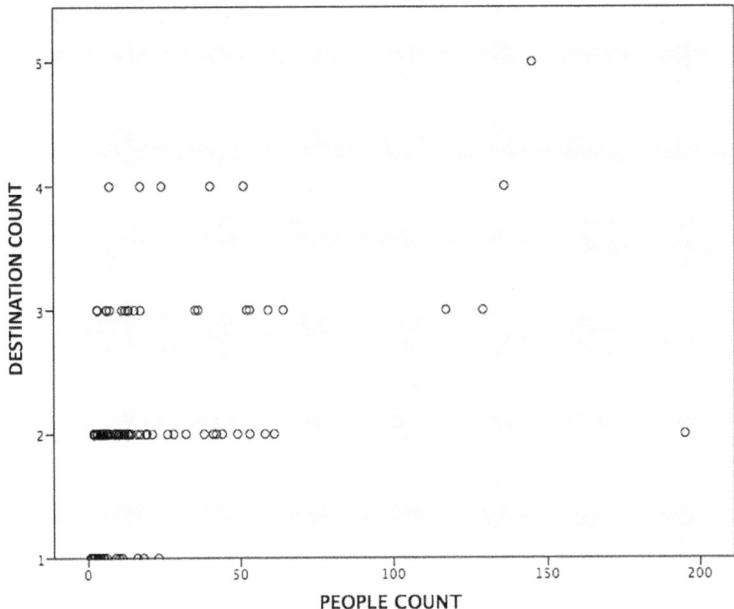

Figure 21.6. The head count and destination count for villages in Zhongshan County, 1910–1949.

Source: Zhang and Cho 2019.

related to the size of a village clan? We used R to statistically study the relationship between these two variables.

A scatter plot is a type of plot that assigns one variable to the horizontal axis and another to the vertical axis; it is useful to visually display if two continuous variables are related to each other and if bivariate outliers exist. We produced a scatter plot in Figure 21.6, using R, showing the relationship between two variables: the number of immigrants who came from a village and the total number of destinations that village is connected to, with each dot being a village in Zhongshan county. The things we can learn from the graph and its implications include the following:

1. For Zhongshan county, the villages are mostly grouped at the lower part of the graph (the number of destinations being only two), including some villages with a large number of immigrants. In other words, the immigrants originated from those villages chose from a very narrow number of destinations,

even for most of the large clans. This is not consistent with Yu and Chan's findings that four major Cantonese clans in Saskatchewan all showed a high geographic distribution in choice of destination.

2. A "Maverick" village (the one at the far right on the graph): this village shows extremely limited choice of destinations despite its extremely large number of immigrants. It is worth further investigation into this big clan to find out what was unique about this village that caused this highly narrowed choice of destinations.

The analysis above is based on just one county—Zhongshan. Certainly, further analysis could be done on the other two counties with the completion of the process of normalizing village names, in order to compare between the counties and see which county shows more correlation between the size of a village clan and the diversity of immigrants' choice of destinations.

In addition to revealing the nuances in the relationship between the diversity of choice of destinations and the size of a village clan, another meaningful research question that can be posed by future studies is a temporal one: How did the diversity of a village's (or family clan's) choice of destination from China to Canada change over time? The emergence of this question is illuminated in Myrdal's theory of "cumulative causation," which describes "the snowball effect in migration patterns, whereby destinations become increasingly attractive to migrants as more of their network settles there" as reiterated by Yu and Chan:

> The best strategy for successful emigration to Canada would be to follow the example of one's relatives in choosing a destination, which a narrow interpretation of the effects of chain migration suggests. Rather than striking out on their own, newly arrived immigrants could take advantage of the connections that their relatives had already put in place, finding work and housing through families. They would have had little incentive to go elsewhere. In this scenario, a small number of origin points would connect to a relatively equal number of destinations. (Yu and Chan 2017, 45–46)

Thus, Yu and Chan concluded that the immigrants' diverse choice of destinations revealed by their research went against Myrdal's theory. However, their Gephi visualization did not adequately touch on the "cumulative" aspect that the name of Myrdal's theory underscores. A possible, and perhaps more complete, speculation would be temporal: to focus on how the number of destinations connected to a particular village changed over time. A curve can be hypothesized—that the number of destinations initially increased and then, after a number of years, it plateaued. This hypothesis may warrant a thorough data analysis to accept or reject it. Most excitingly, to study this hypothesis is computationally and statistically viable given the variable of normalized village names (for the three counties with normalized village names) and the variable that indicate the time of immigrants' arrival.

To conclude, the hidden patterns uncovered by the study proved that normalizing the county names and village names has supported research to work at all levels of granularity—with different scales of study (as a whole, county level, and village level) letting the data tell a different story. Unfortunately, the value of normalized origins has not been systematically exhausted by scholars. The findings in this study reveal much more nuanced patterns that complement and contradict previous studies; thus, these new findings promise to lead to more intriguing research questions to potentially be explored by scholars. As discussed, further studies are warranted to expand the understanding of early Chinese immigrants—spatially and temporally.

TWO LIBRARIANS' APPROACH TO REJUVENATING THIS LEGACY OF RACE AND DATA

Why has this head-tax data been largely underused despite its untapped potential? We offer a few possibilities. First, the Chinese character normalization project so far has attracted little attention; and further, critical information concerning methodology and which counties have been normalized are missing from the dataset. Compounding the issue is the lack of a codebook. For social-sciences data, a codebook is important documentation that contains information intended to be complete and self-explanatory for each variable in a data file. Without a meaningful codebook, it is more difficult to discern variables, especially the origins (such as village names) of migrants. To counter these challenges, we attempted to reduce the barriers of entry to understanding and using the data by demystifying some variables that are critical to understanding how the migrants' origins were recorded and normalized. In addition, one variable, the height of the head taxpayer, was converted to a format that allows for easier manipulation.

However, publishing the results of our research in the traditional sense might not necessarily reach the largest number of individuals, particularly if we want to share interpretations of variables and computing scripts. That would be cumbersome and ineffective. In addition, as librarians, we are deeply concerned with the crisis in scholarly publishing. As John Unsworth puts it, "nobody is reading these books—not even colleagues in the disciplines, much less students, or the general public" (Unsworth 2013, 30). Remarkably, open access and open scholarship are perhaps, so far, the strongest voices advocating for reaching out to a wider audience. Second, as there is currently a reproducibility crisis in scholarship, the open-science movement has sought to counter this by making scientific research (including publications, data, physical samples, and software) and its dissemination accessible and open to all levels of inquiry, be it professional or amateur (Allen and Mehler 2019, 1–14).

These are the considerations that propelled us to publish our project in a venue somewhat unconventional for the humanities—the Open Science Framework (OSF). As an open-source software project originally developed for collaborative and reproducibility research in psychology, the OSF is now used by researchers of all research fields. In our case, the raw materials that are required to replicate the findings have been stored in OSF: the converted dataset, R scripts, and a Readme file that mitigates the lack of a codebook (Zhang and Cho 2019). It ultimately allows for us to communicate with a larger audience in a timely manner.

Is this a type of work that librarians usually undertake? This study may represent a somewhat unique form of librarians engaging the digital humanities in the sense that we examine the data from two perspectives simultaneously. One, as if we are the stewards of the collection—focused on how to increase the usability of the dataset and make it more amenable to computational analysis; the other, that of the researcher—focused on what the potential of the data is and how computational tools can help ask and answer humanistic inquiries. The former point of view may seem more natural in relation to how libraries typically support digital humanities; thus, our project actually echoes a larger movement in the digital libraries community: collections as data (Padilla et al. 2020).

The second stance, the one that speaks more directly to DH scholarship, may appear to be subversive to libraries'/librarians' accustomed roles—providing services to support DH research on campus—and yet resonates deeply with Trevor Muñoz's powerful call to place "digital humanities research as core to the theory and practice of librarianship in its own intellectual terms" (Muñoz 2016). Our own intellectual terms are reflected by the exploration of the interplay between computational tools and humanistic inquiry, which is at the core of DH scholarship, as well as the fact that we still take the neutral stance of librarians—we do not make any historical interpretation but simply reveal the potential in the data, and raise awareness around it and the history of discrimination and exclusion the data represents. Lastly, our intellectual pursuit demonstrates itself by the tools we chose to use—particularly Palladio, a freely available and user-friendly network-analysis tool—to subvert overly obscure scholarly work and thus avoid further marginalizing the data and history. The project will likely pave the way for more researchers who may or may not identify themselves as DH scholars to explore the data and the colonial frameworks behind it.

Roopika Risam has argued that some of the most developed DH work preserves the legacies of "dead white men," specifically individuals unlikely to be forgotten in anglophone literary history even if these projects did not exist (Risam 2015). A similar story can be traced to the histories of digital archives intended to preserve national histories. Adeline Koh warns that as Eurocentric biases continue to exist within current digital work and as the history of the colonized continues to be misrepresented or underrepresented in digital archives, we should be careful that digital humanities does not become a "refuge" from issues of gender, race, class, and sexuality (Koh and Risam 2013). It is because of this fraught yet evolving relationship between digital humanities and diversity that our project on Chinese-Canadian migrants seeks to contribute inclusive representation and a critical approach to the collection and preservation of digital records that had once intended to anonymize those very histories that were recorded by officials to limit the intake of undesirable peoples into the country. It is our hope that by sharing our research findings scholars can continue to unthread the racist project that took place over a century ago and that continues to haunt a nation.

NOTE

1. This image is adapted from the "Mapping the Villages & Towns Recorded in the Register of Chinese Immigration to Canada from 1885 to 1949" produced by UBC Asian Library (University of British Columbia 2012).

REFERENCES

Allen, Chris, and David M. A. Mehler. 2019. "Open Science Challenges, Benefits and Tips in Early Career and Beyond." *PLOS biology* 17, no. 5: 1–14. https://doi.org/10.1371/journal.pbio.3000246.

———. 2018. "Mass Capture Against Memory: Chinese Head Tax Certificates and the Making of Noncitizens." *Citizenship Studies* 22, no. 4: 381–400.

Chiang, Rudy. 2012. "Canada Immigration Chinese Head Tax Record Sun Woy District Part 1: 1885–June, 1903." UBC Asian Library. https://branchasian.sites.olt.ubc.ca/files/2011/09/Head_Tax_Record_SunWoy_RC.pdf.

Der, Gillian. 2018. "Pigs, Pestilence, and Prejudice: The Racialization of Early Chinese Settlers in Vancouver's Chinatown." *Trail Six* 12: 41–49.

Easley, David, and Jon Kleinberg. 2010. *Networks, Crowds, and Markets: Reasoning About a Highly Connected World*. Cambridge, U.K.: Cambridge University Press. https://doi.org/10.1017/CBO9780511761942.

Hermansen, Sally, and Henry Yu. 2014. "The Irony of Discrimination: Mapping Historical Migration Using Chinese Head Tax Data." In *Historical GIS Research in Canada*, edited by Jennifer Bonnell and Marcel Fortin, 225–237. Calgary, Alb.: University of Calgary Press. https://doi.org/10.2307/j.ctv6gqt40.15.

Koh, Adeline, and Roopika Risam. 2013. "Open Thread: The Digital Humanities as a Historical 'Refuge' from Race/Class/Gender/Sexuality/Disability?" *Postcolonial Digital Humanities* (blog). May 10, 2013. https://dhpoco.org/blog/2013/05/10/open-thread-the-digital-humanities-as-a-historical-refuge-from-raceclassgendersexualitydisability/.

Muñoz, Trevor. 2016. "Recovering a Humanist Librarianship through Digital Humanities." In *Laying the Foundation: Digital Humanities in Academic Libraries*, edited by John White and Heather Gilbert, 3–14. West Lafayette, Ind.: Purdue University Press. https://doi.org/10.2307/j.ctt163t7kq.4.

Padilla, Thomas, Laurie Allen, Hannah Frost, Sarah Potvin, Elizabeth R. Roke, and Stewart Varner. 2020. "Always Already Computational: Collections as Data." OSF. July 14. https://doi.org/10.5281/zenodo.3152935.

Risam, Roopika. 2015. "Beyond the Margins: Intersectionality and the Digital Humanities." *DHQ: Digital Humanities Quarterly* 9, no. 2. http://www.digitalhumanities.org/dhq/vol/9/2/000208/000208.html.

University of British Columbia. 2012. "Mapping the Villages & Towns Recorded in the Register of Chinese Immigration to Canada from 1885 to 1949." UBC Asian Library. https://branchasian.sites.olt.ubc.ca/files/2012/01/Head-Tax-brochure2.pdf.

Unsworth, John. 2013. "The Crisis of Audience and the Open-Access Solution." In *Hacking the Academy: New Approaches to Scholarship and Teaching from Digital Humanities*, edited by D. J. Cohen and J. T. Scheinfeldt, 30–34. Ann Arbor: University of Michigan Press. https://doi.org/10.2307/j.ctv65swj3.10.

Ward, W. Peter. 2013. "Stature, Migration and Human Welfare in South China, 1850–1930." *Economics and Human Biology* 11, no. 4: 488–501. https://doi.org/10.1016/j.ehb.2012.10.003.

Ward, W. Peter, and Henry Yu. 2012. "Register of Chinese Immigrants to Canada, 1886–1949." UBC Faculty Research and Publications. https://dx.doi.org/10.14288/1.0075988.

Weingart, Scott B. 2011. "Demystifying Networks, Parts I & II." *Journal of Digital Humanities* 1, no. 1. http://journalofdigitalhumanities.org/1-1/demystifying-networks-by-scott-weingart/.

Yu, Henry, and Stephanie Chan. 2017. "The Cantonese Pacific: Migration Networks and Mobility Across Space and Time." In *Trans-Pacific Mobilities: The Chinese and Canada*, edited by Lloyd Wong, 25–46. Vancouver, B.C.: University of British Columbia Press.

Zhang, Sarah, and Allan Cho. 2019. "Hacking the Historical Data: Register of Chinese Immigrants to Canada, 1886-1949." OSF. June 12. https://doi.org/10.17605/OSF.IO/9ZR6F.

CHAPTER 22

Afterword:
The Landscape and the Horizon

Susan Brown

This substantial volume is a testament to the breadth and quality of digital scholarship, past and present, in the arts and humanities in Canada—some under the banner of digital humanities and some in other contexts. For its size, Canada has achieved an extraordinary level of activity in the field, and the essays here indicate the richness of that landscape. The editors wisely eschew definitional questions, laying out the complexities of situating digital humanities in relation to the concept of Canada, always to some extent under erasure and certainly for decades under increasing scrutiny as a source of deep ambivalence for those who take seriously post-colonialism and, latterly, the calls to action of the Truth and Reconciliation Commission. In keeping with *Future Horizons'* aim to reterritorialize digital humanities, this afterword reflects on the landscape as well as the built human artifices that are the material layers which shape and influence, if not fully determine, the horizon.

The very scope and quantity of digital engagement by Canadian scholars in the humanities and social sciences (since much work straddles that divide) means that no single volume can provide a comprehensive view, for which reason the editors of this volume do not offer a *particular* history of vision of the digital humanities in Canada. However, *Future Horizons* nonetheless presents an array of particular perspectives, grounded in distinct fields, disciplines, or communities of practice. The net effect here leans toward literary studies, even while encompassing a wide range of topics and engaging many broader questions of methodology. Curiously, Voyant Tools (Sinclair), one of the most widely used DH tools globally, especially for literary analysis, is rarely mentioned. This is perhaps due to the collection's emphasis on research rather than hands-on pedagogy, since Voyant puts at the fingertips of novices the kinds of analysis Sandra Djwa used to challenge accounts of Canadian poetry, as described in Djwa's chapter with Sarah Roger, Paul Barrett, and Kiera Obbard as well as in Djwa's own, and so is often used for teaching.

In her reflections from an administrative perspective, Andrea Zeffiro contrasts the creative, divergent, decolonizing potential of DH tools and methods with the hegemonic bent of university courseware, echoing Dani Spinosa and Kendra Cowley's concerns about the complicity of the digital humanities with the neo-liberal corporatization of the university, settler colonialism, and the military-industrial complex. Such concerns have reverberated from the earliest engagements with computers in Canadian poetry and poetics, as detailed by Gregory Betts and Eric Schmaltz. They resurfaced nationally (making international headlines) early in the pandemic, when Proctorio, an exam surveillance proctoring software company, sued University of British Columbia educational technology specialist Ian Linkletter in October 2020 for tweeting criticism based on Proctorio's public videos. The use of such surveillance software has led to widespread student protest via online petitions and social media on the grounds that the technology is discriminatory, ableist, and intrusive (Harwell 2022). The policing of student bodies through such systems by higher-education institutions is a grim contrast to Voyant's attempt to place researchers, often students or those starting to experiment with digital methods, in a position of power in relation to the gaze. Yet it reminds us, as do many pieces here, including Jon Saklofske's advocacy of the transformative world-making potential of digital games, that the virtual has real impacts. Chapter after chapter here reveals contributors probing theories, tools, arts, arguments, and practices for their impacts on lived and embodied experiences as a means of situating their analysis. For situation determines what perspectives are available and returns our focus to the human.

Voyant Tools offers many perspectives on the contents of this collection. Its Cirrus Tool creates a word cloud where the more frequently a word occurs, the larger it appears. Below (see figure 22.1), the left image uses Voyant's default stop-word list, while the one on the right, in tribute to Klara du Plessis's brilliant reflection on stop words, visualizes the top 75 words without exclusion. Word-based visualizations such as this emblematize the text-centricity of much current information visualization as articulated by Julia Polyck-O'Neill. Its simplicity undeniably limits interpretation, but it requires only minimal visual literacy to begin to interpret it.

Voyant[1] confirms the sense that insofar as this volume emphasizes any particular area or domain, it is indicated by *poetry, poem/s, literature, literary,* and perhaps *writing. Archives* and *archival* matters are prominent but go beyond the territory of a single discipline, particularly given the capacious sense of the word in digital-scholarly contexts, pointing to methods and theory. Throughout, the volume emphasizes the conjunction of alternative perspectives and digital scholarly methods as a means of discerning, imagining, or conjuring possible futures.

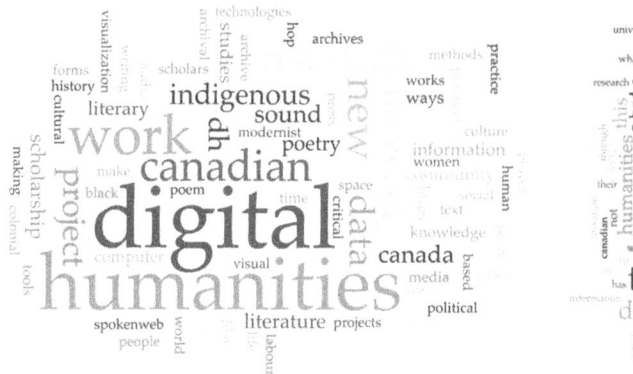

Figure 22.1. Word clouds produced by Voyant's Cirrus Tool.

Source: Voyant Tools and *Future Horizons*, 2022.

Mark V. Campbell's engagement with hip-hop becomes a means of centring Blackness and grappling with anti-Black legacies of digital humanities. By way of Marisa Parham, Simone Brown, and Sylvia Wynter, he reminds us of the extent to which "digital life replicates social binaries and hierarchies" and of the importance of aesthetics in contesting power. Pascale Dangoisse, Constance Crompton, and Michelle Schwartz find their work transformed by the socio-political "perspectives of the activists that we study": they use feminist markup in the service of subaltern queer histories that provide a counterweight to capitalist notions of labour and the silences of the archive.

Considering that DH activity is arguably dominated by textual studies (though these can slide into and intersect with other fields, as Katherine McLeod indicates), it is also important to acknowledge the disciplinary flavours of digital scholarship that do not tend to operate (so much) under the umbrella of the digital humanities. Thus, for instance, the strong tradition of quantitative digital history, which laid the ground for the massive interdisciplinary Canada Century Research Infrastructure census project (Gaffield 2007, 2016), has also provided the foundation of a strong sense of digital history qua history in *The Historian's Macroscope* (Graham et al. 2022) as well as the Archives Unleashed project. The former explicitly addresses historians, while Archives Unleashed grounds itself rhetorically in history—principal investigator Ian Milligan proclaims: "Data is rapidly becoming the building blocks of our histories"—although reaching well beyond the discipline. This observation does not aim to set up understandings of digital methods as expanding disciplinary approaches, on one hand, against an umbrella conception of the digital humanities as a conversation among and beyond disciplines, on the other. Indeed, Allan Cho and Sarah Zhang's analysis of the head-tax registry displays an ease with applying digital approaches as a component of nuanced historical inquiry into institutionalized racism in Canada, even though they approach that work as librarians seeking to reveal the potential in their collections and advance digital scholarship more broadly. Many contributors invoke disciplinary and umbrella approaches as complementary and mutually beneficial, even if the former shore up existing institutional structures, hierarchies, and power relations against which many in this volume position themselves.

This tension between the disciplinary and interdisciplinary should be taken seriously. How are the kinds of rich intellectual inquiry using digital methods described here to be sustained, maintained, and renewed when the digital humanities still seems "supplementary" or even "foolish" (to echo Djwa), if not actively harmful, to many academics? There are advantages to articulating new approaches in relation to familiar content and established disciplinary methods: the possibilities can otherwise be difficult to discern. There is also the ethical question of the best strategy for supporting new scholars in an ever-worsening hiring environment, where the ability to align with a discipline can mean the difference between a tenurable academic job and precarity or the need to retool. We need to be mindful about how best to advance conversations about digital methods. Meanwhile, the ambivalences, warnings, and critiques articulated in this collection demonstrate that value propositions associated with digital methods have been—and continue to be—contested from within, in ways that are both salutary and revealing. Gregory Betts probes the contradictions of "the ongoing hunger to dismantle the power dynamics of Empire by those poets yet caught within its nets" in the responses to early computers in the avant-garde Canadian literary scene of the 1960s and 1970s, whether they rejected or embraced them. Zeffiro links her ambivalence toward the digital humanities to Roopika Risam's articulation of "the local-global quandary within digital humanities" and asks: "Who are we to define digital humanities writ large?" This volume situates the project of territorializing digital humanities in Canada in dialogue with transforming and pluralizing the nation through decolonization, and with a reworking of inherited values and institutions that embraces rather than erases unease and ambivalence.

The nation has already been radically transformed by the digital in a host of ways. As Benjamin H. Bratton argues in *The Stack*, digital systems create contemporary worlds in which software has everything to do with sovereignty. Notably, the nation is absent from the layers Bratton stacks: "If imagined as an emergent nation-state, the Cloud would be today the first largest consumer of electricity," he notes, deeming nation-states "petroglyphs written by law" and obsessed with concretizing "the integrity of virtual boundaries" (2016, 309) that have never held in any meaningful sense, and which are constantly traversed in the digital world. As Risam points out in her chapter here, organizing digital humanities by the nations in which scholars are located, as opposed to the topics on which they work, is illogical and obfuscates diversity within national contexts and networks that cross borders. I do not position the nation as a solution of any kind, but I would argue that there are important specificities—nicely summed up in Risam's allusion to "issues of infrastructure, access, and policy"—that need to be taken into account in thinking about the digital humanities in Canada, even if the eventual aim is to move beyond national categories.

Canada as nation has sidestepped the implications of the digital turn for culture. Although there have been inquiries into more specific matters, the Government of Canada has initiated nothing that resembles in scope the Massey Commission's report on the impacts of broadcasting on Canadian culture. How far globalization has undermined political will—along with the political ability to set a national cultural agenda—is evident in the lack of any coordinated, national attempt to take stock of the enormous impacts (social, cultural, political, economic) of digital technologies on the fabric of Canadian life. What is distinctively Canadian or amenable to government intervention in this transformative process is an open question, especially given the extent to which the impacts are being considered piecemeal or not at all.

As it happens, the first generation of scholars and artists to digitally produce knowledge and cultural productions is now retiring; as they lose institutional footholds, the digital artifacts that they produced—call them documents, files, spreadsheets, databases, e-literature, websites, prototypes, designs, code, or what have you—will, with very few exceptions, evaporate, in contrast to paper-based publications and research archives. Research data management is not on the radar for many first-generation digital-scholarly adopters (Borgman 2015). The efforts of individual universities (even with the new emphasis on research data management) are unlikely to capture a large proportion of this material, even leaving aside the value of digital-only materials collected and curated by scholars for their own purposes, which would be hugely useful to other scholars if effectively shared (Brown 2014). There is, for instance, considerable resistance to data-management plans within the third cohort of Digging into Data, one of the most prestigious, international, multi-agency funding programs for large-scale DH research to date, not because researchers are opposed to the idea but because current funding models do not provide post-award funding to support such work (Poole and Garwood 2020).

Policies, or the lack thereof, can have massive impacts on the shape of scholarship and the types of available cultural knowledge. Challenges posed by the outmoded copyright regime are mentioned here by Campbell in discussing hip-hop, Graham Jensen in relation to modernist periodicals, Deanna Fong and Ryan Fitzpatrick with regard to archiving the works of Fred Wah. Yet the table of contents in this volume would be substantially different if Canada had a different copyright regime. The intellectual-property regime stifles the study of culture in Canada across many fields, given that most cultural production is recent. Certainly, in digital literary studies, stylistical analysis and distant-reading methods are inhibited by the inability to publish the corpora on which analyses, such as Paul Barrett's (2021) on Austin Clarke, are conducted. For a decade now, Canadian copyright law has explicitly committed to balancing owner or creator rights against user rights (Geist 2013), in part in response to the impassioned arguments of legal scholar Michael Geist, who is cited in a Supreme Court decision from 2006 as saying: "The Internet and new technologies have unleashed a remarkable array of new creativity, empowering millions of individuals to do more than just consume our culture, instead enabling them to actively and meaningfully participate in it" (Robertson v. Thomson Corp 2006, para. 79, quoting Geist 2006). In fact, a 2012 Supreme Court decision found that "limiting research to creative purposes would also run counter to the ordinary meaning of 'research,' which can include many activities that do not demand the establishment of new facts or conclusions. It can be piecemeal, informal, exploratory, or confirmatory" (Society of Composers 2012, para. 22). This ruling ought to support the kinds of non-consumptive use advocated by the HathiTrust Digital Library in U.S. contexts (Aaron 2012; Diaz 2013). However, this reassurance that the validity of user rights would be balanced has not unleashed a torrent of digital engagement with in-copyright materials, even though it has the potential, for instance, to transform Canadian literary studies.

Legal decisions affecting the use of content are decidedly part of the stack that enables and constrains how humanities scholars in Canada conduct digital research, which is to say that they are part of the infrastructural conditions of digital humanities in this country. Infrastructure shapes our lives and, in many cases, our lives depend upon it, in both mundane and potentially literal ways, as demonstrated by the massive outage of the Rogers telecom network on July 8, 2022, which knocked out telephone, Internet, and cellular service for about 25 percent of Canadians (Wikipedia 2022). The outage brought glaring attention to the pervasive lack of effective regulation of even essential services, such as the 911 emergency-call system. Questions of jurisdiction regarding the regulation of digital space, which corporate interests aggressively consider non-territorial (Geist 2001), have contributed to inadequate oversight; the consequences are profound. In the wake of an initial laughable hearing held by the Canadian Radio and Telecommunications Commission on the outage, Geist made the sobering observation that we must "prioritize Canada's communications infrastructure and its impact on consumers and business as the single most important policy issue faced by the CRTC [...]. While there has been an emphasis on cultural policy in recent months, CanCon policies don't matter if Canadians can't access the content" (Geist 2022). The mundane mechanisms of digital connectivity are as intimately linked to culture as broadcasting was in the 1950s, when the Massey Commission paved the way for a cultural nationalism that embraced rather than eschewed (at least some forms of) regulation and state intervention.

There are also other forms of regulation, many inherited from earlier media, and this collection stresses the importance of working creatively or subversively against such existing constraints. Above all, David Gaertner reminds us that decolonizing means rethinking core assumptions about, for instance, universal access as an unquestionable good within digital humanities. Building on work by Kimberly Christen, Jennifer Wemigwans, and others, Gaertner advocates boundary setting and closure to facilitate relationships and "culturally specific pathways for data" rather than blanket mandates of openness (Christen 2012; Wemigwans 2018). Campbell notes the extent to which the digital humanities can learn from hip-hop to move beyond the stultifying reification of older knowledge systems into new digital forms. The analysis of the dynamics of sampling and remix offers as much to thinking through the affordances of the digital as have medieval manuscripts, but with the added challenge of grappling with the limiting logic of the copyright system. Campbell also demonstrates the power of community-based counter-archives to support intersectional engagement and analysis grounded in subversion, and a refusal of institutionalized frameworks of legitimation. Martin and Kaur, however, note the potential of even relatively recent community-led efforts—that is, the creation and running of maker-spaces—to reproduce patterns of exclusion embedded in social and economic relations, which in turn reinforce ableist, gender, and race privilege. They also note the many ways in which the infrastructure of a physical space impacts what can happen and how. Thinking through infrastructure, then, whether we are talking about the built environment or matters of policy, provides a means of specifying conditions of possibility for digital humanities.

Digital scholarship in the humanities is massively inflected by the division of governmental powers within Canada, wherein the provinces and territories have authority over education but telecommunications, including broadcasting and the internet, are federal responsibilities. How this division works with respect to research has shifted significantly in the past few decades. The Social Sciences and Humanities Research Council of Canada (SSHRC) was founded in 1978 by a federal statute to "promote and assist research and scholarship in the social sciences and the humanities," which had previously been funded by the Canada Council; it serves the largest group of research faculty and graduate students in the country (Doern 2009). In transforming itself (starting in 2004) from a granting council to a "knowledge council" (SSHRC 2006), SSHRC became increasingly policy-driven in a range of ways, including allocating funding to specific initiatives based on themes deemed to be of high social value (LaPointe 2006; Doern 2009). Some programs resulting from this shift have undoubtedly been of great benefit in advancing DH research. The modest-but-flexible, multidisciplinary, and team-oriented Image, Text, Sound and Technology (ITST) program was a major catalyst for DH conferences, prototyping, research, and other initiatives (SSHRC 2009). One advantage of its focus was that peer-review adjudication panels were drawn from humanities scholars and social scientists with experience of digital research, with the result that even if one member had a conflict of interest there would still be others with technological expertise to help with the decision. By contrast, even interdisciplinary panels for other SSHRC adjudications often have only one person with DH expertise, if any; if that person has a conflict of interest, which is not uncommon given the highly collaborative nature of the DH community in Canada, then no one is left to speak to the technological elements of a project, which are often central to its methodology. In other words, true peer review happened thanks to the thematic focus of the now defunct ITST program.

SSHRC has also moved increasingly in the direction of stressing the importance of partnerships, replacing Major Collaborative Research Initiatives with Partnership Grants, most of which require contributions from partners. This can be a real challenge for the humanities, and consequently pushes projects in directions that allow applicants to procure matching funds. A significant difference from other contexts (such as in the United States) is that the tri-councils do not provide overhead costs, nor are researcher-salary costs eligible expenses. These factors can impede international collaboration and, even when they do not, can result in prolonged negotiations over inter-institutional agreements that significantly impede projects. More significantly, the ineligibility of salaries significantly disadvantages precariously employed early-career scholars whose positions do not include research time. The ever-worsening academic job market in Canada has normalized precarity and increased competition for positions to an extent that, Jensen observes, makes questions about the status of DH scholarship in relation to traditional research and publication more than academic. SSHRC's policy means that after the postdoctoral period, precariously employed academics cannot receive material support to work on independent research, no matter how meritorious. Such policies reduce the range of options for researchers already suffering precarity; combined with the paucity of dedicated DH positions, they threaten the continuity of the field. Digital humanities has created and worked to valorize "alt-ac" or "alternative academic" positions and career paths, often in combination with large projects or centers, though employment is itself too often precarious. McLeod flags the importance of the SpokenWeb project's recognition "in the current climate of academic employment—significant research contributions can happen with one foot in and one foot out of academic institutions." Often those in alt-ac positions play crucial roles in developing new methods, tools, or infrastructure. The ineligibility of those in alt-ac positions to apply for external grants (typically due to university policies) can create real problems for career development and project leadership succession.

Digital humanities is an unusual field in the extent to which researchers themselves develop tools or infrastructure either as a necessary methodological component of their work or as the primary goal of their research. Sustainability is a major problem for digital tools developed through scholarship. Many tools are created as a component of research grants, and they therefore have no sustainability funding beyond the scope of the initial grant; this makes it particularly hard to meet the requirements for "findability, scaleability and usability" (Poole and Garwood 2020, 87–88). Websites disseminating combinations of project information, research data, and results (often fuelled by SSHRC's knowledge-mobilization policies) undoubtedly make up the largest category of tools funded by research grants, and the websites or the knowledge they disseminate should be preserved. There are also other, more generic tools—often produced as a means of thinking through specific problems, and also often divorced from specific content—that have been developed through research grants and taken up by other scholars. In other words, research and infrastructure bleed into each other. There are pros and cons of separating them (as compared to other jurisdictions where they are often combined within funding programs), but this separation is a defining feature of the DH landscape in Canada. Only a few programs—such as the Canada Research Chairs (Polster 2002) and the international Digging into

Data (Poole and Garwood 2020) programs—have made modest attempts to coordinate research and infrastructure-funding processes.

One of the most distinctive features of the DH landscape is the Canada Foundation for Innovation (CFI), an arm's-length organization founded by the federal government in 1997 to fund research infrastructure with an initial budget of $800 million. CFI's funding model leverages matching funds from the provinces equivalent to the amount committed by CFI, and it requires that researchers secure half that amount from partners, typically industry or institutions, resulting in a 40/40/20 split. Since CFI is far and away the largest and most stable source of research-infrastructure funding, it in effect acts not only as a patron but also as a regulator, since its policies shape research planning, activities, and outputs; many of CFI's objectives flowing from its funding agreement with the federal government include economic growth and job creation (Lopreite and Murphy 2009). Guppy, Grabb, and Mollica argue that the CFI is one of several indicators of the "re-engineering" of Canadian universities away from a traditional liberal model associated with the pursuit of pure knowledge and substantial government funding, but little government control, to a "technical efficiency" model that has resulted in some quite significant changes over the past half century. They note that, within the Canadian university system,

> By the early 2000s, for example, 65 percent of the operating funds came from government [...], compared to 90 percent in 1960. The shortfall has been covered largely by students and their families[...]. A second change is that the federal government now plays a more direct role in providing money for universities, as part of its expanded emphasis on science policy. (Guppy, Grabb, and Mollica 2013, 2)

From the beginning there was criticism that the arts, humanities, and social sciences would be disadvantaged by the CFI model. Indeed this has turned out to be the case: at most (given some misclassification in CFI's data), "[t]he arts, literature, humanities, and social sciences, which represent 55 percent of all university research appointments (AUCC 2007, 4), have received just 5 percent" of CFI funding from 1998 to 2009 (Guppy, Grabb, and Mollica 2013, 4). Among many factors that contribute to this inequity are the emphasis on the procurement of equipment and physical infrastructure; the greater difficulty of articulating "benefits to Canada" for projects from the arts and humanities when compared to STEM (science, technology, engineering, math) projects; the demands of preparing large-scale grant applications, given differential teaching loads and research assistance; and an emphasis on innovation that is often understood as a proxy for commercialization and spinoffs, a sense reinforced by CFI reporting requirements that look for results such as patents (Doern 2009).

All these challenges are bound up in who even gets to apply for CFI funding, since the amount for which any eligible institution can apply, called the "CFI envelope," is pegged to the amount of research funding that the institution has received from the tri-councils. Many institutions apportion the envelope internally in proportion to the amount of funding awarded to the university from each funding council, which further disadvantages applications from the humanities, since SSHRC has less than 18 percent of the total tri-council budget.[2] In addition, many institutions conduct internal competitions for a share of their CFI envelope, in which those projects that seem to be most likely to succeed—in other words, the best fit for the CFI model—are the ones chosen to apply. Matching funding requirements disadvantage researchers in the humanities and social sciences, compared to those in STEM and health (Polster 2002, 291n[14]); some provinces have made it policy not to match CFI grants in the humanities, or not to match CFI grants at all. The envelope system makes inter-institutional collaborative grants more challenging, since researchers need to secure an envelope at each institution, which means fitting in with multiple sets of local priorities, while also competing against the fact that projects that are actually led by a university raise its institutional profile more than ones in which that university participates as a partner, no matter how desirable the outcome. The CFI's now defunct Cyberinfrastructure program countered this problem by not requiring envelope from participating institutions in its attempt to foster truly national infrastructure, and it is highly unlikely that the Linked Infrastructure for Networked Cultural Scholarship (LINCS, discussed in chapter 18) could have even reached the application stage had there not been this significant variance from usual CFI policy.

The Cyberinfrastructure program was meant to be iterative to address the sustainability challenge baked into project-based funding. Indeed, the notion of a "project" is at odds with the idea of infrastructure, which should be funded for as long as it is useful. Sustainability for research infrastructure, and in particular research software, is a major challenge across all fields, and an even greater one in the digital humanities, where there are fewer funding opportunities and the user-fee model employed by some infrastructure in STEM and medicine is anathema to a community that values openness, equity, and accessibility.

The case of Artmob, an innovative and sophisticated but short-lived online content management system for arts and cultural organizations, exemplifies the loss involved when infrastructure is not sustained. An innovative, Drupal-based platform designed to support open access and creative commons, collaborative knowledge production, exhibits, and annotation, Artmob was developed at the Centre for Digital Policy and Cultural Rights Initiatives within the New Media Collaboration Centre at York University to help "press for Canadian cultural policy changes while tabling fair technological solutions that fairly balance the interests of creators, owners, citizens, and institutions" (Meurer 2009). Artmob was the infrastructure for the first Fred Wah Digital Archive developed under Susan Rudy. In chapter 15, Fong and Fitzpatrick describe the complexity of the Wah archive, the diverse forms of labour, and the number of people required to reboot this initiative, providing an eloquent account of what is lost when infrastructure and its content are not adequately maintained, as well as a recognition of the continuity between cultural production in different forms. Massive intellectual expenditures of design thinking and applied theory, of the kind also articulated by Dani Spinosa in her discussion of the work involved in the *Electronic Literature Collection* (chapter 12), evaporate when digital content infrastructure goes dark.

The Artmob example illustrates the extent to which researchers should work with preservation specialists (more available to scholars now that research data management has been mandated by the tri-councils) from the project-design stage—to ensure that project data and metadata can be preserved for use by others. However, archiving for long-term preservation should not be mistaken for the fuller kind of access needed for cultural-data reuse by most scholars and citizens, when compatible with Indigenous-Knowledge protocols. True accessibility in the humanities requires interfaces that deliver not datasets but digital content, as Fong and Fitzpatrick recognize. Their achievement is not only an innovative approach to representing the relationality of literary communities, but in establishing a model for what they call "loose sustainability" based on "reciprocality." Their work contributes to strategies for addressing ethical tensions in the complex if often rewarding relationships operative in many large DH projects involving student or precarious labour (Anderson et al. 2016; Di Pressi 2015; Mukamal et al. 2021). Another achievement of the rebooted Wah project is its partnership with, and institutional embeddedness in, a university library to enhance its prospects of long-term sustainability. The expectation by CFI that universities will have the resources to maintain infrastructure funded by CFI has not proven reasonable, and libraries that have pivoted to meet the challenges of new forms of knowledge production are the institutions best situated in Canada to help sustain infrastructure centred on digital cultural content.

Voyant, mentioned above, is perhaps Canada's greatest generalized DH infrastructure success story, and a case study in the difficulty of transforming researcher-driven tool building to true infrastructure. Developed by scholar-coder Stéfan Sinclair with his long-time interlocutor and collaborator Geoffrey Rockwell (2016), and promoted and supported tirelessly by them in a host of ways, including through irregular grant funding from both SSHRC and CFI, it has become one of the best-known and most widely used DH tools in the world, and the most-used tool on the Compute Canada, and now the Digital Research Alliance of Canada, Cloud. Yet Stéfan's death in 2020 has left Voyant in a precarious position, which Rockwell is seeking to address by forming a consortium (Rockwell 2022). Voyant's vulnerability, despite being led by well-funded, tenured professors at major Canadian universities, puts a further spin on Zeffiro's probing questions about the relationship between service and servitude, labour and precarity. She voices concern for how "tools and methods exacerbate inequities in the production of new knowledge about what sorts of research attract funding, about the inequitable distribution of resources across departments, and about how questions of power, equity, and race are sidelined for tools training across multiple facets of collective life." The inquiries into equity, ethics, and social justice that permeate so many of the arguments collected here are required to arrive at new and less precarious models for how we move together, collectively, as communities of digital practice.

This overview of factors shaping the conditions of the collective life of scholars demonstrates some of the specific features and complex processes that give shape to the DH landscape in Canada, and consequently its horizons. Many other factors that contribute to the conditions of possibility for DH Canada go undiscussed here for want of space, including interprovincial differences in funding, the complex relationship between Canada Council and electronic-creative practices, the impacts of Canadian Society for Digital Humanities / Societé canadienne des humanités numériques and expanding training networks, and how the Federation of the Social Sciences and Humanities enables cross-fertilization through its annual multidisciplinary Congresses. The chapters in this volume offer, in rather different ways, counterpoints to any naïve sense of technology as a panacea for intellectual, economic, or social-justice challenges. They serve as reminders of the particularities and materialities at play in a wide range of contexts, and they point to the messy, creative, and pointedly human elements of scholarly and artistic engagements with the digital in relation to the contested space and nation called Canada. Singling out points of connection and continuity within and beyond Canada as much as they do resistance and rupture, they offer a rich, diverse, and promising vision for how future digital work will continue to intervene in and reshape our worlds and our relationships.

NOTES

1. Voyant is an interactive system that allows for live links to visualizations, but the links are not persistent, so none are included here. However, since this volume is Open Access thanks to the University of Ottawa Press, one may download the text, upload to Voyant, and explore the volume further using Voyant's array of tools.

2. Based on the 2021–2022 departmental reports for each of the tri-council funders, the SSHRC projected budgetary spending of $560,220,669 (SSHRC 2021), the Natural Sciences and Engineering Research Council of Canada projected spending of $1,356,837,786 (NSERC 2021), and the Canadian Institutes of Health Research projected spending of $1,220,744,865 (CIHR 2021), for a total budget of $3,137,803,320. As share of the total of tri-council funding, this works out to 17.85 percent for SSHRC, 43.24 percent for NSERC, and 38.90 percent for CIHR.

REFERENCES

Aaron, James. 2012. "The Authors Guild v. HathiTrust: A Way Forward for Digital Access to Neglected Works in Libraries." *Lewis & Clark Law Review* 16, no. 4: 1317–47. http://dx.doi.org/10.2139/ssrn.2205374.

Anderson, Katrina, Lindsey Bannister, Janey Dodd, Deanna Fong, Michelle Levy, and Lindsey Seatter. 2016. "Student Labour and Training in Digital Humanities." *Digital Humanities Quarterly* 10, no. 1. http://www.digitalhumanities.org/dhq/vol/10/1/000233/000233.html.

AUCC (Association of Universities and Colleges of Canada). 2007. *Trends in Higher Education: Volume 2: Faculty.*

———. 2008. *Trends in Higher Education: Volume 3: Finance.* https://globalhighered.files.wordpress.com/2008/06/trends_2008_vol3_e.pdf.

Barrett, Paul. 2021. "Austin Clarke's Digital Crossings." In *The Digital Black Atlantic,* edited by Roopika Risam and Kelly Baker Josephs, 84–94. Minneapolis: University of Minnesota Press. https://doi.org/10.5749/j.ctv1kchp41.11.

Borgman, Christine L. 2015. *Big Data, Little Data, No Data: Scholarship in the Networked World.* Cambridge, Mass.: The MIT Press. https://doi.org/10.7551/mitpress/9963.001.0001.

Bratton, Benjamin H. 2016. *The Stack: On Software and Sovereignty.* Cambridge, Mass.: The MIT Press. https://doi.org/10.7551/mitpress/9780262029575.001.0001.

Brown, Susan. 2014. "Scaling Up Collaboration Online: Toward a Collaboratory for Research on Canadian Writing." *International Journal of Canadian Studies* 48, no. 1: 233–51. https://doi.org/10.3138/ijcs.48.233.

Christen, Kimberly A. 2012. "Does Information Really Want To Be Free? Indigenous Knowledge Systems and the Question of Openness." *International Journal of Communication* 6: 2870–93. https://ijoc.org/index.php/ijoc/article/view/1618/828.

CIHR (Canadian Institutes of Health Research). 2021. "2021–22 Departmental Plan." Last modified February 25, 2021. https://cihr-irsc.gc.ca/e/52272.html.

Diaz, Angel Siegfried. 2013. "Fair Use & Mass Digitization: The Future of Copy-Dependent Technologies After Authors Guild v. HathiTrust." *Berkeley Technology Law Journal* 23: 683–713. https://www.jstor.org/stable/24122033.

Di Pressi, Haley, Stephanie Gorman, Miriam Posner, Raphael Sasayama, and Tori Schmitt, with contributions from Roderic Crooks, Megan Driscoll, Amy Earhart, Spencer Keralis, Tiffany Naiman, and Todd Presner. 2015. "A Student Collaborators' Bill of Rights." HumTech, June 8, 2015. https://humtech.ucla.edu/news/a-student-collaborators-bill-of-rights/.

Doern, G. Bruce. 2009. "The Granting Councils and the Research Granting Process: Core Values in Federal Government-University Interactions." In *Research and Innovation Policy: Changing Federal Government-University Relations*, edited by G. Bruce Doern and Christopher Stoney, 89–122. Toronto, Ont.: University of Toronto Press. https://doi.org/10.3138/9781442697478.

Gaffield, Chad. 2007. "Conceptualizing and Constructing the Canadian Century Research Infrastructure." *Historical Methods: A Journal of Quantitative and Interdisciplinary History* 40, no 2: 54–64. https://doi.org/10.3200/HMTS.40.2.54-64.

———. 2016. "Mindset and Guidelines: Insights to Enhance Collaborative, Campus-wide, Cross-sectoral Digital Humanities Initiatives." *International Journal of Humanities and Arts Computing* 10, no. 1: 8–21. https://doi.org/10.3366/ijhac.2016.0156.

Geist, Michael A. 2001. "Is There a There? Toward Greater Certainty for Internet Jurisdiction." *Berkeley Technology Law Journal* 16, no. 3: 1345–406. https://doi.org/10.2139/ssrn.266932.

———. 2006. "Our Own Creative Land: Cultural Monopoly & the Trouble with Copyright." *Marquette Intellectual Property Law Review* 10, no. 3: 411–31. https://scholarship.law.marquette.edu/cgi/viewcontent.cgi?article=1088&context=iplr.

———. 2013. *The Copyright Pentalogy: How the Supreme Court of Canada Shook the Foundations of Canadian Copyright Law*. Ottawa: University of Ottawa Press.

———. 2022. "The CRTC Shrugged." July 26, 2022. https://www.michaelgeist.ca/2022/07/the-crtc-shrugged/.

Graham, Shawn, Ian Milligan, Scott B. Weingart, and Kim Martin. 2022. *Exploring Big Historical Data: The Historian's Macroscope*. 2nd ed. Hackensack, N.J.: World Scientific Publishing. https://doi.org/10.1142/12435.

Guppy, Neil, Edward Grabb, and Clayton Mollica. 2013. "The Canada Foundation for Innovation, Sociology of Knowledge, and the Re-engineering of the

University." *Canadian Public Policy* 39, no. 1: 1–19. https://doi.org/10.3138/CPP.39.1.1.

Harwell, Drew. 2022. "Cheating-detection Companies Made Millions During the Pandemic. Now Students Are Fighting Back." In *Ethics of Data and Analytics*, edited by Kirsten Martin, 410–17. New York: Auerbach Publications. https://doi.org/10.1201/9781003278290-60.

LaPointe, Russell. 2006. "The Social Sciences and Humanities Research Council: From a Granting Council to a Knowledge Council?" In *Innovation, Science, Environment: Canadian Policies and Performance, 2006-2007*, edited by G. Bruce Doern, 127–48. Montréal, Que., and Kingston, Ont.: McGill-Queen's University Press.

Lopreite, Débora, and Joan Murphy. 2009. "The Canada Foundation for Innovation as Patron and Regulator." In *Research and Innovation Policy: Changing Federal Government-University Relations*, edited by G. Bruce Doern and Christopher Stoney, 123–47. Toronto, Ont.: University of Toronto Press. https://doi.org/10.3138/9781442697478-007.

Meurer, David M. 2009. "Artmob CMS: Arts-Oriented Digital Archiving Software." Poster presented with Bill Kennedy at the Association of Canadian Archivists 2009 Conference, Calgary, May 14–17, 2009. https://www.academia.edu/5064822/Artmob_CMS_Arts_Oriented_Digital_Archiving_Software_Poster_Presentation_with_Bill_Kennedy_.

Mukamal, Anna, Kate Moffatt, Kandice Sharren, and Claire Battershill. 2021. "Student Labour and Major Research Projects." *Digital Studies / Le champ numérique* 11, no. 1: 1–33. https://doi.org/10.16995/dscn.375.

NSERC (Natural Sciences and Engineering Research Council of Canada). 2021. "NSERC—Departmental Plan—2021–22." Last modified February 25, 2021. https://www.nserc-crsng.gc.ca/NSERC-CRSNG/Reports-Rapports/DP/2021-2022/.

Polster, Claire. 2002. "A Break from the Past: Impacts and Implications of the Canada Foundation for Innovation and the Canada Research Chairs Initiatives." *Canadian Review of Sociology/Revue canadienne de sociologie* 39, no. 3: 275–99. https://doi.org/10.1111/j.1755-618X.2002.tb00621.x.

Poole, Alex H., and Deborah A. Garwood. 2020. "Digging into Data Management in Public-funded, International Research in Digital Humanities." *Journal of the Association for Information Science and Technology* 71, no. 1: 84–97. https://doi.org/10.1002/asi.24213.

Robertson v. Thomson Corp. 2006. Supreme Court of Canada 43, [2006] 2 S.C.R. 363. https://decisions.scc-csc.ca/scc-csc/scc-csc/en/item/2317/index.do.

Rockwell, Geoffrey. 2022. "Voyant Tools." DH2022: Antonio Zampolli Prize Lecture, July 26, 2022. https://dh2022.dhii.asia/dh2022bookofabsts.pdf.

Rockwell, Geoffrey, and Stéfan Sinclair. 2016. *Hermeneutica: Computer-assisted Interpretation in the Humanities*. Cambridge, Mass.: The MIT Press. https://doi. org/10.7551/mitpress/9780262034357.001.0001.

Sinclair, Stéfan, and Geoffrey Rockwell. Voyant Tools. https://voyant-tools.org/.

Society of Composers, Authors and Music Publishers of Canada v. Bell Canada. 2012. Supreme Court of Canada 36, [2012] 2 S.C.R. 326. http://scc.lexum.org/ decisia-scc-csc/scc-csc/scc-csc/en/item/9996/index.do.

SSHRC (Social Sciences and Humanities Research Council of Canada). 2006. "Moving Forward as a Knowledge Council: Canada's Place in a Competitive World." https://www.sshrc-crsh.gc.ca/about-au_sujet/publications/finance_ committee_e.pdf.

———. 2009. "Image, Text, Sound and Technology." https://www.sshrc-crsh. gc.ca/funding-financement/programs-programmes/itst/research_grants- subventions_recherche-eng.aspx.

———. 2021. "Social Sciences and Humanities Research Council 2021–22 Departmental Plan." https://www.sshrc-crsh.gc.ca/about-au_sujet/publications/dp/ 2021-2022/dp-eng.aspx.

Wemigwans, Jennifer. 2018. *A Digital Bundle: Protecting and Promoting Indigenous Knowledge Online*. Regina, Sask.: University of Regina Press.

Wikipedia. 2022. "2022 Rogers Communications outage." Wikimedia Foundation. Accessed August 5, 2022. https://en.wikipedia.org/wiki/2022_Rogers_ Communications_outage.

Contributors

Paul Barrett is Associate Professor in the School of English and Theatre Studies and Culture and Technology Studies at the University of Guelph. He is the author of *Blackening Canada: Diaspora, Race, Multiculturalism* (University of Toronto Press, 2015) and the editor of *'Membering Austin Clarke* (Wilfred Laurier University Press, 2020). His research is at the intersection of Canadian literature, digital humanities, diaspora, and critical race theory. He is the co-investigator, with Sarah Roger, on two SSHRC-funded research projects that study how Canadian literary discourse is transformed in online spaces.

Gregory Betts is a scholar, editor, and experimental poet with collections published in Canada, the United States, Australia, and Ireland. He is most acknowledged for *If Language* (Book*hug, 2005), a collection of paragraph-length anagrams, and *The Others Raisd in Me* (Pedlar, 2009), 150 poems carved out of Shakespeare's Sonnet 150. His other books explore conceptual, collaborative, and concrete poetics. He has lectured and performed internationally, including at the Sorbonne Université, the Johannes Gutenberg Universität Mainz, the National Library of Ireland, and the 2010 Vancouver Olympic Games as part of the "Cultural Olympiad," among others. He is a professor of Canadian and avant-garde literature at Brock University, where he has produced two of the most exhaustive academic studies of avant-garde writing in Canada, *Avant-Garde Canadian Literature: The Early Manifestations* (2013) and *Finding Nothing: The VanGardes, 1959–1975* (2020), both published with University of Toronto Press. He has served as the President of the Association of Canadian College and University Teachers of English (ACCUTE), the Craig Dobbin Professor of Canadian Studies at University College Dublin, and the Chancellor's Chair for Research Excellence at Brock University. He is currently the curator of the bpNichol.ca Digital Archive and Associate Director of the Social Justice Research Initiative.

Susan Brown is Professor of English and Canada Research Chair in Collaborative Digital Scholarship at the University of Guelph. Her research explores intersectional feminism, literary history, and online modes of collaborative knowledge production. She directs the Orlando Project in British women's writing, the Canadian Writing Research Collaboratory, and the Linked Infrastructure for Networked Cultural Scholarship. She collaborates with colleagues at Guelph in running The Humanities Interdisciplinary Collaboration (THINC) Lab, the DH@Guelph Summer Workshops, and the major in Culture and Technology Studies.

Mark V. Campbell is a DJ, scholar, and curator. His research explores the relationships between Afrosonic innovations, hip-hop archives, and notions of the human. Mark is currently the principal investigator in the SSHRC-funded research project *Hip-Hop Archives: The Poetics and Potentials of Knowledge Production* and founder at Northside Hip-Hop Archives. His recent books include the monograph *AfroSonic Life* (Bloomsbury, 2022), the co-edited collection of essays *We Still Here: Hip Hop in North of the 49th Parallel* (McGill-Queen's University Press, 2020), and his collection *Hip-Hop Archives: The Politics and Poetics of Knowledge Production* (University of Chicago Press, 2023), co-edited with Murray Forman. He is Assistant Professor of Music and Culture at the University of Toronto Scarborough and holds research fellow positions with the Laboratory for Artistic Intelligence and the Research Centre for Music, Sound and Society in Canada.

Allan Cho has an MLIS in Library and Information Studies, an MA in History, and an MA in Educational Technology, all from the University of British Columbia. His previous professional roles include Research Commons librarian and digital humanities liaison librarian. Allan's research interests are in Asian Canadian history, literature, and culture, and outside of work he volunteers his time for several community organizations with anti-racism and solidarity building. Allan's work includes supporting ongoing community initiatives and leading new ones, focusing on community engagement with historically underrepresented groups, subject liaison librarian with the School of Information, and developing an Equity, Diversity, and Inclusion Scholars-in-Residence program with the support of the Peña Family Foundation.

Kendra Cowley is a public librarian and forever-researcher based in Tkaronto.

Constance Crompton is Assistant Professor in the Department of Communication at the University of Ottawa and Canada Research Chair in Digital Humanities. She directs the University of Ottawa's Labo de données en sciences humaines/ The Humanities Data Lab and is a member of the Lesbian and Gay Liberation in Canada (LGLC), Linked Infrastructure for Networked Cultural Scholarship, and Implementing New Knowledge Environments Partnership research teams. She serves as Associate Director of the Digital Humanities Summer Institute, North America's largest digital humanities training institute. She is the co-editor of two volumes, *Doing Digital Humanities* and *Doing More Digital Humanities* (Routledge 2016, 2019). She lives and works on unceded Algonquin land.

Pascale Dangoisse is a PhD candidate at the University of Ottawa and a research assistant on the Lesbian and Gay Liberation in Canada (LGLC) project. Her research focuses on the study of liberal political discourses on the topic of feminism and women's rights in Canada. Her research is particularly interested in understanding how systemic discrimination persists under liberal or progressive governments.

Sandra Djwa is a scholar of Canadian Literature and author of ten books, including *The Politics of the Imagination: A Life of F.R. Scott* (McClelland and Stewart, 1987), *Professing English: A Life of Roy Daniells* (University of Toronto Press, 2002), and *Journey with No Maps: A Life of P.K. Page* (McGill-Queen's University Press, 2012; winner of the 2013 Governor General Award for Non-fiction). She co-founded the Association of Canadian and Quebec Literatures in 1973. She has been a member of the Royal Society of Canada since 1994.

Klara du Plessis is a FRQSC-funded, final-year PhD candidate at Concordia University, and is affiliated with the SpokenWeb research network. An interdisciplinary project straddling English literature, curatorial studies, and performance, her doctoral work aims to schematize different modes of literary event curation and to think critically about the often-neglected labour that goes into shaping poetry reading series, whether live or in the audio archive. Her research focuses on twentieth century and contemporary Canadian poetry, and develops a research creation component called Deep Curation, an approach that places poets' work in deliberate dialogue with each other and heightens the curator's agency toward the poetic product. In this capacity, she has worked with an amazing array of poets, including Alexei Perry Cox and Kama La Mackerel. Klara is the author of *Ekke* (Palimpsest, 2018; winner of the 2019 Pat Lowther Memorial Award) and *Hell Light Flesh* (Palimpsest, 2020) and has also edited a book of experimental criticism based on transcription and citation with SpokenWeb and in collaboration with Emma Telaro called *Quotes: Transcriptions on Listening, Sound, Agency*.

Ryan Fitzpatrick is a poet and researcher living in Toronto/Tkaronto. His research focuses on contemporary poetics and questions of space and intimacy. He has recently published academic articles in *Studies in Canadian Literature* and *Canadian*

Literature. He is the author of four books of poetry, including *Sunny Ways* (Invisible 2023) and *Coast Mountain Foot* (Talonbooks 2021). With Deanna Fong, Janey Dodd, and others, he worked on the second iteration of the Fred Wah Digital Archive (fredwah.ca).

Deanna Fong is a SSHRC-funded researcher at Concordia University, where she directs the digital archive of Canadian poet Fred Wah (fredwah.ca). With a team of student researchers and Systems Librarian Tomasz Neugebauer, she is working on visualizing the site's social metadata, which represents the roles and activities that go into literary production. With Cole Mash, she is the co-editor of a forthcoming collection of essays, interviews, and art titled *Resistant Practices in Communities of Sound* (McGill-Queen's University Press, 2023). Her book of interviews, *Concern and Commitment: Seven Oral Histories with Innovative Vancouver Women*, is forthcoming with Talonbooks (2024). She is the literary editor at *The Capilano Review*.

David Gaertner is a settler scholar and Assistant Professor in the Institute for Critical Indigenous Studies at the University of British Columbia and the co-director of the CEDaR Space, a community-oriented new media and digital storytelling lab. He is the author of *The Theatre of Regret: Literature, Art, and the Politics of Reconciliation in Canada* (University of British Columbia Press, 2020), the editor of *Sôhkêyihta: The Poetry of Sky Dancer Louise Bernice Halfe* (Wilfred Laurier University Press, 2018), and co-editor of *Read, Listen, Tell: Indigenous Stories from Turtle Island* (Wilfred Laurier University Press, 2017).

Asen Ivanov holds a PhD in Information Studies from the University of Toronto and an MA in Heritage Studies from the University of Amsterdam. His research and teaching expertise is in the technologies and practices through which cultural heritage and media organizations collect, organize, preserve, and assign value to cultural works. Most recently, Asen was Michael Ridley Postdoctoral Fellow in Digital Humanities at the University of Guelph.

Graham H. Jensen is the principal investigator of the Canadian Modernist Magazines Project (modernistmags.ca) and a Mitacs accelerate and INKE Partnership postdoctoral fellow in the Electronic Textual Cultures Lab at the University of Victoria. In the latter role, he helps manage research, development, and user testing for the Canadian Humanities and Social Sciences Commons (beta version at hsscommons.ca). Previously, at the University of Victoria, he was a SSHRC postdoctoral fellow in English and an associate fellow in the Centre for Studies in Religion and Society. His wide-ranging research interests include Canadian literature, modernism, twentieth- and twenty-first-century literature and religion, critical infrastructure studies, critical digital humanities, and open scholarship. His research is published or forthcoming in *The Edinburgh Companion to Modernism, Myth and Religion*; *Interdisciplinary Digital Engagement in Arts & Humanities*; *Open Scholarship Press Collections: Connection*; *Pop! Public. Open. Participatory*; *English Studies in Canada*; *University of Toronto Quarterly*; *William James Studies*; *Canadian Poetry*; and *The Sound and the Fury: A Hypertext Edition*.

Rashmeet Kaur completed her Bachelor of Science degree at the University of Guelph and is currently a Master of Public Health candidate at Drexel University's Dornsife School of Public Health. Rashmeet loves to merge her passion for both the sciences and the humanities with poetry and mixed media artwork. She believes all forms of art have transformative power and this has sparked her passion for facilitating community workshops where she encourages participants to use various art forms as storytelling and social justice tools. Rashmeet's artwork and poetry have been published in local and international publications, including *Kaleidoscope*, *Margins Magazine*, and *Nature is a Human Right*. You can visit her online at https://dissectionoftheself.wordpress.com/ and follow her @_rashmeet.k on Instagram.

Kim Martin is Assistant Professor of History and Culture and Technology Studies at the University of Guelph. She is Associate Director of THINC Lab and Research Board Chair for the Linked Infrastructure for Networked Cultural Scholarship (LINCS). Her research interests include serendipity in digital environments; the information behaviour of humanities scholars; and local, community-focused oral history.

Katherine McLeod is Affiliate Assistant Professor in the Department of English at Concordia University. She is writing a book that is a feminist listening to recordings of women poets on the radio, and she is the principal investigator of her SSHRC-funded project "Literary Radio: Developing New Methods of Audio Research." She has co-edited, with Jason Camlot, *CanLit Across Media: Unarchiving the Literary Event* (McGill-Queen's University Press, 2019) and has published on poetry, performance, and archives in journals such as *Canadian Literature* and *Mosaic*. She produces *ShortCuts*—a monthly series about archival audio—for *The SpokenWeb Podcast*.

Kiera Obbard is a poet and PhD candidate in the School of English and Theatre Studies at the University of Guelph. Her SSHRC-funded project, "The Instagram Effect: Contemporary Canadian Poetry Online," examines the complex social, cultural, technological, and economic conditions that have enabled the success of social media poetry in Canada, how the technological affordances of social media platforms mediate reading and writing, and the relationship between social media poetry and data mining practices. She completed her MA in Cultural Studies and Critical Theory at McMaster University and her honours BA with a joint major in English and Communication at the University of Ottawa. She is currently a graduate research assistant for the Translating Digital Canadas project, a fellow at The Humanities Interdisciplinary Collaboration (THINC) Lab, and an editorial board member for the Centre for Media and Celebrity Studies.

Julia Polyck-O'Neill is an artist, curator, critic, poet, and writer. A former visiting scholar at University of the Arts London (Chelsea College of Arts), lecturer at the Obama Institute at Johannes Gutenberg Universität Mainz (2017–2018), and international fellow of the Electronic Literature Organization, she is currently a SSHRC postdoctoral fellow in the Department of Visual Art and Art History and the Sensorium Centre for Digital Arts and Technology at York University (Toronto) where she studies digital, feminist approaches to interdisciplinary artists' archives. Her writing has been published in *Zeitschrift für Ästhetik und Allgemeine Kunstwissenschaft* (*The Journal for Aesthetics and General Art History*), *English Studies in Canada*, *DeGruyter Open Cultural Studies*, *BC Studies*, *Canadian Literature*, and other places.

Roopika Risam is Associate Professor of Film and Media Studies and of Comparative Literature and Faculty of Digital Humanities and Social Engagement at Dartmouth College. She is the author of *New Digital Worlds: Postcolonial Digital Humanities in Theory, Praxis, and Pedagogy* (Northwestern University Press, 2018). Among her edited collections, *The Digital Black Atlantic*, part of the Debates in the Digital Humanities series, was published by University of Minnesota Press in 2021. Risam is the co-editor of *Reviews in Digital Humanities*, a journal offering peer review of digital scholarship, and director of the Digital Ethnic Futures Consortium, a Mellon Foundation-funded initiative to support teaching and research at the intersections of ethnic studies and digital humanities. More information is available at http://roopikarisam.com.

Sarah Roger is the project manager for the Linked Infrastructure for Networked Cultural Scholarship (LINCS) and Adjunct Professor in the School of English and Theatre Studies at the University of Guelph. She is the author of *Borges and Kafka:*

Sons and Writers (Oxford University Press, 2017). Sarah is the co-investigator, with Paul Barrett, on two SSHRC-funded research projects that study how Canadian literary discourse is transformed in online spaces.

Jon Saklofske, Literature Professor at Acadia University, is insatiably curious about intersections between media forms and cultural perceptions. In addition to experimenting with virtual environments and games as tools for academic research, communication, and pedagogy, Jon's other research and research-creation interests include environmental storytelling in theme parks, values-based game design, alternative platforms for open social scholarship, and the critical potential of feminist war games.

Eric Schmaltz is an academic, poet, and editor. He holds a PhD from York University, where he studied Canadian and avant-garde literature. He is the author of *Surfaces* (Invisible Publishing, 2018) and several shorter creative works, including *Language in Hues* (Timglaset, 2021). He is also co-editor of *I Want to Tell You Love* by bill bissett and Milton Acorn (University of Calgary Press, 2021). His writing has appeared in *Canadian Literature, English Studies in Canada, Jacket2, Bomb, The Capilano Review*, and other places. A former SSHRC postdoctoral fellow at the University of Pennsylvania, he is currently Writer-on-the-Grounds at Glendon College.

Michelle Schwartz is an educational developer at Toronto Metropolitan University's Centre for Excellence in Learning & Teaching where they focus on inclusive and accessible teaching and learning. They co-direct Lesbian and Gay Liberation in Canada (LGLC; lglc.ca), a SSHRC-funded digital humanities research project that is building an interactive digital resource for the study of LGBTQ history in Canada, and serve on the Board of Directors of the ArQuives: Canada's LGBTQ2+ Archives.

Dani Spinosa is a poet of digital and print media, an on-again-off-again precarious professor, the managing editor of the Electronic Literature Directory, and a co-founding editor of Gap Riot Press. She has published several chapbooks of poetry, several more peer-reviewed journal articles on poetry, one long scholarly book, and one pink poetry book.

Andrea Zeffiro is Assistant Professor in critical technology studies in the Department of Communication Studies and Media Arts and Academic Director for the Lewis & Ruth Sherman Centre for Digital Scholarship at McMaster

University. Her work has appeared in *Cultural Analytics, the Canadian Journal of Communications, Journalism & Mass Communication Quarterly, Convergence, Studies in Social Justice*, and many edited collections.

Sarah Zhang is the Librarian for Geography, GIS, and Maps at Simon Fraser University. Sarah holds master's degrees in Library and Information Studies and Ecology. As an immigrant, she is constantly inspired by the cultures around her. Her current research interests include spatial literacy and open scholarship.

Index

CANADIAN LITERATURE COLLECTION /
COLLECTION DE LITTÉRATURE CANADIENNE

Series editor: Dean Irvine

The *Canadian Literature Collection* (CLC) is a series of nineteenth- to mid-twentieth-century literary texts produced in new critical editions. All texts selected for the series were either out of print or previously unpublished. Each text appears in a print edition with a basic apparatus (critical introduction, explanatory notes, textual notes, and statement of editorial principles) together with an expanded web-based apparatus (which may include alternate versions, previous editions, correspondence, photographs, source materials, and other related texts by the author).

Recent titles in the *Canadian Literature Collection*

Barney Allen, *They Have Bodies, by Barney Allen: A Critical Edition*, 2020.

Scott K. Duchesne, ed., *Creative Theatre, by Roy Mitchell: A Critical Edition*, 2020.

Heather Macfarlane, *Divided Highways: Road Narrative and Nationhood in Canada*, 2019.

Colin Hill, ed., *Man Should Rejoice, by Hugh MacLennan: A Critical Edition*, 2019.

Robert D. Denham, *Northrop Frye and Others. Volume III: Interpenetrating Visions*, 2018.

Bruce Nesbitt, ed., *Conversations with Trotsky: Earle Birney and the Radical 1930s*, 2017.

Robert D. Denham, *Northrop Frye and Others. Volume II: The Order of Words*, 2017.

Richard J. Lane and Miguel Mota, eds., *Malcom Lowry's Poetics of Space*, 2016.

Emily Ballantyne, Marta Dvořák, and Dean Irvine, eds., *Translocated Modernisms: Paris and Other Lost Generations*, 2016.

Lee Skallerup Bessette, *A Journey in Translation: Anne Hébert's Poetry in English*, 2016.

For a complete list of titles published by the University of Ottawa Press, please visit:
www.Press.uOttawa.ca